Quantum Dots - Recent Advances, New Perspectives and Contemporary Applications

Edited by Jagannathan Thirumalai

Published in London, United Kingdom

Quantum Dots - Recent Advances, New Perspectives and Contemporary Applications
http://dx.doi.org/10.5772/intechopen.102101
Edited by Jagannathan Thirumalai

Contributors
Kristin S. Williams, Geeta Bhatt, Pawan Kumar, Shweta Dua, Balaram Pani, Karunanithi Rajamanickam, Minakshi Chakraborty, Sandip Sen, Paweł Kwaśnicki, Ncediwe Tsolekile, Nozikumbuzo Anati Vitshima, Noluvuyo Mngcutsha, Rachid Houça, Abdelhadi Belouad, Abdellatif Kamal, El Bouâzzaoui Choubabi, Mohammed El Bouziani, Penio Lebamovski, Murtadha Jasim Hasan AlTimimi, Vikram P. Bhalekar, Hajar Shekarchizadeh, Shabnam Sistani, Salar Khaledian, Mohadese Abdoli, Reza Fatahian, Saleh Salehi Zahabi, Meysam Safari, Jagannathan Thirumalai

First published in London, United Kingdom, 2023 by IntechOpen
IntechOpen is the global imprint of INTECHOPEN LIMITED, registered in England and Wales, registration number: 11086078, 5 Princes Gate Court, London, SW7 2QJ, United Kingdom

British Library Cataloguing-in-Publication Data
A catalogue record for this book is available from the British Library

Additional hard and PDF copies can be obtained from orders@intechopen.com

Quantum Dots - Recent Advances, New Perspectives and Contemporary Applications
Edited by Jagannathan Thirumalai
p. cm.
Print ISBN 978-1-80356-593-4
Online ISBN 978-1-80356-594-1
eBook (PDF) ISBN 978-1-80356-595-8

For EU product safety concerns:
IN TECH d.o.o., Prolaz Marije Krucifikse Kozulić 3, 51000 Rijeka, Croatia,
info@intechopen.com or visit our website at intechopen.com.

Meet the editor

Jagannathan Thirumalai received his Ph.D. from Alagappa University, Karaikudi in 2010 and was awarded a post-doctoral fellowship at Pohang University of Science and Technology (POSTECH), Republic of Korea, in 2013. He was an assistant professor of physics at B.S. Abdur Rahman University, Chennai, India from 2011 to 2016. Currently, he is a Senior Assistant Professor and Head of the Physics Department, at SASTRA Deemed University, Kumbakonam (T.N.), India. His research interests focus on luminescence, self-assembled nanomaterials, thin film optoelectronic devices, and spintronics applications. He has published more than 65 SCOPUS/SCI-indexed papers and ten book chapters and has edited seven books, six for IntechOpen. His research activities are funded by the Science and Engineering Research Board (SERB) of India.

Contents

Preface

Quantum dots (QDs) were first theoretically proposed by Alexander Efros in 1982, and the term came into widespread use in 1986, although nanoparticles of semiconductor QDs had already been theorized in the 1970s. The attributes of quantum dots have led to their use in a number of research areas such as QD light emitting diodes (LEDs), QD solar photovoltaics, and quantum computing (qubits), as well as bio- and food technologies. This book focuses on a range of important real-time applications for QDs.

The book is in three sections. Section 1 (Quantum Dots in Solar and Other Energy Sector) reviews important advances in QD-based solar energy systems. The introduction of semiconductor quantum dots into photoelectric material can be an important and effective way to increase the efficiency of photoelectric devices and photovoltaic cells. Chapter 2 reports efficient energy harvesting mechanisms. Chapter 3 discusses two performance-enhancing applications of carbon quantum dots (CQDs). In solar cells, CQDs have been used as active absorbing layers, electron-transporting layers (ETL), hole-transporting layers (HTL), donor/acceptor or dopant. In white LEDs, CQDs have been used as a phosphor material in the presence and absence of composite material to produce white emission. In Chapter 4, different types of nanomaterial synthesis using QD physical and chemical properties are discussed. The chapter also covers the structure and synthesis of PbS quantum-dot sensitized solar cells (QDSSCs) using simple, safe and chemical-free ultrasonication techniques and without a capping agent. Finally, Chapter 5 proposes a simple, cost-effective and efficient solar photovoltaic system.

Section 2 (Quantum Dots in Bio and Food Technology) reviews important advances in the application of QDs in the biomedical field. Chapter 6 examines bio-sensing, bio-imaging, drug loading capacity, targeted drug delivery, anti-stacking limit hostile to bacterial activity, photo-thermal treatment, and photodynamic treatment. In Chapter 7, surface modification of QDs for biosensor applications with high sensitivity and selectivity is discussed. QDs are ideal fluorescent labels for lateral flow assays (LFA) due to their unique optical properties and resistance to chemical and photo-degradation, and the prospects and challenges of QDs in LFA are reported in Chapter 8. Chapter 9 introduces in-vitro and in-vivo bio-imaging applications and synthesis, modification and cytotoxicity expression methods. The application of QDs in the food industry is discussed in Chapter 10. Quantum dot-based sensors have promising prospects for food analysis and water treatment as nanomembranes, nanosensors, and absorbent photocatalysts. The many benefits of QDs for food packaging, including improved mechanical properties, better thermal stability, increased water resistance, UV barrier properties, and antimicrobial and antioxidant activity, are reported in Chapter 10.

Section 3 (Quantum Dots in Other Important Applications) contains a further four chapters. Chapter 11 reports basic findings of a theoretical investigation of QD scattering in monolayer molybdenum disulfide (MoS2) electron propagation wave plane in a circular quantum dot of MoS2, and the scattering efficiency, square modulus of the scattering coefficients, the radial component of the far-field scattered current, and electron density. In Chapter 12, the determination of Qubit entanglement in one-step double photoionization of a helium atom is reported. Chapter 13 reviews QLED applications of Photobiomodulation (PBM) as a photodynamic therapy in medicine and examines the quantum mechanics of Transcranial Photobiomodulation (tPBM) and its ability to affect electrophysiological activity. Chapter 14 focuses on the application of 3D technology in teaching the discipline of stereometry.

The aim of this book is to provide readers with a detailed overview of the most recent advances in the fascinating world of quantum dots, as well as to outline some possible future directions. The book is addressed to students and scientists working in the field of quantum dots and advanced technologies.

We would like to express our thanks to all the contributors to this book and our gratitude to Ms. Marica Novakovic for her valuable assistance and constant support during the writing and final preparation of the book.

Dr. Jagannathan Thirumalai
Department of Physics,
Srinivasa Ramanujan Centre,
SASTRA Deemed University,
Kumbakonam, Tamil Nadu, India

Section 1

Quantum Dots in Solar and Other Energy Sector

Chapter 1

Introductory Chapter: The Fame of Quantum Dots in Space-age Improvements for Multifunctional Application

Jagannathan Thirumalai

1. Introduction

The word "'Quantum dot (QD)" is a precisely trifling structure (ranging between 1 and 10 nm wide), for instance, a semiconductor-made nanocrystal implanted in alternative semiconductor-based materials, in which the electrons or other charge carriers can be confined in all the three dimensions along with reference to the respective electronic physiognomies contingent on its shape and size. Over the past three decades, quantum dots (QDs) had well established in connection with numerous remarkable applications. Amid the supreme contemporary machineries are typically built on their attractive optical and electronic properties and their part in light absorption, emission, conversion, and detection have a massive volume of innovative day-to-day applications, and gratifying ever more controllable and accustomed to the societal nature. In 1981, Alexey I. Ekimov is a Russian solid-state physicist invented the semiconductor nanocrystals well known as quantum dots in glass matrix, while he was working at the Vavilov State Optical Institute [1]. Further, systematic progress in the remarkable science and technology of the QDs was determined in 1985, when Louis E. Burs at Columbia University come up with a relationship between size and a degree of bandgap relies on the semiconductor base nanoparticles which relates a particle of kind in a sphere model approximations to the respective wave function ideologies for the aforesaid bulk semiconductors [2–5] and the QDs were discovered from the colloidal mixture of semi-conductor nanocrystals [6]. Also, spin qubits in semiconductor quantum dots signify a prominent family of solid-state qubits, which provides greater efforts to build a quantum computer [7].

Murray et al. have efficaciously synthesized the colloidal state CdX (X = S, Se, Te) QDs along with a tunable size of band-edge absorption as well as emissions and it took more than a decade for the preparation of new-fangled QD material [4]. Owing to its outstanding optical and electrochemical physiognomies, the CdX QDs had widely investigated. In the infancy of core-shell QD research, CdSe/ZnS and CdSe/CdS are the utmost exhaustively examined materials [5, 8–10]. A decade back, increasingly new "core-shell" QDs were synthesized, such as CdSe/ZnSe [11], CdTe/CdS [12], CdSe/ZnS, and CdTe/ZnS [13], and at even multilayer CdTe/CdS/ZnS "core/shell/shell" QDs [14]. By employing zinc stearate as a zinc source, bright luminescent and low toxic, CdSeTe@ZnS-SiO2 QDs would be made with ZnS-like clusters

Field	Application with examples
Engineering/ Processing	*scale-up engineering development:* • Trap engineering in PbSe quantum dots for photodetectors [19] • Wave function engineering—Type—I/II Excitons [20] • Large-scale production of quantum dots for energy storage applications [21] • Efficient calibration of crosstalk in a quadruple quantum dot array [22]
Photonic devices	• Quantum dot—Si photonic integrated circuits [23] • High-quality quantum dot laser/waveguide [24] • QDs and nano-photonic waveguides [25]
Optoelectronics	• Semiconductor quantum dot for photovoltaic applications [26] • Quantum dot optoelectronics [27] • Quantum dot semiconductor optical amplifiers (QD-SOA) [28] • Flexible quantum dot light-emitting diodes [29]
Electronics	• Single-electron transistors with quantum dots [30] • High-Performance quantum dot thin-Film transistors [31] • Integrated circuits (VLSI, hybrid integrated) [32] • Single quantum emitter detection with amateur CCD [33]
Electricity (without semiconductors)	• Insulating matrices [34] - e.g., for resistors, capacitors • Piezoelectric energy harvester [35]
Cryotechnics	• Cryogenic-temperature thermodynamically Suppressed and strongly confined quantum dots [36] • SQUIDS (superconducting quantum interference devices) [37]
Mechanics	• Quantum-dot array with self-aligned electrodes [38] • Reversible adhesion *via* quantum dots [39] • Reduction of friction [40]
Magnetics	• Quantum-dot cellular automata [41] • Multimodal imaging [42]
Sensorics	• Data acquisition in antagonistic surroundings and media [43] • Telemetry monitoring via PDA [44] • Optical and electrochemical (bio)analytical sensors [45]
Chemistry	• Drift-diffusion [46] • Carbon dots are in contrast to corrosion inhibitors [47] • Resistive-based gas sensors [48]
Biomedicine	• Biocompatible implant coating [49] • Neurological sensors [50]
New materials	• Metastable phases/metallic glasses: Mn^{2+}- doped CdS quantum dots in a silicate glass [51] Metal-doped PbSe quantum dots in silicate glasses [52] • Low-noise GaAs quantum photonics [53] • Ultra-stable carbon quantum dots [54]
(Alternative) Energies	• Digital light processing to inkjet 3D printing [55] • Quantum Dots as potential electrode materials for Supercapacitors application [56]

Table 1.
Reviews some major state-of-the-art development in technology connected to the applications of quantum dots in a diverse spectrum.

packed into the SiO2 shell through a microwave-assisted process [15]. Hypersensitive type photosensors were prepared with respect to the cesium lead bromide (CsPbBr3) perovskite quantum dots (QDs) with ample amount of higher sensitivity for chemiluminescence based immunoassays [16]. Vastly luminescent all-inorganic cesium lead bromide (CsPbBr3-QDs/CuPc) heterostructures of perovskite natured quantum dots (QDs) had been comprehensively in the usage as a photosensitizer in the known optoelectronic devices, while p-type small-organic-molecule comprised of copper phthalocyanine (CuPc) would greatly use as a photoactive material in solar cells and organic field-effect transistors (OFETs) [17].

In the year 1988, Mark Reed investigated electronic transport via a three-dimensional confined semiconductor quantum well ("quantum dot"). In 2013, the first commercial deliverance of a product employing the quantum dots would be the Sony XBR X900A series of flat panel televisions. In the year 2013, the Kindle Fire HDX is released covering quantum dot technology, and a blue glow bleeding in from the edge of the display. Followed by in the year 2015, the quantum dots would be a featured innovation at the consumer electronics show, projects that, the quantum dots have been flaunted to be the next breakthrough visual technology to enhance the LED TV picture quality. In recent years, the focus is based on the development of effectual and stable lead-free wise perovskite solar cells (PSCs) and also the prospect of reaching 20% power conversion efficiency (PCE) for tin PSCs. These issues concern the enlargement of the cell size and apprehending scalable production in the future [18] (**Table 1**).

2. Recent automated advancements in quantum dots for multifunctional application

With due diligence to the nano-size regime, quantum dots (QDs) had demonstrated unique physicochemical properties which are prosperous for multifunctional applications. Exclusively, owing to the respective quantum size effect, remarkable optoelectronic properties have been perceived. These substantial growths have not only elevated the systematic practice of QDs but also fortified to formulate numerous hybrid manners of materials to accomplish higher efficacy by eradicating definite inadequacies. Such problems can be overwhelmed by amalgamating QDs with potential applications, including single-electron transistors, solar cells, LEDs, lasers, single-photon sources, quantum computing, second-harmonic generation, cell biology research, medical imaging, microscopy. In addition, the next-Generation QDs can be lingering further into developing technologies, such as (i) quantum dots of very high-quality optical applications, light guide plate (LGP) with QDs, (ii) quantum dot white light-emitting diodes (QD-WLED), (iii) quantum dot-based photodetectors (QDPs), perovskite QDs, (iv) quantum dot solar photovoltaics, (v) biological applications (to study *in-vivo* & *in-vitro* observation of subcellular trafficking, drug delivery, and high-resolution cellular imaging), (vi) quantum computers (qubits), and (vii) the future of quantum dots (magnetic quantum dots & graphene quantum dots) owing to the unique physicochemical characteristics. A typical QD color enrichment scheme by including fluorescent QDs into a polycarbonate (PC) LGP, has been also established to broadcast quite homogeneous planar lighting and might be likely for the new-generation backlighting as well as display devices [57]. Outlook on stability of Quantum dots, patterning, and deposition which yield dare of micro-LED displays provide progressive development in the field of QD-based micro-LED displays, exhibiting the optimistic future prospects of the same kind of spelled technology [58]. The device on blue LED exhibited that efficacy levels of various 230 lm/W are achievable using ultra-revolutionized blue LED pumps. By employing liquid injection fabrication, high performance of white LEDs using QD liquids indicates a high potential for future prospects on lighting devices [59]. Perovskite quantum dots (PQDs) have captured an innovative impression that is exploited to improve the photovoltaic performance of numerous kinds of solar cells and summarized some challenges and perspectives concerning of PQDs [60]. The latest innovations in perovskite technology via diverse schemes, such as the amplify its functionality of methodological procedures, surface amendments, and solid QD inks near the device manner engineering for accomplishing sophisticated PV efficiency and enhanced longstanding steadiness [61]. Quantum dots (QDs) make available a multipurpose component to aid the most precise investigative tools and fluoroimmunoassays, dual imaging, multiplexed imaging, therapeutic platforms, real-time *in vivo* and cellular process imaging, and so on. With the extensive augmented interest in researchers to report these challenges and endure to move forward toward the clinical translation of QDs [62]. The team Krause P et al., have successfully modeled the simulations exposed in what way the quantum dot duos absorb, exchange, and store light energy. Similarly, it is of interest how size and geometry variations of three Ge/Si nanocrystals impact the transfer times and thus the efficacy of laser-driven populations of the said electron-hole pair states paid to a wide range of applications for quantum dots including quantum bits (qubits) [63]. Tajik S, et al., reviewed the applications of carbon-based QDs (GQDs and CQDs) in biological and sensing areas, and then covers sensing features of key neurotransmitters [64]. In conclusion, in recent years, quantum dots play a major role in multifunctional applications to make acquainted with the novel techniques of semiconductor nanocrystals, and advanced devices which favor the outcomes from translational research, a convenient collaboration lined by the research scientists of extreme necessity for the societal benevolent.

3. Conclusion

The endorsement of quantum dots (QDs) has led to swift developments in multifunctional QDs applications over the early few years throughout the world amidst the advanced scientific improvements, though confirmations, moreover, validate that it has been more extensive and extremely significant in the way of innovative examination in contradiction of novel applications for the societal prerequisite. Amongst the major noteworthy progress in diverse fields of QDs the roadmap for next-generation QDs points to the advent of single-electron transistors, single-photon sources, solar cells, μ-LEDs, lasers, quantum computing, second-harmonic generation, microscopy, cell biology research, and medical imaging, etc., highly necessary for the societal benign. However, a probable drawback when utilized in biological applications is the fact that owing to their large physical size, they could not prolix across cellular membranes. Also, the delivery process may essentially be hazardous for the cell and even affect in destroying it. In other cases, a QD may be venomous for the cell and not suitable for any biological application. The dispute on the degradation of quantum dots inside the living organism has to be explored in the near future. Thus, studies on superior QDs hi-tech vicinities might pay way to atoms/molecules that might perchance turn out to be the upcoming policies of multifunctional prospects in the forthcoming outline.

Acknowledgements

All authors contributed toward data analysis, drafting, and revising the paper and agree to be accountable for all aspects of the work.

The authors apologize for inadvertent omission of any pertinent references.

Conflict of interest

The authors declare that there is no conflict of interests regarding the publication of this paper.

Author details

Jagannathan Thirumalai
Department of Physics, School of Electrical and Electronics Engineering, SASTRA Deemed University, Srinivasa Ramanujan Centre, Kumbakonam, Tamil Nadu, India

*Address all correspondence to: thirumalaijg@gmail.com; thirumalai@src.sastra.edu

References

[1] http://navillum.com/history-of-quantum-dots/#synved-sections-1-4

[2] Ekimov AI, Onushchenko AA. Quantum size effect in three-dimensional microscopic semiconductor crystals. JETP Letters. 1981;**34**(6):345-349

[3] Reed MA, Randall JN, Aggarwal RJ, Matyi RJ, Moore TM, Wetsel AE. Observation of discrete electronic states in a zero-dimensional semiconductor nanostructure. Physical Review Letters. 1988;**60**:535. DOI: 10.1103/PhysRevLett.60.535

[4] Murray CB, Norris DJ, Bawendi MG. Synthesis and characterization of nearly monodisperse CdE (E = sulfur, selenium, tellurium) semiconductor nanocrystallites. Journal of the American Chemical Society. 1993;**115**(19):8706-8715

[5] AAM W, AEG H, Saleh M, Hoda G, Hala H, Hanan M, et al. Quantum dots synthetization and future prospect applications. Nanotechnology Reviews. 2021;**10**:1926-1940

[6] De Arquer PG, Dmitri VT, Victor IK, Yasuhiko A, Manfred B, Edward HS. Semiconductor quantum dots: Technological progress and future challenges. Science. 2021;**373**:6555. DOI: 10.1126/science.aaz8541

[7] Vandersypen LMK, Bluhm H, Clarke JS, Dzurak AS, Ishihara R, Morello A, Reilly DJ, Schreiber LR, Veldhorst M. Interfacing spin qubits in quantum dots and donor,—Hot, dense, and coherent. npj Quantum Inference. 2017;**3**:34. DOI: https://doi.org/10.1038/s41534-017-0038-y

[8] Hines MA, Guyot-Sionnest P. Synthesis and characterization of strongly luminescing ZnS-capped CdSe nanocrystals. Journal of Physical Chemistry. 1996;**100**(2):468-471

[9] Dabbousi BO, Rodriguez-Viejo J, Mikulec FV, Heine JR, Mattoussi H, Ober R, et al. (CdSe)ZnS core-shell quantum dots: Synthesis and characterization of a size series of highly luminescent nanocrystallites. Journal of Physical Chemistry B. 1997;**101**(46):9463-9475

[10] Peng X, Schlamp MC, Kadavanich AV, Alivisatos AP. Epitaxial growth of highly luminescent CdSe/CdS core/shell nanocrystals with photostability and electronic accessibility. Journal of the American Chemical Society. 1997;**119**(30):7019-7029

[11] Reiss P, Bleuse J, Pron A. Highly luminescent CdSe/ZnSe core/shell nanocrystals of low size dispersion. Nano Letters. 2002;**2**(7):781-784

[12] He Y, Lu HT, Sai LM. Microwave-assisted growth and characterization of water dispersed CdTe/CdS core-shell nanocrystals with high photoluminescence. Journal of Physical Chemistry. 2006;**110**(27):13370-13374

[13] Sreeram C, Zhiqiang Y, Christopher MS, Kenneth JK. Synthesis of CdSe/ZnS and CdTe/ZnS quantum dots: Refined digestive ripening. Journal of Nanomaterials. 2012;**2012**:12

[14] He Y, Lu HT, Sai LM, Su YY, Hu M, Fan CH, et al. Microwave synthesis of water-dispersed CdTe/CdS/ZnS core-shell-shell quantum dots with excellent photostability and

biocompatibility. Advanced Materials. 2008;**20**(18):3416-3421

[15] Shen YY, Li LL, Lu Q, Ji J, Fei R, Zhang JR, et al. Microwaveassisted synthesis of highly luminescent CdSeTe@ZnS-SiO2 quantum dots and their application in the detection of Cu(II). Chemical Communications. 2012;**48**:2222-2224

[16] Kim H-R, Bong J-H, Park J-H, Song Z, Kang M-J, Son DH, et al. Cesium lead bromide (CsPbBr3) perovskite quantum dot-based photosensor for chemiluminescence immunoassays. ACS Applied Materials & Interfaces. 2021;**13**(25):29392-29405

[17] Preeti S, Sarita Y, Patel MS, Pramendra K, Naresh K, Lokendra K. The effects of cesium lead bromide quantum dots on the performance of copper phthalocyanine-based organic field-effect transistors. Nanotechnology. 2021;**32**(19):195208

[18] Tianhao W, Xiao L, Xinhui L, Xuesong L, Danyu C, Yanbo W, et al. Lead-free tin perovskite solar cells. Joule. 2021;**5**(21):863-886

[19] Mahboubeh D, Ali R, Sanjay M, Axel K. Trap engineering in solution processed PbSe quantum dots for high-speed MID-infrared photodetectors. Journal of Materials Chemistry C. 2019;7(4):5658-5669. DOI: 10.1039/C8TC06093B

[20] Nandan Y, Mehata MS. Wavefunction engineering of type-I/type-II excitons of CdSe/CdS Core-Shell quantum dots. Scientific Reports. 2019;**9**(2):1-11. DOI: 10.1038/s41598-018-37676-3

[21] Quan X, Yingchun N, Jiapeng L, Ziji Y, Jiajia G, Lan D, et al. Recent progress of quantum dots for energy storage applications. Carb Neutrality. 2022;**1**(13):1-18. DOI: 10.1007/s43979-022-00002-y

[22] Hsiao TK, van Diepen CJ, Mukhopadhyay U, Reichl C, Wegscheider W, Vandersypen LMK. Efficient orthogonal control of tunnel couplings in a quantum dot Array. Physical Review Applied. 2020;**13**(5):054018. DOI: 10.1103/PhysRevApplied.13.054018

[23] Chen S, Yating W, Jennifer S, Eamonn H, Robert H, Kunal M, et al. Perspectives on advances in quantum dot lasers and integration with Si photonic integrated circuits. ACS Photonics. 2021;**8**(9):2555-2566. DOI: 10.1021/acsphotonics.1c00707

[24] Ravitej U, Hans TE, Henri T, Aslı DU, Ying W, Sven S, et al. On-chip deterministic operation of quantum dots in dual-mode waveguides for a plug-and-play single-photon source. Nature Communications. 2020;**11**:3782

[25] Pregnolato T, Chu XL, Schröder T, Schott R, Wieck AD, Ludwig A, et al. Deterministic positioning of nanophotonic waveguides around single self-assembled quantum dots. APL Photonics. 2020;**5**(7):086101. DOI: 10.1063/1.5117888

[26] Chistyakov AA, Zvaigzne MA, Nikitenko VR, Tameev AR, Martynov IL, Prezhdo OV. Optoelectronic properties of semiconductor quantum dot solids for photovoltaic applications. Journal of Physical Chemistry Letters. 2017;**8**(17):4129-4139. DOI: 10.1021/acs.jpclett.7b00671

[27] Ramiro I, Özdemir O, Christodoulou S, Gupta S, Dalmases M, Torre I, et al. Mid- and long-wave infrared optoelectronics via Intraband transitions in PbS colloidal quantum

dots. Nano Letters. 2020;**20**:1003-1008. DOI: 10.1021/acs.nanolett.9b04130

[28] Izadyar SM, Razaghi M, Hassanzadeh A. Quantum dot semiconductor optical amplifier: Investigation of ultra-fast cross gain modulation in the presence of a second excited state. Journal of Physics D: Applied Physics. 2020;**53**:35

[29] Bang SY, Suh Y-H, Fan X-B, Shin D-W, Lee S, Choi HW, et al. Technology progress on quantum dot light-emitting diodes for next-generation displays. Nanoscale Horizontal. 2021;**6**(2):68-77. DOI: 10.1039/D0NH00556H

[30] Asgari M, Coquillat D, Menichetti G, Zannier V, Diakonova N, Knap W, et al. Quantum-dot single-Electron transistors as thermoelectric quantum detectors at terahertz frequencies. Nano Letters. 2021;**21**(20):8587-8594. DOI: 10.1021/acs.nanolett.1c02022

[31] Yen TJ, Chin A, Gritsenko V. High-performance top-gate thin-film transistor with an ultra-Thin Channel layer. Nanomaterials. 2020;**10**(11):2145. DOI: 10.3390/nano10112145

[32] Elshaari AW, Pernice W, Srinivasan K, Benson O, Zwiller V. Hybrid integrated quantum photonic circuits. Nature Photonics. 2020;**14**:5

[33] Gritchenko AS, Eremchev IY, Naumov AV, Melentiev PN, Balykin VI. Single quantum emitters detection with amateur CCD: Comparison to a scientific-grade camera. Optics & Laser Tech. 2021;**143**(11):107301. DOI: 10.1016/j.optlastec.2021.107301

[34] Rubino A, Caliò L, Calvo ME, Míguez H. Ligand-free MAPbI3 quantum dot solar cells based on nanostructured insulating matrices. Solid. 2021;**5**:2100204. DOI: 10.1002/solr.202100204

[35] Li L. Rectifying the output of vibrational piezoelectric energy harvester using quantum dots. Scientific Reports. 2017;**7**(3):44859. DOI: 10.1038/srep44859

[36] Cao J, Yan C, Luo C, Li W, Zeng X, Xu Z, et al. Cryogenic-temperature thermodynamically suppressed and strongly confined CsPbBr3 quantum dots for deeply blue light-emitting diodes. 2021;**9**(17):2100300

[37] Schupp FJ, Vigneau F, Wen Y, Mavalankar A, Griffiths J, Jones GAC, et al. Sensitive radiofrequency readout of quantum dots using an ultra-low-noise SQUID amplifier editors-pick. Journal of Applied Physics. 2020;**127**(2):244503. DOI: 10.1063/5.0005886

[38] Wang I-H, Hong P-Y, Peng K-P, Lin H-C, George T, Li P-W. Germanium quantum-dot Array with self-aligned electrodes for quantum electronic devices. Nanomaterials. 2021;**11**:2743. DOI: 10.3390/nano11102743

[39] Mi Y, Niu Y, Ni H, Zhang Y, Wang L, Liu Y, et al. Gecko inspired reversible adhesion via quantum dots enabled photo-detachment. Chemical Engineering Journal. 2022;**431**(2):134081. DOI: 10.1016/j.cej.2021.134081

[40] Zhang W, Li T, An R, Wang J, Tian Y. Delivering quantum dots to lubricants: Current status and prospect. Friction. 2022;**10**(4):1-21. DOI: 10.1007/s40544-021-0591-4

[41] Sivasubramani S, Mattela V, Pal C, Acharyya A. Dipole coupled magnetic quantum-dot cellular automata-based efficient approximate nanomagnetic subtractor and adder design approach.

Nanotechnology. 2020;**31**(2):025202. DOI: 10.1088/1361-6528/ab475c

[42] Koole R, Mulder WJM, van Schooneveld MM, Strijkers GJ, Meijerink A, Nicolay K. Magnetic quantum dots for multimodal imaging. Wiley Interdisciplinary Reviews. Nanomedicine and Nanobiotechnology. 2009;**1**(5):475-491. DOI: 10.1002/wnan.14

[43] Greve KD, Yu L, McMahon PL, Pelc JS, Natarajan CM, Kim NY, et al. Quantum-dot spin-photon entanglement via frequency downconversion to telecom wavelength. Nature. 2012;**491**(7424): 421-425. DOI: 10.1038/nature11577

[44] Tweedie M, Subramanian R, Craig I, McAdams ET, McLaughlin JA. Fabrication of impedimetric sensors for label-free point-of-care immunoassay cardiac marker systems, with microfluidic blood flow delivery, and results telemetry to PDA. Conference Proceedings: Annual International Conference of the IEEE Engineering in Medicine and Biology Society. 2006;**1**:4610-4614

[45] Kalkal A, Kadian S, Pradhan R, Manik G, Packirisamy G. Recent advances in graphene quantum dot-based optical and electrochemical (bio) analytical sensors. Materials Advance. 2021;**2**(7):5513-5541. DOI: 10.1039/D1MA00251A

[46] Lin WMM, Yazdani N, Yarema O, Yarema M, Liu M, Sargent EH, et al. Recombination dynamics in PbS nanocrystal quantum dot solar cells studied through drift–diffusion simulations. ACS Applied Electronic Materials. 2021;**3**(11):4977-4989. DOI: 10.1021/acsaelm.1c00787

[47] Berdimurodov E, Verma DK, Kholikov A, Akbarov K, Guo L. The recent development of carbon dots as powerful green corrosion inhibitors: A prospective review. Journal of Molecular Liquids. 2022;**349**(3):118124. DOI: 10.1016/j.molliq.2021.118124

[48] Mirzaei A, Kordrostami Z, Shahbaz M, Kim J-Y, Kim HW, Kim SS. Resistive-based gas sensors using quantum dots: A review. Sensors (Basel). 2022;**22**(12):4369. DOI: 10.3390/s22124369

[49] Auciello O, Renou S, Kang K, Tasat D, Olmedo DA. Biocompatible Ultrananocrystalline diamond (UNCD) coating for a new generation of dental implants. Nanomaterials. 2022;**12**:(02):782. DOI: 10.3390/nano12050782

[50] Hasannejadas B, Janbaz FP, Choupani E, Fadaie M, Hamidinejad MA, Ahmadvand D. Quantum dots application in neurodegenerative diseases. Thrita. 2021;**9**(2):e100105. DOI: 10.5812/thrita.100105

[51] Panmand RP, Tekale SP, Daware KD, Gosavi SW, Jha A, Kale BB. Characterisation of spectroscopic and magneto-optical faraday rotation in Mn^{2+}- doped CdS quantum dots in a silicate glass. Journal of Alloys and Compounds. 2020;**817**(3):152696. DOI: 10.1016/j.jallcom.2019.152696

[52] Saito K, Inoue C, Ikegawa J, Yamazaki K, Goto S, Takamura M, et al. Effects of size and distribution of Spheroidized cementite on void initiation in the punched surface of medium-carbon steel. Metallurgical and Materials Transactions A. 2020;**51**(9):4499-4510. DOI: 10.1007/s11661-020-05854-6

[53] Zhai L, Löbl MC, Nguyen GN, Alisa Javadi RJ, Spinnler C, Wieck AD, et al. Low-noise GaAs quantum dots for quantum photonics. Nature

Communications. 2020;**11**(09):4745. DOI: 10.1038/s41467-020-18625-z

[54] Chen X, Xiong Z, Chen M, Zhou P. Ultra-stable carbon quantum dot nanofluids for direct absorption solar collectors. Solar Energy Materials and Solar Cells. 2022;**240**(15):111720. DOI: 10.1016/j.solmat.2022.111720

[55] Klein M, Steenhusen S, Löbmann P. Inorganic-organic hybrid polymers for printing of optical components: From digital light processing to inkjet 3D-printing. Science and Technology. 2022;**101**(3):649-654. DOI: 10.1007/s10971-022-05748-6

[56] Das HT, Barai P, Dutta S, Das N, Das P, Roy M, et al. Polymer composites with quantum dots as potential electrode materials for supercapacitors application: A review. Polymers (Basel). 2022;**14**(5):1053. DOI: 10.3390/polym14051053

[57] Chen G-H, Yang S-H, Yeh C-W, Ho S-J, Liu M-C, Chen H-S. Polycarbonate light guide plates with embedded quantum dots fabricated by large-scale injection moulding for wide colour gamut displays. Materials & Design. 2021;**201**(3):109504. DOI: 10.1016/j.matdes.2021.109504

[58] Liu ZJ, Lin C-H, Hyun B-R, Sher C-W, Lv Z, Luo B, et al. Micro-light-emitting diodes with quantum dots in display technology. Light: Science & Applications. 2020;**9**:83. DOI: 10.1038/s41377-020-0268-1

[59] Onal A, Eren GO, Sadeghi S, Melikov R, Han M, Karatum O, et al. High-performance white light-emitting diodes over 150 lm/W using near-Unity-emitting quantum dots in a liquid matrix. ACS Photonics. 2022;**9**(4):1304-1314. DOI: 10.1021/acsphotonics.1c0180

[60] Liu L, Najar A, Wang K, Du M, Liu S. Perovskite Quantum Dots in Solar Cells. 2022;**9**(7):2104577. DOI: 10.1002/advs.202104577

[61] Khan J, Ullah I, Yuan J. CsPbI3 perovskite quantum dot solar cells: Opportunities, progress and challenges. Materials Advance. 2022;**3**(12):1931-1952. DOI: 10.1039/D1MA01075A

[62] Wagner AM, Knipe JM, Orive G, Peppas NA. Quantum dots in biomedical applications. Acta Biomaterialia. 2019;**94**(8):44-63. DOI: 10.1016/j.actbio.2019.05.022

[63] Krause P, Tremblay JC, Bande A. Atomistic simulations of laser-controlled exciton transfer and stabilization in symmetric double quantum dots. The Journal of Physical Chemistry. A. 2021;**125**(22):4793-4804. DOI: 10.1021/acs.jpca.1c02501

[64] Tajik S, Dourandish Z, Zhang K, Beitollahi H, Le QV, Jang HW, et al. Carbon and graphene quantum dots: A review on syntheses, characterization, biological and sensing applications for neurotransmitter determination. RSC Advances. 2020;**10**(04):15406-15429. DOI: 10.1039/D0RA00799D

Chapter 2

Quantum Dots as Material for Efficient Energy Harvesting

Paweł Kwaśnicki

Abstract

The essence of the photovoltaic effect is the generation of electric current with the help of light. Absorption of a quantum of the energy of light (photon) generates the appearance of an electron in the conduction band and holes in the valence band. The illumination of the material, in general, is not uniform, which leads to the appearance of spatially inhomogeneous charge in the band valence and conductivity. Besides, electrons and holes generally diffuse with different velocities, which leads to the creation of a separated space charge and generation of an electric field (sometimes called the Dember field). This field inhibits further separation of cargo. The reverse processes also take place in the system, i.e. electron recombination and holes. These processes are destructive from the point of view of photovoltaics and should be minimized, which is achieved; thanks to the spatial separation of electrons and holes. The point is that electrons and holes were carried away from the area where they formed as quickly as possible, yes to prevent their spontaneous recombination. The use of semiconductor quantum dots introduced into the photoelectric material is currently a very important and effective way to increase the efficiency of photoelectric devices and photovoltaic cells. This is due to the fact that in semiconductor photoelectric materials with no quantum dots, there is always some upper limit of the wavelength $\lambda_{gr}[gr] \simeq 1, 24/E_g[eV]$ for absorbed light, above which the light is not absorbed.

Keywords: quantum dots, photovoltaic, QDSC, transparent PV

1. Introduction

The maximum coefficient of light to electric conversion in semiconductor systems is determined by the so-called Shockley-Queisser limit, which is approximately 32% with the optimal width of the band gap of 1.2–1.3 eV. However, performance can be improved by using solutions based on semiconductor nanostructures [1, 2]. Such cells are called Third Generation Photovoltaic Cells (KFTG). One type of nanostructure that plays an important role is the so-called quantum dots, that is, small semiconductor grains with sizes in the order of nm. As bounded systems in all three dimensions, quantum dots are characterized by a discrete energy spectrum. The advantage of such systems is the fact that due to the lack of translational symmetry, the limitation on quantum transitions resulting from the behaviour of the wave vector (equality of the photon and electron wave vector) is removed. Exclusion of this limitation leads to the

IntechOpen

fact that absorption can arise from deep discrete levels to high energy states. As a result, the useful range of the light spectrum expands considerably towards violet light. However, the width of the energy gap of the output semiconductor limits the possibility of using the spectral part from the low-energy light side [3].

Another beneficial element resulting from the location of an electron and a hole (exciton) in the area of a quantum dot is the slowing down of thermalization processes and increasing the effective number of generated carriers through transitions in which the exciton while relaxing to lower energy, generates another exciton (the phenomenon of multiple exciton generation) [1]. Thermalisation occurs when carriers transfer their excess energies to the crystal lattice through interaction with phonons. In this way, large populations of non-equilibrium (hot) charges are created and the carriers lose their energy to the lattice vibrations. Thermalisation times depend on many factors such as carrier concentration and lattice temperature but are usually in the range below 100 ps. The multiple generations of an exciton can only occur for high-energy photons, due to the conservation of energy principle. This effectively means that one photon can generate two (or more) carriers. Of course, there are also recombination processes in which the exciton generated in the photon absorption process disappears. All these factors mean that the performance of such systems may exceed the Shockley-Queisser limit.

Numerous types of quantum dots can be used in photovoltaics: semiconductor polycrystalline and granular materials, quantum dots obtained by epitaxial methods or from colloidal solutions, nanoparticles of organic dyes, etc. There are also many possibilities for the architecture of photovoltaic cells. Their common feature is that the phenomenon of multiple exciton excitation in dots is used, and the generated charges (electrons and holes) are discharged in various ways to the electrodes while ensuring their spatial separation. One possibility is to use dots dispersed in the conductive material (e.g. in organic polymers). With the appropriate concentration of the dots, the discharge of the charge from the dots to the electrodes can take place due to the coupling between the quantum dots. In the case of regular networks of dots (one, two or three-dimensional), discrete dot states are formed into mini-electron bands, ensuring charge transport. This problem has been and is still widely studied in the literature [4, 5].

Photovoltaic cells using regular quantum dot networks and their electronic mini-band structure (also called intermediate bands) have become one of the significant directions of photovoltaics development [6]. The essence of this type of solution is the fact that in the area between the electrodes in the p-n junction there is a layer with quantum dots between which the distance is so small that an intermediate band is created in this area. This allows the use of low-energy photons (with energy lower than the width of the output semiconductor gap) to generate electrons in the conduction band and holes in the valence band. This is due to the optical transitions from the valence band to the intermediate band and from the intermediate band to the conduction band. An important element is also the fact that recombination processes are much less likely in the case of the intermediate band than in the case of isolated quantum dots. In this case, it is enough for the wave functions of the dots to be quite delocalized. This can be achieved in systems with complexes of quantum dots instead of regular lattices [7].

2. A single quantum dot

First, consider a single semiconductor quantum dot. In general, the quantization associated with the limited size of the quantum dot leads to discrete energy values,

DOI: http://dx.doi.org/10.5772/intechopen.106579

ε_n (n = 1,2,3 ...). Each eigenvalues value corresponds to an orbital wave function ψ_n (**r**). For the sake of simplicity, we ignore the magnetic and spin-orbital interactions in this consideration. Then, we deal with a two-fold spin degeneration of all discrete levels of a quantum dot. However, if there are spin-orbital or other magnetic forces in the system, then this degeneration is generally cancelled out. A diagram of the discrete levels of a semiconductor quantum dot in some other semiconductor matrix is shown in **Figure 1**. The energy gap, both in the matrix and in the material of which the dot is made of, is marked with a dashed line (the bottom of the conductivity band and the upper limit of the valence band).

Both the own energy values and the corresponding wave functions fulfil the Schrodinger equation:

$$\left[-\frac{\hbar^2}{2m}\left(\frac{\partial^2}{\partial x^2}+\frac{\partial^2}{\partial y^2}+\frac{\partial^2}{\partial z^2}\right)+V(\boldsymbol{r})\right]\psi_n\,(\boldsymbol{r})\equiv\hat{H}\psi_n(\boldsymbol{r})=\varepsilon_n\varnothing_n(\boldsymbol{r}) \qquad (1)$$

where V (**r**) is the limiting potential of the quantum dot, and $\hat{H}$ is the operator of the total energy, that is kinetic and potential energy.

2.1 Two interacting quantum dots

Let us now consider two quantum dots at some distance from each other. Let us assume that the separation of levels on the dots is large enough and consider only one discrete level in a selected dot and the closest discrete level in the second dot. Let us assume that the other levels are sufficiently distant on the energetic scale. Let us denote this level in the first dot as ε_1 and in the second dot as ε_2. If the dots were identical, then $\varepsilon_1 = \varepsilon_2$, but in general these energies may be different. We denote the corresponding normalized wave functions as $\psi_1(\mathbf{r})$ and $\psi_2(\mathbf{r})$. If these dots are close enough and their wave functions partially overlap, then the wave functions of a system composed of two dots can be taken in the form

$$\psi(\mathbf{r}) = a\psi_1(\mathbf{r}) + b\psi_2(\mathbf{r}) \qquad (2)$$

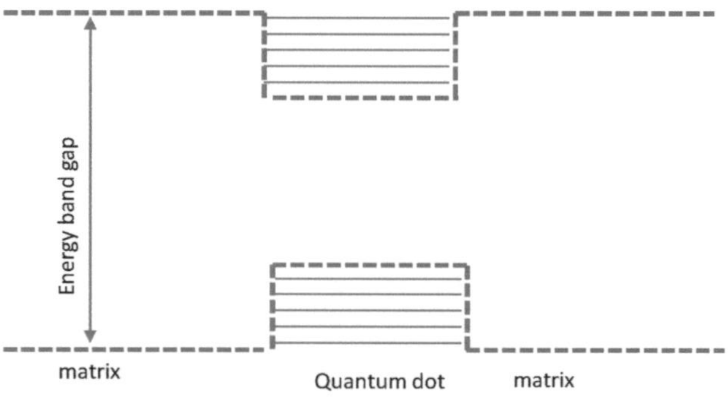

Figure 1.
A diagram of the energy structure of a single semiconductor quantum dot placed in a matrix made of another semiconductor material. The dashed line corresponds to the location of the bottom of the conduction band and the upper limit of the valence band in the semiconductor matrix and in the material of which the semiconductor quantum dot is made. The solid lines represent the discrete energy levels of the quantum dot.

where *a* and *b* are constants. When the wave functions are not too large, one can put the normalization condition $|a|^2 + |b|^2 = 1$. The energies of the system, E= $\langle \hat{H} \rangle$, can then be written in the form

$$E = \varepsilon_1 + \varepsilon_2 + a * b \int dr \psi_1^*(r) V_1(r) \psi_2(r) + ab^* \int dr \psi_2^*(r) V_2(r) \psi_1(r) \tag{3}$$

where $V_1(r)$ and $V_2(r)$ are the limiting potentials for both quantum dots.

If the dots are the same, so they have the same limiting potential, $V_1(\mathbf{r}) = V_2(\mathbf{r}\text{-}\mathbf{d}) \equiv V(\mathbf{r})$, where d is the vector connecting the dots 1 and 2, and the same levels energy $\varepsilon_1 = \varepsilon_2 \equiv \varepsilon$, and the same corresponding wave functions, $\psi_1(\mathbf{r}) = \psi_2(\mathbf{r}\text{-}\mathbf{d}) \equiv \psi(\mathbf{r})$, then the energies of the system can be written in the form

$$E = 2\varepsilon + (a * b + ab *) \int dr \psi^*(r) V(r) \psi(r - \boldsymbol{d}) \tag{4}$$

The state corresponding to a = b ($|a| = |b|$ $1/\sqrt{2}$) is a symmetric state with energy ε_2- t and the antisymmetric state, a = −b, is a state with energy ε_2 + t, where

$$t = -\int dr \psi^*(r) V(r) \psi(r - \boldsymbol{d}) \tag{5}$$

The hoping parameter t is positive because V (r) as the attracting potential is negative. The symmetric state is the binding (lower energy) state, while the antisymmetric state is the anti-binding (higher energy) state.

2.2 Periodic networks of quantum dots

Now, let us consider the periodic networks of quantum dots and first consider the one-dimensional periodic chain of equal quantum dots. The wave function then satisfies Bloch's theorem, $\psi(\mathbf{r} + \mathbf{d}) = \psi(\mathbf{r})$ exp. iQ•d where Q is a wave vector, Q = (Q, 0, 0) and d = (d, 0, 0). Consider a chain with N dots and assume periodic Born-Karman conditions. From the considerations, so far, it follows that the energies of such a chain can be written in the form

$$E = \sum_l \varepsilon_l + \frac{1}{2} \sum_l (a_l a_{l+1} t_{l,l+1} + a_l a_{l-1} t_{l,l-1}) \tag{6}$$

where a_l is the amplitude of the wave function at node l, l = 1, ... N, and summing over l means summing over all N nodes. Factor 1/2 in the second term eliminates double counting of hoping terms. In turn $t_{l,\, l \pm 1}$ means the element hoping to the nearest neighbours. It follows from Bloch's theorem that $a_{l \pm 1} = a_l$ exp. ± iQd. Considering the constant value of the hoping term, $t_{l,\, l \pm 1}$ = t, and the constant value of the discrete energies, $\varepsilon_l = \varepsilon$, the total energy can be written as

$$E = N\varepsilon + \mathrm{tcosQd} \tag{7}$$

The one-dimensional wave vector takes the following values: Q = (2π/d) l for l = 0, ± 1, · ± N. The electronic states thus form an energy band with a width of 2 t. This can be interpreted in such a way that, in the case of the interacting dots, the discrete levels

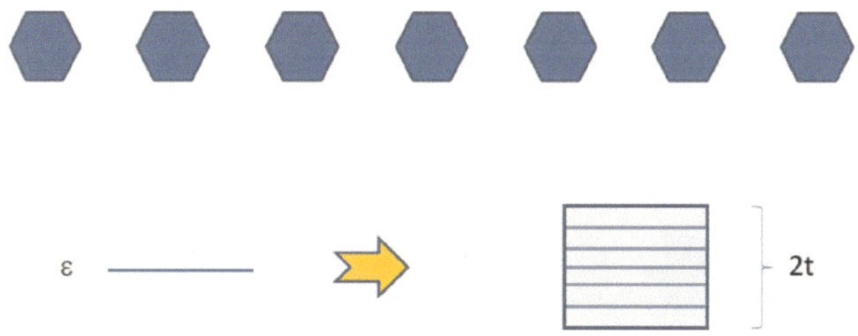

Figure 2.
Diagram of the energy structure of a one-dimensional network of interconnected quantum dots. In the case of a large number of dots in the chain, a continuous energy band with a width of 2 t is created, corresponding to the shaded area.

blur to form an energy band, as shown schematically in **Figure 2**. This figure shows a finite (and relatively small) number of quantum dots, so the energy levels of the entire system are a discrete structure. The original degeneracy is abolished. However, in the case of a long chain, the states resulting from the interaction between the dots create a continuous (at the limit of an infinitely long one-dimensional network) energy band with a width of 2 t. In **Figure 2**, this band is marked with a shaded area in which discrete levels are visible.

In the case of a two-dimensional rectangular quantum dot network, where the interaction between the dots is the same for all four closest neighbours, the bandwidth resulting from the interaction is greater due to the greater number of the nearest neighbours and amounts to 4 t. This is shown schematically in **Figure 3**. Similar to **Figure 2**, the energy structure diagram is shown for a finite number of quantum dots (49 to be exact). The energy band resulting from the discrete levels corresponds to the shaded area. The discrete structure resulting from a finite number of dots is now virtually indistinguishable in this figure. In the case of a network with $N \times N$ dots, the band becomes continuous for large values of N (more precisely for $N \rightarrow \infty$).

Similarly, the three-dimensional network of quantum dots can be described. As in the above-discussed cases of the one-dimensional and two-dimensional dot network, the interaction between the dots leads to the formation of an energy band from the discrete level of individual dots. However, the width of this band in the three-dimensional case is greater and amounts to 6 t for the cubic quantum dot network.

In summary, discrete levels in periodic quantum dot networks blur to form an energy band. The width of this band depends on the value of the hoping parameter t, and it decreases to zero with $t \rightarrow 0$. Besides, the bandwidth also depends on the number of the nearest neighbours, which is 2 t for a one-dimensional chain, 4 t for a two-dimensional square lattice, and 6 t for a three-dimensional cubic lattice.

An important aspect of the resulting band structure is the fact that the electronic states are stretched as opposed to the localized states of a single dot. Electrons and holes are therefore mobile in these bands and can be led relatively easily to the appropriate electrode, even if their generation takes place at a considerable distance. Moreover, the value of the parameter t depends on the wave function of the given energy level of the quantum dot. Therefore, the widths of the mini-bands resulting from the levels of a quantum dot can vary significantly from level to level.

2.3 Non-periodic systems

Let us now consider non-periodic systems in which quantum dots of various shapes and sizes are arranged randomly. In such systems, the connection between the

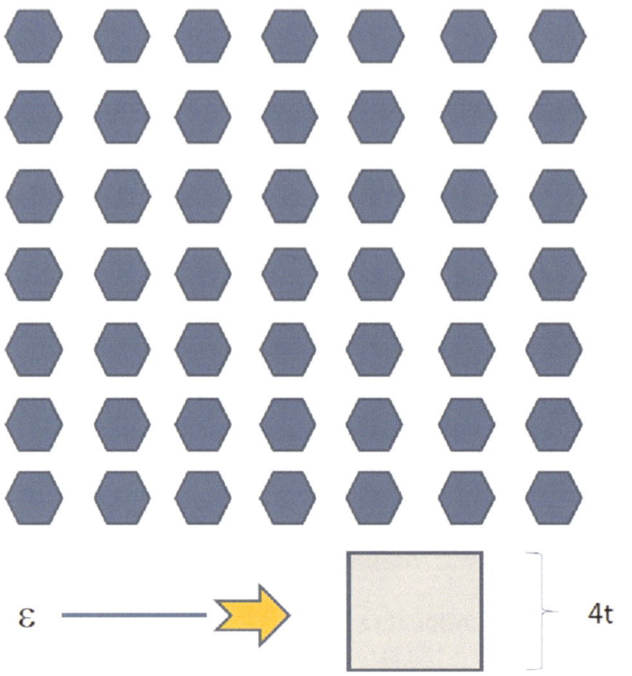

Figure 3.
Diagram of the electronic structure in the case of a two-dimensional network of interconnected quantum dots. In the case of a large (unlimited) two-dimensional network, a continuous 4 t wide energy band is created (discrete structure from a finite number of dots is now virtually indistinguishable).

dots is also described by random hoping parameters. Moreover, if the dots are arranged quite rarely, then only some of them are connected together. In the simplest case, such a system includes complexes of two or three interconnected quantum dots. In the simplest model system, the complexes of double dots are arranged randomly, and it can be assumed that the coupling parameter between two dots and the eigen energies of individual dots in the complex are statistically different. If the energies of states in dot 1 and 2 are ε_1 and ε_2, respectively, and the coupling parameter is t, then the eigen energies of the interacting complex, ε_- for the bonding state and ε_+ for the anti-bonding state, are respectively

$$\varepsilon_{\pm} = \frac{1}{2}(\varepsilon_1 + \varepsilon_2) \pm \frac{1}{2}\sqrt{(\varepsilon_1 - \varepsilon_2)^2 + 4t^2} \tag{8}$$

The appropriate wave functions are located on the dots and in the space between the dots. If the colon complexes are different, then a spectrum of discrete levels located in different areas of the matrix is effectively obtained.

In the case of triple complexes, in which three dots are connected together to form one complex, the coupling is described by two hoping parameters, and the interaction result produces three different discrete levels located on the complex and in the space between the dots.

In the case of networks with relatively small deviations from the translational symmetry, the interaction between the dots leads to the formation of a continuous band. However, the corresponding wave functions may be localized. The length of the location, that is, figuratively speaking of the number of quantum dots in the area of which the wave function related to a given electron state extends, may be large enough, comparable to the size of the structure, so effectively such states meet the mobility conditions needed for pairing electrons and holes.

3. Two-dimensional cells with chains of quantum dots

Let us now consider examples of possible planar photovoltaic cells containing one-dimensional networks of quantum cells. The simplest such structure is shown in **Figure 4**. Quantum dots are embedded in a thin layer of optically active semiconductor, which form one-dimensional chains. The distance between the strings is so long that the interaction between dots from different strings can be ignored. This interaction is non-zero only for the adjacent dots in a given string. If the quantum dots within one chain are the same, then the interaction between the dots generates energy bands from discrete levels of quantum dots. The electrons and holes generated in a certain quantum dot can then freely flow to the appropriate electrode (collecting electrons or holes). This is possible because the states in the electron bands made of discrete dot levels are conductive. If both electrodes were the same, then there would be no such separation. Note that the condition is the formation of bands from discrete dot levels, however, these bands may be different within different chains. Thus, a more general version of the photovoltaic cell shown in **Figure 4** would be a cell in which each of the strings would be composed of other quantum dots (however, the same in the given chain).

However, even in this case, the same electrodes can be achieved by means of appropriate modulation of quantum dots. Besides, such a modified arrangement of dots should also strengthen the separation in the case of two different electrodes (as in **Figure 4**). Well, if quantum dots in a given chain will gradually change their sizes (in any direction), as shown schematically in **Figure 5**, then the energies of a given dot level will gradually increase for electrons and decrease for holes (or vice versa). In **Figure 5**, shown quantum dots are stretched in one direction but one can consider different direction. The change of dot parameters is so small that the shift of a given level is much smaller than the distance between the nearest discrete levels. Then the

Figure 4.
Diagram of a planar photovoltaic cell with single-winding chains of the same quantum dots. Red- electron transporting electrode, green- hole transporting electrode.

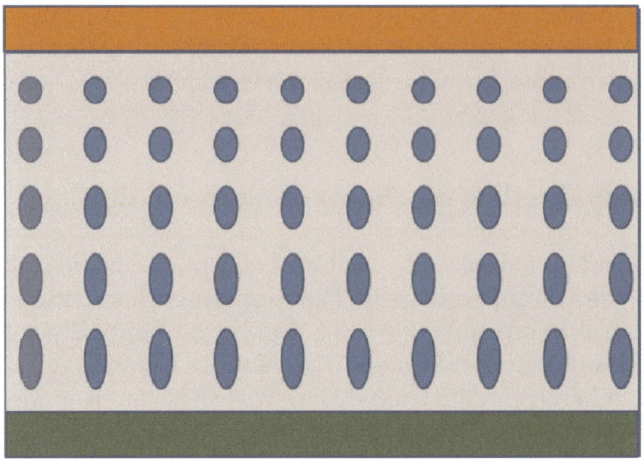

Figure 5.
Diagram of a planar photovoltaic cell with single-axis chains of quantum dots, the parameters of which (e.g. sizes) gradually change. Red- electron transporting electrode, green- hole transporting electrode.

electrons will go to the dot with a lower energy level (losing excess energy in thermalization processes), while the holes will be transported from the given dot to the dot with higher energy, which effectively leads to electron-hole separation. As before, a more general situation can be considered in which individual one-dimensional strings are different.

4. Relaxation processes for energy and spin in quantum dots important for photovoltaics applications

Possible applications of quantum dots in photovoltaics [6] call for the studies and understanding of their optical and transport properties. These properties are either controlled or strongly related to the energy and spin relaxation processes in these systems. The energy relaxation is important since it determines transport properties of the photovoltaic devices such as the probability of electron escape from the quantum dot to the glass layer and the transparent conductive oxide and, as a result, the integral photovoltaic efficiency. The spin relaxation is important since it is directly related to the coupling mechanisms responsible for the energy relaxation and allows one to find the efficient mechanisms of both the energy and the spin relaxation. In addition, the spin states are critically important for the formation of the spectra of quantum dots with several carriers (electrons and/or holes) and, thus, spin relaxation can determine the photovoltaic efficiency of the charged quantum dots in the regime of strong light irradiation, where the light intensity is sufficiently strong to produce more than one carrier per quantum dot [8].

There are different types of quantum dots applicable in photovoltaics. Let us consider ellipsoidal quantum dots serving as optical absorbers and, subsequently, source of the photo excited electrons upon their escape from the dot. The main source of the energy relaxation in single-electron quantum dots is due to various types of electron-phonon coupling, which will be analysed below. In addition, we will analyse the mechanisms of the energy relaxation in multi-electron quantum dots taking into account the electron-electron interaction. We will demonstrate that a novel mechanism of spin relaxation, different from the conventional admixing mechanism, can be relevant for the quantum dots based on III-V group materials. Let us begin with a model of a semiconductor quantum dot applicable in photovoltaics. These dots have a variety of different properties related to their sizes and shapes. Although this variety extends the ability to use the quantum dots for nonmonochromatic light sources such as sunlight, on the other hand, it decreases the controllability of their applications. Direct system-dependent analysis of properties of quantum dots is difficult and can basically be done with the effective mass approximation as included in the Hamiltonian

$$H_0 = \frac{p^2}{2m^*} + U_e(r). \tag{9}$$

Where $\frac{p^2}{2m^*}$, is the electron kinetic energy, p is the electron momentum, m∗ is the electron effective mass, and U (r) is the effective confining potential of the quantum dot. Frequently, for model calculations the confinement is taken in the form of an anisotropic oscillator:

$$U_e(r) = 2m^* \left[\Omega_{\|}^2 (x^2 + y^2) + \Omega_z^2 z^2\right]/2, \tag{10}$$

where $\Omega_{\|}$ and Ω_z, are the corresponding frequencies of the anisotropic oscillator. However, this potential is well-suitable only for the description of the low-energy states of the photoexcited electrons, where wavefunction is well-localized near the potential minimum. Another form of the potential is given by $U_e(r)$ = 0 for r inside the quantum dot, $U_e(r)$ = U for r outside the quantum dot, where the inside/outside boundary determines the quantum dot shape. Usual model shapes of quantum dot are ellipsoidal with an example presented in **Figure 6**. The typical scale of the quantum kinetic energy is determined by the z − axis extension of the quantum dot ω as $\hbar^2 m^* \omega^2$ with the corresponding quantity of the order of 10 meV for $\omega \sim 10$ nm and m^* of the order of 0.1 of the free electron mass. For highly excited states, with the energies above the excitation threshold of the order of 100 meV, but less than the confinement potential U, a classical description in terms of the electron trajectories becomes possible. Here emission of phonons leads to the relaxation to the lower-energy quantum states. To study the relaxation, we present electron-phonon coupling Hamiltonian (assuming the crystal volume ≡ 1) in the form:

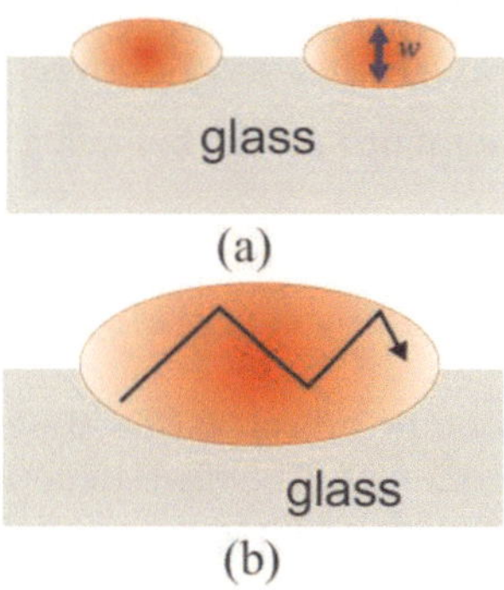

Figure 6.
(a) Two model ellipsoidal quantum dot on the surface of a glass. The vertical size of the quantum dot is denoted as w, typically of the order of 10 nm. (b) Classical trajectory of highly photoexcited electron in the quantum dot.

$$V_{e-ph} = \frac{D\sqrt{\hbar}}{\sqrt{2\zeta c}} \sum_q \sqrt{q}(de)\left(e^{-iqr} a_q^\dagger + H.c\right) \tag{11}$$

Where $a_q^\dagger$ is the creation operator for the phonon with momentum q, D is the deformation potential (D = −5.5 eV in GaAs), ζ is the crystal density, d = q/q is the phonon propagation direction, e is the phonon polarization, c is the speed of the longitudinal sound mode, and summation is taken over the momenta. We have chosen a single longitudinal phonon branch with (de) = 1, since for the transverse branches (de) = 0, taking into account only the strongest interaction with the deformational potential of the acoustic phonons.

There are different regimes of energy relaxation. The first regime is relevant for the highly-excited semiclassical states, leading to their energy loss and subsequent quantization with localization in the quantum states. The second regime is relevant for the localized states, where phonon-induced transitions occur between the low-energy quantum states in the dots. We begin with the most important for the photovoltaics semiclassical regime, where the electron states can be presented as plane waves. For the relaxation of these states shown in **Figure 6(b)** one can use the classical Boltzmann equation based on Fermi's golden rule. The energy-dependent phonon emission rate 1/τ (E) per single phonon mode with wavevector q leading to the energy relaxation is given by this rule as [9]:

$$\frac{1}{\tau(E)} = \frac{2\pi D^2 \hbar}{\hbar 2\zeta c} \int q\delta\left[E_f - E_i + \hbar\Omega(q)\right] \frac{d^3 k_f}{(2\pi)^3} \tag{12}$$

where the initial and final energy is $E_i = \frac{\hbar^2 k_i^2}{2m^*}$ and $E_f = \frac{\hbar^2 k_f^2}{2m^*}$ respectively, final electron wavevector $k_f = k_i - q$, and δ [..] – function corresponds to the energy conservation with (q) = cq. Taking into account that the characteristic wave vector k_i of a photoexcited electron is of the order of 106 cm^{-1} or higher, at the corresponding phonon momentum its energy $\hbar ck_i$ is of the order of less than one meV, that is 10 K, much less than the electron energy. As a result, the electron scattering is quasi-elastic and the occupation number of the phonons $n_B = \left[\exp\left(\frac{\hbar\Omega(q)}{T} - 1\right]^{-1}$, equal to T/ħΩ at the room temperature T, is large. Therefore, the emission probability is greatly enhanced by the factor nB + 1. However, this occupation factor increases the phonon absorption probability as well and these processes partially compensate each other. This process leads to the energy dependence of $\frac{1}{\tau(E)}$ determined by the electron density of states, proportional to $\sqrt{E}$ and by the absolute value of the phonon momentum, also behaving as $\sqrt{E}$ and resulting in:

$$\frac{1}{\tau(E)} \approx \frac{1}{\tau_D} \frac{E}{\hbar\Omega_D} \left(\frac{m^* c^2}{\hbar\Omega_D}\right)^{1/2} \tag{13}$$

where τ_D is the nominal momentum relaxation time Ref. [9] (τ_D = 2.5 ps in GaAs), and D is the Debye phonon frequency. The time-dependence of the energy due to nearly balanced quasi-elastic emission and absorption of phonon is given by: $\frac{dE(t)}{dt} = \frac{\hbar\Omega(E)}{\tau(E)}$, where the energy goes to phonon subsystem. The numerical value of eτ is rather long, being of the order of 10^{-7} s as a result. The resulting time dependence of energy has the form:

$$E(t) = \frac{E(0)}{\left(\sqrt{\frac{E(0)}{2\sqrt{\hbar\tau}}}t + 1\right)^2} \tag{14}$$

For the initial energy E(0) = 100 meV this estimate yields the value $\tau \sim 10$ ps, demonstrating that energy relaxation of photoexcited electrons is fast and can set up the initial condition for the diffusion through the glass layer. Taking into account that the electron velocity is of the order of 10^8 cm/s, electron during this time hits the quantum dot boundary at least 100 times. As a result, the electron can escape from the quantum dot if the escape probability per single boundary collision is higher than 10^{-2}, thus, requiring for a relatively transparent boundary between the quantum dot and the glass (see **Figure 6(a)**).

Quantum dots are often single-electron charged since they can absorb electrons from the bulk of the semiconductors. Therefore, complex states built by a photoexcited electron-hole pair inside the dot and the resident electron, usually called 'trions' Ref. [9], can be formed, as shown in **Figure 7(a)**. The formation of trions extends the absorption range of quantum dots. The spectrum and the structure of a trion strongly depends on electron-electron and electron-hole coulomb interaction between electrons. The Hamiltonian of the system becomes:

$$H_0 = \frac{p_1^2}{2m^*} + \frac{p_2^2}{2m^*} + \frac{P^2}{2m_h^*} + U_e(r_1) + U_e(r_2) + U_h(R) + \frac{e^2}{\epsilon|r_1 - r_2|} + \left[\frac{e^2}{\epsilon|r_1 - R|} + \frac{e^2}{\epsilon|r_2 - R|}\right] \tag{15}$$

where r_1 and r_2 are positions of electrons (with momenta p_1 and p_2, respectively), R is the position of the hole (with mass m_h^* and the momentum P with the confinement potential Uh (R)), and ϱ is the quantum dot material dielectric constant. The process leading to the overall decrease in the escape probability is shown in **Figure 7(b)**. Note that this process reduces the energy of the upper electron and increases the energy of the lower one.

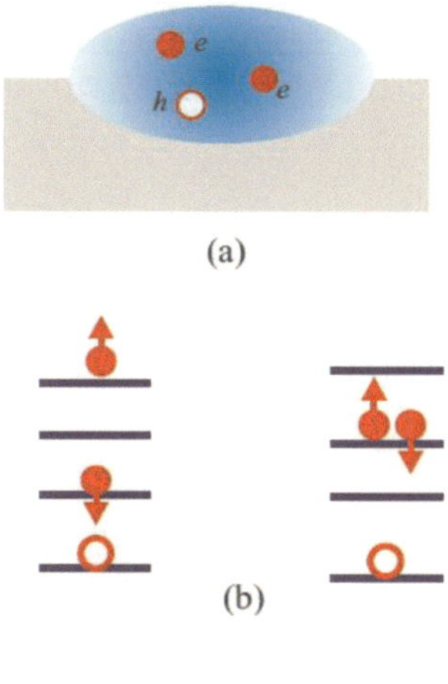

Figure 7.
(a) Trion in a photoexcited quantum dot. Electron positions are r_1 and r_2, and the hole position is R. (b) Interaction-induced transition between initial (left) and final (right) states in a quantum dot. Note that, due to the Pauli principle, this transition is possible only for the singlet electron spin state.

Despite this partial increase, the escape probability, strongly dependent on the electron energy, decreases. The corresponding energy and time scales are given by the following matrix element, where we neglected the exchange effects take only the direct interactions as: $V_C = \int \psi_f^1(r_1)\psi_f^2(r_2)\psi_f^h(R)H_C\psi_i^1(r_1)\left(r_{2)}\psi_i^h(R)d^3r_1d^3r_2d^3R\right.$

where the coulomb term

$$H_C = \frac{e^2}{\epsilon|r_1 - r_2|} + \left[\frac{e^2}{\epsilon|r_1 - R|} + \frac{e^2}{\epsilon|r_2 - R|}\right] \tag{16}$$

The probability of this process is given by the time scale corresponding to the energy scale of the Hamiltonian (16), typically of the order of $\frac{e^2}{\epsilon\omega}$, that is higher than 1 meV. Therefore, these processes are very fast, being, in general, energy-conserving and dependent on the total spin of participating electrons. This energy value corresponds to short time scales of these processes, of the order of a picosecond. Then, on the top of these fast transitions, a relatively slow energy relaxation due to phonons occurs, complicated by the transitions between the localized states in the quantum dots. Yet, the energy relaxation weakly depends on these fast electron transitions since these states are localized and, therefore, the estimate of long $\tau[\mathrm{loc}]$ given in the previous subsection is valid here as well.

Thus, understanding of energy and spin relaxation processes in quantum dots suitable for applications in photovoltaics is crucial. The energy relaxation is due to the phonon emission. It was found the relaxation times of semiclassical states of the order of 1 ps, sufficient for electron escape from the quantum dot and its contribution to the photovoltaic effects. In addition, the redistribution of electrons among the energy levels and subsequent energy evolution can be caused by the electron-electron interactions. Spin relaxation in quantum dots is caused by direct spin-phonon coupling and leads to the spin relaxation times of the order of $1-10^2$ microsecond, being the longest relaxation process in quantum dots.

Acknowledgements

This work was supported by the National Centre for Research and Development under the project No. POIR.01.02.00-00-0265/17-00.

Author details

Paweł Kwaśnicki[1,2]

1 Department of Physical Chemistry and Physicochemical Basis of Environmental Engineering, John Paul II Catholic University of Lublin, Institute of Environmental Engineering in Stalowa Wola, Stalowa Wola, Poland

2 Research and Development Centre for Photovoltaics, ML System S.A., Zaczernie, Poland

*Address all correspondence to: pawel.kwasnicki@mlsystem.pl

References

[1] Lopez ABC, Vega AM, Lopez AL. Next Generation of Photovoltaics: New Concepts. Springer, Berlin: Eds Springer Series in Optical Sciences; 2012. ISBN 139783642233692

[2] Duan L, Hu L, Guan X, Lin C, Wu T. Quantum Dots for Photovoltaics: A Tale of Two Materials. Advanced Energy Materials. 2021;**11**(20):2100354. DOI: 10.1002/aenm.202100354

[3] Nozik AJ. Quantum dot solar cells. Physica E: Low-dimensional Systems and Nanostructures. 2002;**14**:115. DOI: 10.1016/S1386-9477(02)00374-0

[4] Tomic S. Intermediate-band solar cells: Influence of band formation on dynamical processes in InAs/GaAs quantum dot arrays. Physical Rewiev B. 2010;**82**:195321. DOI: 10.1103/PhysRevB.82.195321

[5] Kłos JW, Krawczyk M. Two-dimensional GaAs/AlGaAs superlattice structures for solar cell applications: Ultimate efficiency estimation. Journal of Applied Physics. 2009;**106**:093703. DOI: 10.1063/1.3253584

[6] Luque A, Marti A, Antolin E, Tablero C. Intermediate bands versus levels in non-radiative recombination. Physica B. 2006;**382**:320. DOI: 10.1016/j.physb.2006.03.006

[7] Nozik AJ, Beard MC, Luther JM, Law M, Ellingson RJ, Johnson JC. Semiconductor quantum dots and quantum dot arrays and applications of multiple exciton generation to third-generation photovoltaic solar cells. Chemical Reviews. 2010;**110**:6873. DOI: 10.1021/cr900289f

[8] Gantmakher VF, Levinson YB. Carrier Scattering in Metals and Semiconductors. Amsterdam: North-Holland; 1987. DOI: 10.1002/crat.2170230219

[9] Laurent S, Eble B, Krebs O, Lemaitre A, Urbaszek B, Marie X, et al. Electrical control of hole spin relaxation in charge tunable InAs/GaAs quantum dots. Physical Review Letters. 2005;**94**:147401. DOI: 10.1103/PhysRevLett.94.147401

Chapter 3

Application of Fluorescent CQDs for Enhancing the Performance of Solar Cells and WLEDs

Pawan Kumar, Shweta Dua, Balaram Pani and Geeta Bhatt

Abstract

Carbon quantum dots (CQDs) are emerging as promising materials for applications like flexible or transparent solar cell, white light emitting diodes (WLEDs), etc. due to their low cost, eco-friendliness, substantial absorption coefficient, wide absorption spectrum, tuneable optical properties, good charge transfer/separation ability, good quantum yield and large two-photon absorption cross-section. They have been employed in solar cells as active absorbing layers, electron acceptors/donors, electron sinks, electron transporting layers (ETL), hole transporting layers (HTLs), dopants, and interlayer spacing. Consequently, such solar cells have exhibited enhanced performance. In contrast to commercial rare-earth phosphors and traditional semiconductor quantum dots (SQDs) (usually toxic), CQDs exhibit wide range of emission characteristics [full width half maxima (FWHM) > 80 nm]. Interestingly, the emission characteristics of these nanomaterials are tuneable which makes them suitable for WLEDs applications. Red-CQDs are gaining importance as they are required to realize the warm WLEDs. Though a lot of work has been done to modulate the properties of CQDs in order to enhance the performance of solar cells and WLEDs, there are immense possibilities to further exploit the potential of CQDs in these applications.

Keywords: carbon quantum dots (CQDs), solar cell, white light emitting diodes (WLEDs)

1. Introduction

Carbon quantum dots (CQDs) are tiny (2–10 nm) quasi-spherical fluorescent nanoparticles having a core-shell structure. It consists of sp^2 and/or sp^3 carbon domains (core) and oxygen and nitrogen-rich functional groups (shell) [1]. CQDs are synthesised by either bottom-up or top-down approaches; however, the bottom-up approach dominates due to being cost-effective and environmentally friendly.

CQDs were accidentally discovered by Xu et al. in 2004 [2] while segregating single-wall carbon nanotubes from carbon soot using gel electrophoresis. CQDs are very popular for their size, fluorescence, water solubility, biocompatibility, chemical inertness, photo-stability, eco-friendliness, and easy and fast synthesis. They have

IntechOpen

been widely used in numerous applications like solar cells, WLEDs, drug delivery, bio-imaging, sensing, data encryption, etc. [1, 3, 4].

In addition, CQDs possess excellent optical properties that can be easily tuned or altered by (i) passivating agents, (ii) functional groups, and (iii) doping/co-doping with heteroatoms [1].

They are a perfect choice for application in PL-based sensing systems because they display photoluminescence (PL) behavior that is excitation wavelength-dependent or -independent, have high quantum yield, and has promising analyte binding capabilities. However, majority of the reports show their excitation wavelength-dependent behavior [1, 4].

CQDs have various functional groups on their surface (–OH, –COOH, $-NH_2$, etc.) resulting from a variety of chemical treatments. Consequently, CQDs have a strong capacity to connect with both organic and inorganic molecules. Chemi-luminescence and electrochemical luminescence are caused by the remarkable electronic properties of CQDs [1, 3].

Solar cell is an electronic device that converts light energy into electricity through the photovoltaic (PV) effect. Low-weight, flexible and transparent solar cells have gained the interest of researchers in past few years due to their charming commercial applications like pliable chargers, transparent door curtains, vehicle glasses, window shades, etc. Various strategies have been adopted to enhance the performance of these solar cells amongst which employment of semiconductor quantum dots (SQDs) is a very common and effective technique. But SQDs have an adverse effect on environment and health of living beings because it contains carcinogenic materials like Cd, Pb, etc. [5]. Therefore, cost-effective and eco-friendly fluorescent CQDs were thought to be a possible alternative to the SQDs. Interestingly, it was observed that CQDs exhibit large absorption coefficients, broad absorption spectra, [6] and good charge transfer/separation ability, good quantum yield, and a large two-photon absorption cross-section [7]. These characteristics suggest that CQDs have good potential to enhance the performance of optoelectronics devices like solar cells and white light emitting diodes (WLEDs). As a consequence, attention toward CQDs for these applications have boomed in recent years.

WLEDs have become a strong contender as the future solid-state lighting sources owing to their advantages of high luminous efficiency, high luminance, and energy conservation. The characteristic broad emission property makes CQDs widely studied in the field of WLEDs [8]. Firstly, CQD-based WLED was prepared simply by coating yellow CQDs on blue GaN chips but these WLEDs exhibited (1) poor color rendering index (CRI), (2) cold emission [correlated color temperature (CCT) > 5000 K], and (3) excessive blue light which are harmful to human retina. The limitations associated with this WLED were due to the lack of a sufficient red component [9]. Therefore, for warm WLEDs (CCT < 4000 K) efficient red emissive CQDs were extensively researched. Alternately, single-component white emission CQDs (SCWE-CQDs) have gained importance for WLEDs applications because of various advantages such as (1) simple device fabrication procedures, (2) no color fading over time, and (3) no phase separation [10, 11]. In addition, the fluorescence-phosphorescence dual emissive CQDs have been reported which could achieve both SCWE and higher theoretical maximum efficiency in electroluminescent WLEDs simultaneously [11].

Present chapter mainly focuses on applications of CQDs in enhancing the performance of solar cells and WLEDs. In solar cells, CQDs have been used as active absorbing layers, electron-transporting layer (ETL), hole-transporting layer (HTL), donor/acceptor or dopant. In WLEDs, CQDs have been used as a phosphor material in the presence and absence of composite material to produce white emission.

2. Application of CQDs in solar cells

The CQDs have been employed as energy downshift (EDS) material, sensitizer or co-sensitizer, ETL, hole transport layer, donor/acceptor material and dopant in dye-sensitized solar cells (DSSCs), organic solar cells (OSC), perovskite solar cells (PSCs), and other PV cell configurations. Reports show that the addition of CQDs into the various solar cell components has reduced electron recombination, increased charge density, and boosted electron mobility, improved the performance of the PV cells. Enhancing the power conversion efficiency (PCE) of PV devices is essential in propagating green energy technology. Thus, CQDs offer an affordable, safe, and environmentally friendly method to advance PV performance.

2.1 Energy down shift material

96.3% of total solar radiation reaches the earth's surface whose spectrum lies between 250 and 2500 nm. In order to utilize solar energy to its fullest, it is mandatory for the solar cell to absorb the maximum possible solar spectrum. However, it is strenuous to practically accomplish it like in the case of DSSCs, the sensitizer deteriorates by the UV radiation coming from the sun hence adversely affecting the long-term stability of solar cell especially when Ru-free dyes (indoline dyes D149, etc.) are used [12]. There is an increasing demand for Ru-free dyes for low-cost and environmentally friendly PV systems.

Therefore, various strategies have been adopted to address these concerns and one frequently used technique is the use of EDS materials. EDS material absorbs UV photons and produces low-energy photons. CQDs are among the emerging EDS materials due to their low cost, ease of synthesis, eco-friendliness, and excellent absorbance in the UV region and emission in the visible region. In CQD-based DSSCs, CQDs convert the incident UV radiation into visible light, which is then absorbed by the sensitizer layer. This not only protects the DSSCs from UV rays but also improves the external quantum efficiency (EQE) in the 300–400 nm range [1, 13].

Han et al. prepared highly luminescent CQDs (QY = 84.8%, size = 3.7 nm) from citric acid (carbon source) and ethylenediamine (nitrogen source) in the presence of moderate ammonia water (AW) to attain proper inner structure and excellent N-passivation. They used hydrothermal method for synthesis of CQDs. The CQDs were then mixed with polyvinyl alcohol (PVA) aqueous solution (5 wt%) to obtain the CQDs/PVA composite solution. This solution was spin-coated on the silicon nanowire (SiNW) solar cells and heated (80°C for 20 min). This resultant layer serves as the EDS layer. This leads to the enhancement of the PCE from 10.85% to 10.96% which was attributed to the competing result of the deterioration of the surface reflectance and the optical absorption redistribution [14].

Riaz et al. synthesized N-doped CQDs (QY = 70%, size = 5 nm) from citric acid and 2,3-diaminonaphthalene (DAN) by a one-pot hydrothermal method and applied them as EDS in DSSC. EDS layer converted the incident UV radiation (from sun) into green light which was also absorbed by the active layer of DSSC in addition to the already absorbed spectrum. Since more photon absorption took place hence more carriers were generated and improved short circuit current density (J_{SC}) was recorded. EQE

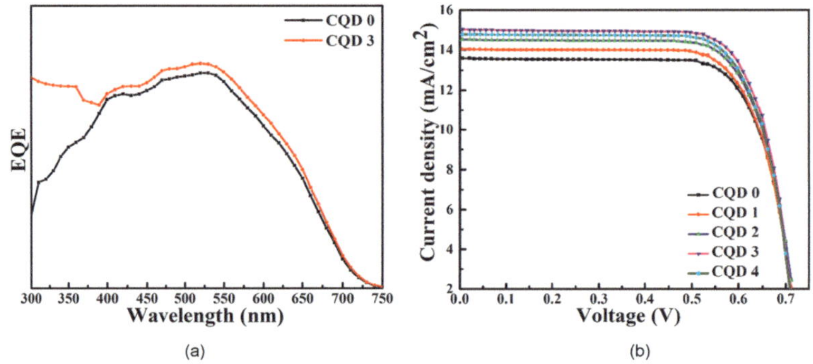

Figure 1.
(a) Enhancement of External quantum efficiency (EQE) in the UV region by the incorporation of CQDs, (b) I–V curve of DSSC at different numbers of coatings (CQD X; X is the number of N-CQDs solution coatings.

test confirmed the enhanced absorption of photons as shown in **Figure 1a**. On the whole, the PCE of the DSSC was amplified in the presence of CQDs as shown in **Figure 1b** and the highest recorded PCE was 8.2% for CQD-3 in comparison to 7.3% for CQD-0. It was observed that the PCEs of CQD-3 and CQD-0 were 6.3% and 2.1%, respectively after 3 weeks which denotes that CQDs layer significantly resists the photo-degradation of sensitizers. Result claims that CQD based EDS layer not only improved the efficiency but also enhanced the stability of the solar cell [15].

Ali et al. synthesised red emissive hollow nitrogen-doped CQDs (NR-CQDs) (QY = 61%, size = 11 nm) from 1, 3-dihydroxy benzene (DHB) and hydrazine monohydrate in the presence of ethanol and NH_4OH by hydrothermal method. They coated NR-CQDs as an EDS layer on crystalline silicon solar cells (c-Si SC). This layer of NR-CQDs absorbed the UV radiation from sun and emitted the red light which was absorbed by the cell in addition to the already coming spectrum and caused the PCE to increase by 5.8% [16].

Lu et al. synthesised CQDs from o-phenylenediamine (oPD) and acetic acid by hydrothermal method. They introduced a luminescent and anti-reflective (AR) silica layer on the top glass of a solar panel (with and without CQDs) for solar panel encapsulation. They found that the improved short-circuit current (I_{SC}) increased the PCE of solar panels from 20.76% (AR coating only) to 20.94% (AR-CQD coating). The improved I_{SC} was attributed to the higher EQE by the AR-CQD coating in the UV range (300–400 nm) in contrast to only AR-coated layer. On contrary, the role of the CQDs was almost insignificant in the remaining spectrum and EQE was slightly lower at longer wavelengths for AR-CQD film. This may be due to the surface scattering caused by the aggregated CQDs in porous silica. The inclusion of CQDs was compatible with the standard AR process of PV glass with no additional cost and it provided a better means for solar panel encapsulation along with enhanced performance [17].

2.2 Sensitizer or co-sensitizer

Sensitizers are photosensitive materials that absorb incoming light and convey the excited electrons to the nearby molecules. Various inorganic and organic/metal-free dyes/natural dyes like N3, N719, N749 (black dye), K19, CYC-B11, C101, K8, D102, SQ, Y123, Z907, etc. have been used as sensitizers in DSSCs.

Among the various dyes available, Ru-based sensitizers (dyes) are commonly used in the most advanced DSSCs. This is because these dyes have carboxylate ligands that anchor the dyes to TiO_2, thus offering excellent stability when adsorbed on TiO_2 surfaces [6, 12]. Although Ru-based sensitizers generally offer excellent performance and have the ability to provide highly efficient DSSCs, their synthesis relies on the rare and expensive Ru metal center, usually, time-consuming and the range of Ru dyes is limited to DSSCs [12, 18]. Thus, Ru-based sensitizers are unfit for economical and environmentally safe PV systems and researchers looked for an alternative. They found that metal-free organic dyes are economical, environmentally safe and have higher absorption coefficient (one order of magnitude higher than Ru complexes) and these dyes started developing at a fast pace.

However, the efficiency based on these dyes is lower than that based on Ru dyes, also these dyes are less stable at high temperatures [18].

As a result, researchers are constantly looking for sensitizers or co-sensitizers that are more appropriate and efficient for solar cells. Due to low cost, sustainability and environmental friendliness, many researchers have tried CQDs as sensitizers or co-sensitizers as new materials and they found superb results [19].

Mirtchev et al. synthesized CQDs (QY = 0.5%, height = 9 ± 6 nm) from g-butyrolactone by chemical ablation (sp^2 hybridized core with hydroxyl, carboxyl, and sulfonate groups) and used them as sensitizer on the photo-anode of nanocrystalline TiO_2 solar cells. The surface functional groups of CQDs enabled the coordination of CQDs to TiO_2 in a similar manner as done by the carboxylate ligands of the dye. Consequently, the PCE of DSSC was enhanced to 0.13% in comparison to 0.03% (non-sensitized nano-crystalline TiO_2). The enhancement is attributed to the broad absorption spectrum throughout the visible region. However, the J_{SC} reported in their study was much lower than the value routinely obtained through Ru sensitizer-based solar cells. The proposed reason for lower J_{SC} was the presence of various emission trap sites on the CQDs surface that serve as the recombination centers [6].

Wang et al. synthesized the NR-CQDs (QY = 36%, size = 10.8 nm) from citric acid and ammonia for different mass ratios through pyrolysis method. The absorption reaches to maximum value for the mass ratio of 1:4 (ammonia:citric acid). These CQDs were incorporated as sensitizer in the TiO_2-based solar cells and recorded the highest PCE of 0.79% under 1 sun illumination (AM 1.5) at that time. This improved performance of solar cell was due to the rich absorption of CQD/TiO_2 in the visible region and the increased transfer of electron from CQD to the TiO_2 conduction band (CB) [20].

Mistry et al. synthesised un-doped CQDs (QY = 5.2%) and NR-CQDs (QY = 9.8%) and incorporated them as green sensitizers in the TiO_2-based quantum dot (QD) solar cells. CQDs divulged a narrow particle size of 9.5 ± 1.9 nm. The highest recorded PCE was 0.56% (for 0.25% NR-CQDs) which was almost double that of the un-doped CQDs solar cell (PCE = 0.30%) under 1 sun illumination. Under a low light intensity of 0.1 sun illumination (AM 1.5), the PCE of nitrogen-doped (0.25% doping) CQDs-based solar cell was further increased to 1.2%. The presence of nitrogen in CQDs facilitated the injection of photo-carriers into the TiO_2 layer, improving the photo-current and hence the PCE of the solar cell device [21].

Yang et al. synthesized N-CQDs (size = 2 nm) using natural chitosan powders from hydrothermal method for different periods (0.5, 1, 2, 4, 8, 16 hours), as shown in **Figure 2a**. Each of these N-CQDs (having different properties) (**Figure 2b**) was used independently as a co-sensitizer for the DSSC (N719 dye) and consequently performance of all the cells get enhanced. The highest recorded cell PCE was up to 9.15% (with N-CQDs synthesized for 2 hours) which was much higher than 8.5% for controlled devices (excluding N-CQDs) under 1 standard sun (AM 1.5). This improvement was due to the following reasons: (i) Since, the energy level of N-CQD is in good agreement with the energy level of TiO_2 and the reduction potential of I^-/I_3^-, electrons can be injected into TiO_2 but not into the electrolyte thereby acting as an electron blocking layer (a layer which inhibits the recombination of injected electrons with red-ox pair in the electrolyte). (ii) Under photoexcitation, N-CQD functions as a charge transfer antenna that receives and transfers energy to the N719 dye via Förster or Fluorescence Resonance Energy Transfer Mechanism (FRET). As a result, this mechanism has significantly improved the light absorption capacity of the device. (iii) The optimized N-CQD showed excellent up-conversion emission which broadened the absorbance of the device and improved the capacity to collect long wavelengths [22].

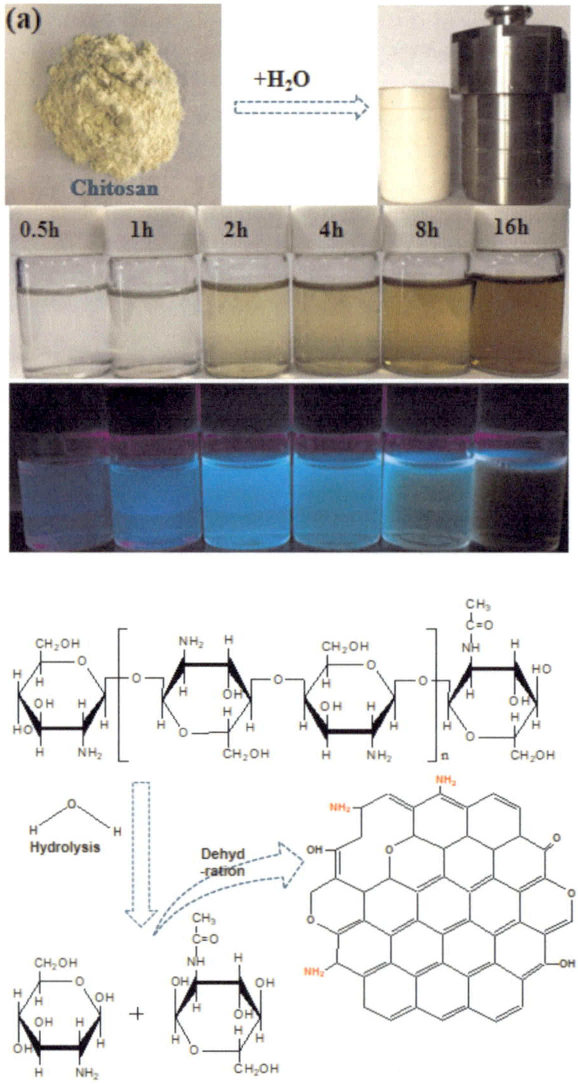

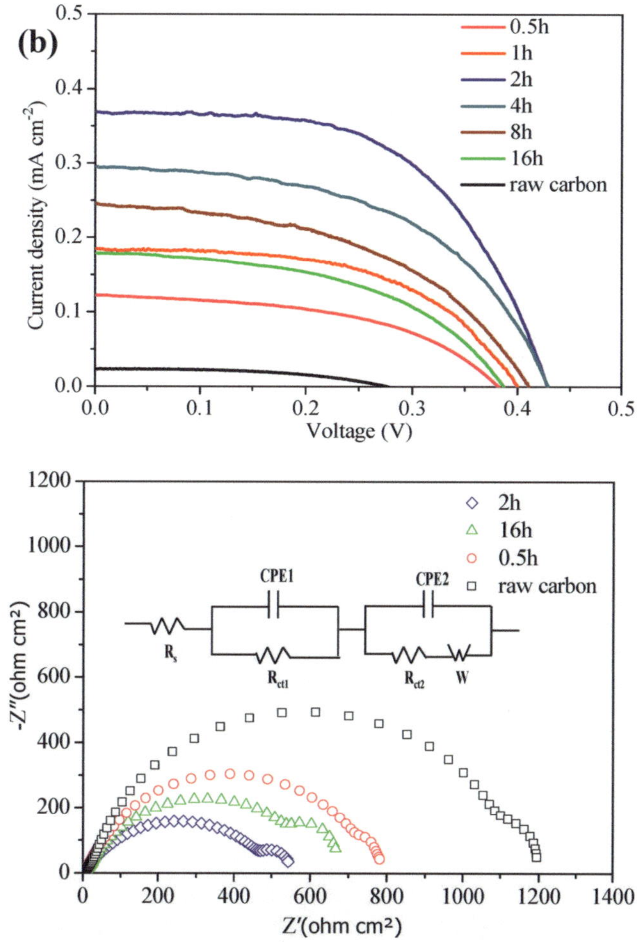

Figure 2.
(a) Conversion from chitosan powders to CQDs by a hydrothermal method and images of nitrogen-doped CQDs aqueous solutions for different periods (0.5, 1, 2, 4, 8, 16 hours) under ambient and UV exposure. Each of these N-CQDs possess different properties and (b) Photocurrent density–voltage (J–V) curves for N-CQDs sensitized solar cells under simulated sunlight (AM1.5, 100 mW cm^{-2}) and impedance different of solar cells.

2.3 Electron transport layer

The purpose of the ETL (between the active layer and the cathode) is to extract electrons from the active layer towards the cathode and prevent recombination of free electrons and holes at defects present at the interface (active layer to cathode).

Presence of the ETL improves the number of electron transfer from active layer to cathode. TiO_2 and ZnO films are metal oxide ETLs widely used in PSCs, of which TiO_2 is the most reported ETL not only in PSCs but also in DSSCs.

TiO_2 has a HOMO/LUMO position (−7.63 eV/−4.4 eV) suitable for the extraction and injection of charge carriers, easy to fabricate, and economical. But low electron mobility, high trap-state density below the CB, and relatively high processing temperatures are serious problems.

On contrary, ZnO has much higher conductivity than TiO_2 and thus exhibits low recombination loss and does not demand high-temperature therapy.

But like TiO_2, ZnO too has a degradation issue, which turns down the stability of device [18, 23]. It seems, that another ETL material, namely tin oxide (SnO_2), may replace both ZnO and TiO_2 in certain aspects. SnO_2 does not require high-temperature processing, possesses high electron mobility, reasonable HOMO/LUMO position, excellent stability, and high transparency in the visible and near-infrared regions.

Although metal oxide ETLs (TiO_2, ZnO, and SnO_2) are chemically and physically more stable than organic ETLs, they nonetheless contain a disproportionately high number of surface defects and traps when compared to organic ETLs [23].

Among organic ETLs, CQDs have been studied and found to be an emerging contestant in a variety of organic ETL materials. This is due to its excellent photo-stability, low toxicity, excellent electron extraction ability, easy bandgap tuning, etc. CQD as an ETL is deployed by researchers in two ways: (i) replacing existing ETL entirely, or (ii) blending with existing ETL [1].

Yan et al. synthesized the CQDs (size = 3.5 nm) by chemical vapor deposition (CVD) in an Ar atmosphere using C_2H_2 (carbon source). Their group used synthesised CQD as an ETL in fabrication of solution-treated OSCs. According to the results, the optimized PCE of P3HT:PC61BM, PTB7:PC61BM, and PTB7-TH:PC71BM devices (consisting of CQDs as ETLs) reached 3.11%, 6.85% and 8.23%, respectively. These results are comparable with the Lithium fluoride (LiF, one of the ETL) based devices. Due to lower resistance of CQDs-ETL, the interfacial connection between the Al electrode and the active layer was improved. This lower interfacial resistance and

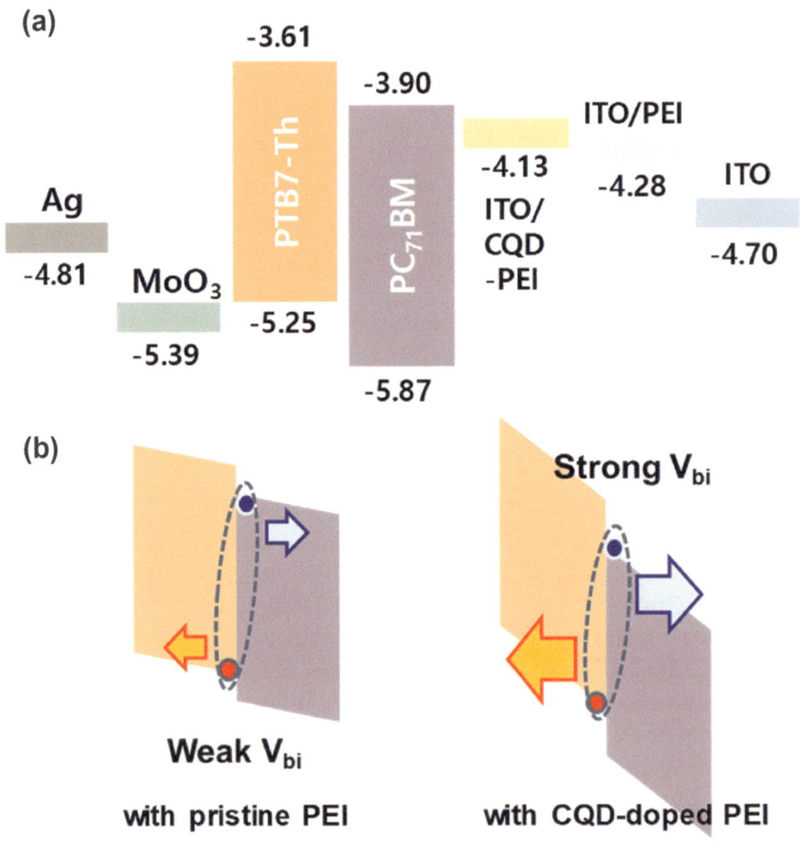

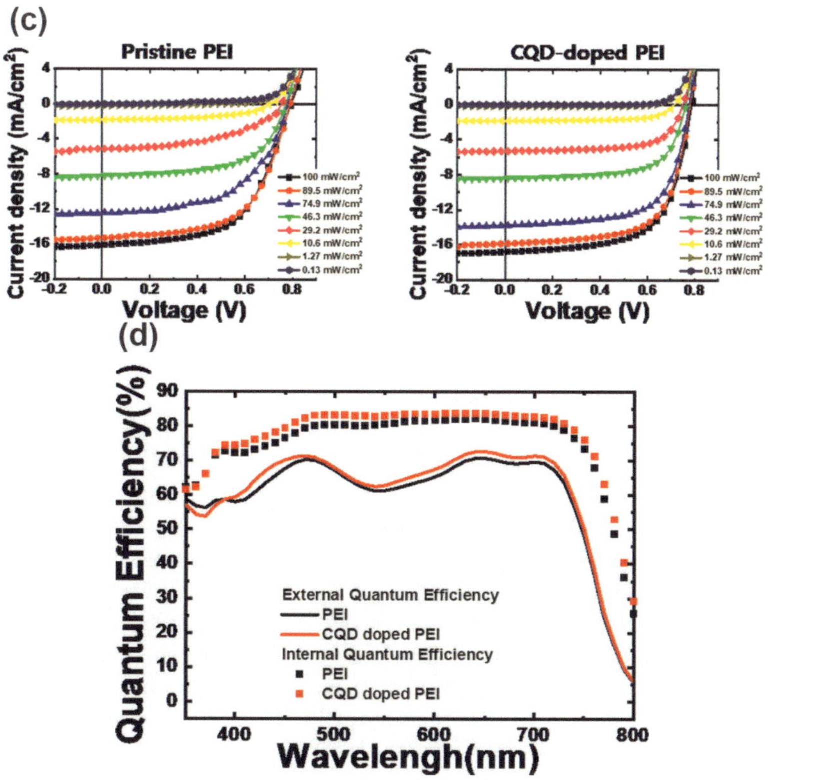

Figure 3.
(a) Energy level diagram of PTB7-Th: PC71BM solar cells, (b) CQD-doped PEI induced a stronger internal field due to the lower work-function. This strengthened internal field induced better exciton dissociation efficiency, (c) Current density-voltage characteristics of Pristine CQDs and CQD doped PEI, (d) Quantum Efficiency of Pristine CQDs and CQD doped PEI.

better electron injection resulted in superior device performance. The long-term thermal stability of CQDs-ETLs based devices was also enhanced significantly due to minimised diffusion of CQDs in the active layer [24].

Li et al. synthesized CQDs from glucose via one-step alkali-assisted ultrasonic chemical method. They observed that the PCE of planar n–i–p hetero-junction PSCs were boosted to 18.89% upon incorporation of CQDs/TiO_2 composite layer as ETL.

CQDs improved the electronic coupling between the $CH_3NH_3PbI_{3x}Cl_x$ and TiO_2 ETL interface, which enhanced the electron transport property (charge extraction and injection) at TiO_2/perovskite (from perovskite to TiO_2). Consequently, enhanced short-circuit current density (J_{SC}), open circuit voltages (V_{OC}), and hence PCE was achieved [25].

Zhu et al. synthesised CQDs (size = 4 nm) via microwave-assisted hydrothermal. They built a planar heterojunction (PHJ) $CH_3NH_3PbI_3$ PSC using CQDs-doped PCBM as ETL. Doping with CQDs resulted in a PCE of 18.1% (39.2% increase), compared to a PCE of 13% for pure PCBM-based devices. Moreover, CQD doping improved the long-term stability of PSCs by increasing the diffusion of I^- [26].

Park et al. synthesized CQDs (size = 3.6 ± 1.2) from neutral red powder and ethylene-diamine via microwave irradiation. These CQDs have NH_2 ligands, these CQDs were incorporated in different ratios in the polyethyleneimine (PEI) layer of fabricated solar cells. In solar cell, the PEI layer modifies (reduce) the work function of ITO, and CQDs in PEI extend this effect as shown in **Figure 3a** and **b**.

Consequently, all the solar cells performed better after incorporation of CQDs (different concentrations).

The impact of CQD concentration on J_{SC} and % EQE are shown in **Figure 3c** and **d**. The highest recorded PCE was ~9.5% (for 2% CQDs) in comparison to PCE of ~8.6% pristine (for 0% CQDs) PEI layer-based solar cells. The enhancement of the performance of CQD-based solar cell was attributed to (i) better electron transport and (ii) improved exciton dissociation probability due to the strengthened internal field [27].

2.4 Hole transport layer

The objective of an HTL (between the anode and active layer) is to extract holes from the active layer towards the anode and to minimize the recombination of free electrons and holes at defects states that exist on the interface (anode and active layer). Presence of the HTL improves the number of holes transferred from active layer to anode. PEDOT:PSS is one of the most widely used HTLs but its strong acidity and hygroscopicity reduce the device stability. In addition, its insulating nature reduces the electrical conductivity (low hole mobility) reduces the device performance, and high cost limits its application [18]. This forced the researchers to explore low-cost and stable material. This brought inorganic p-type materials based HTMs such as CuI, NiO, Cu2O, CuO, MoOx, VOx, WOx, etc. into PSCs [28]. These inorganic materials exhibit a wider band gap, high conductivity and also prevent the device from exposure to the ambient atmosphere. So, employing such inorganic materials in PSCs could help to achieve higher PCE, yet their sensitivity towards oxygen and moisture seems to be a hindrance. To overcome such sensitivity towards ambient condition, low-cost carbon-based materials like CQDs have been employed as HTLs in solar cells and better performance have been reported. Therefore, in comparison to the aforementioned materials, CQDs emerge as one of the better alternatives [18, 29].

Paulo et al. synthesized CQDs from citric acid and p-phenylenediamine via hydrothermal approach. The cyclic voltammetry (C–V) measurements revealed that the prepared CQD's HOMO and LUMO energy levels are suitable for hole transfer (or electron blocking) from perovskite to CQDs. They incorporated CQDs as the HTL for the very first time in methylammonium lead iodide (MAPI) solar cells and observed a PCE of 3%. This efficiency was quite poor in comparison to another similar device's PCE of 8.06%, which was observed utilizing spiro-oMeTAD as the HTL. According to ESEM analysis of the devices, the poor perovskite coverage across the mp-TiO2 surface was the cause of the poor performance of CQD-HTL-based device [30].

Nguyen et al. synthesized the N-CQDs (size = 3.32 nm) using fumaronitrile by solvothermal process and O-CQDs (oxidised) using carbon nanofiber via chemical oxidation method. Their idea was to utilize the strength of both PEDOT: PSS and CQDs, consequently they studied the impact of both varieties of CQDs (as HTL) in the presence of PEDOT:PSS on the performance of fabricated solar cells.

They recorded the highest PCE of 8.57% (with N-CQDs), 8.17% (with O-CQDs) and 7.26% (No CQDs) as shown in **Figure 4**. The performance of device was improved because the incorporation of CQDs in PEDOT:PSS drastically (i) decreased the resistance, (ii) raised the work function, (iii) altered the surface energy, and (iv) smoothed the surface morphology of PEDOT:PSS. In addition, it decreased the electrostatic interaction between PEDOT and PSS, which resulted in the π–π

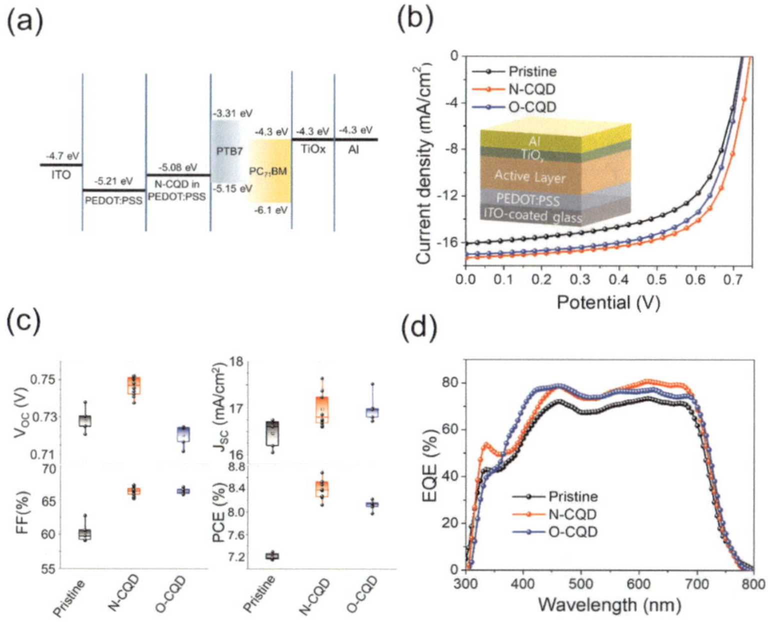

Figure 4.
OPV device performance and characterization. (a) Energy band diagram (b) Current density voltage (J–V) curves (inset: OPV device with the structure ITO/PEDOT:PSS/active layer/TiOx/Al). (c) Photovoltaic response, and (d) EQE spectra of OPV devices.

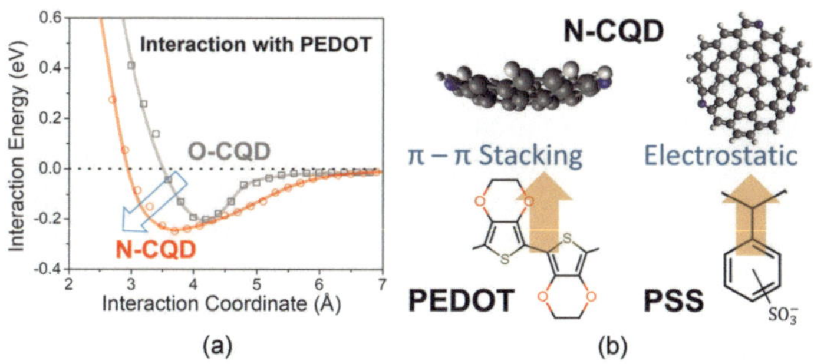

Figure 5.
(a) Potential energy scan for each pair of N-CQD/PEDOT and O-CQD/PEDOT; (b) π–π stacking of PEDOT with N-CQDs and electrostatic interactions of PSS with N-CQDs.

stacking of PEDOT and the hydrophilic interaction of PSS with N-CQDs as shown in **Figure 5**. The interactions in the presence of N-CQDs were quite strong in contrast to O-CQDs. The combination of PEDOT:PSS and N-CQDs improved the charge extraction and charge injection, consequently boosting the improved performance of the solar cells [31].

2.5 Donor/acceptor

In numerous reports, fullerene derivatives have been fully supplanted by CQDs as donor or acceptor material in PV devices [29]. Some research groups have considered the "CQDs as acceptors" when CQDs are incorporated into the active layers. But in this chapter, we have categorised it separately where CQDs will be considered as "dopant" (CQDs as dopant may either accept or donate electrons).

Feng et al. synthesized CQDs (QY = 12.1%, size = 6.2 nm) from L-ascorbic acid (carbon source) and ethanediamine (catalyst) by a hydrothermal method. They found that integration of as prepared CQDs in P3HT (mass ratio 1:1) drastically quenched the PL of P3HT. Since the energy level of the CQDs was well aligned with those of the P3HT consequently, majority of photo-excited electrons are transferred from P3HT to CQDs instead of undergoing radiative recombination. Hence, quenched PL is observed in the presence of the CQDs [32].

Cui et al. synthesized C-CQDs (size = 3.5 nm) by CVD in an Ar atmosphere using C_2H_2 as a carbon source. They fabricated the inverted OSCs with P3HT as donor and C-CQDs as an acceptor (for different amount of C-CQDs: 0, 2.5, 5.0, and 10 wt% ratio to P3HT) and compared the performances of all the solar cells.

PL quenching was seen in the P3HT:C-CQDs composite film, indicating that photo-induced charge transfer between P3HT and C-CQDs had taken place. V_{OC} and FF were found to increase with the C-CQDs concentration and reached the maximum value for 10 wt% C-CQDs. Similarly, J_{SC} rose with C-CQDs concentration but peaked at 5 wt% C-CQDs before falling.

Consequently, the highest PCE of 0.29% was recorded for 5 wt% C-CQDs based solar cells which were almost 2.5 times higher than device without C-CQD (0.08%) [33].

The enhanced performance of C-CQDs-based solar cell is attributed to the improved absorption by C-CQD:P3HT composite film and effective charge transfer from P3HT (donor) to C-CQDs (acceptor), as shown in **Figure 6**.

They also studied the impact of the hydrothermally produced CQDs (H-CQDs) on the performance of solar cell and found that the PCE was just 0.1% for 5 wt% H-CQDs. The aggregation (poor dispersion) of H-CQDs in the P3HT matrix, [33] which was observed using the FETEM, was the cause of the lower performance with H-CQDs.

Privitera et al. synthesized N-CQDs (size = 2.4 nm) from chlorohydrated arginine and ethylendiamine via assisted hydrothermal methods and functionalized them with various thiophene-containing groups. The purpose of CQD's functionalization was to (i) improve their electron-donating capability and (ii) enhance their solubility in non-polar solvents. They showed that the PL of a thin film of a CQD:PCBM mixture dramatically quenched as a result of the transfer of electrons from the CQDs (donor) to the PCBM (acceptor) which was confirmed electron paramagnetic resonance (EPR). They also noted that various functionalization derivatives on CQDs changed the charge transfer efficiency of the CQD: PCBM blend using time-resolved EPR spectroscopy [34]. There

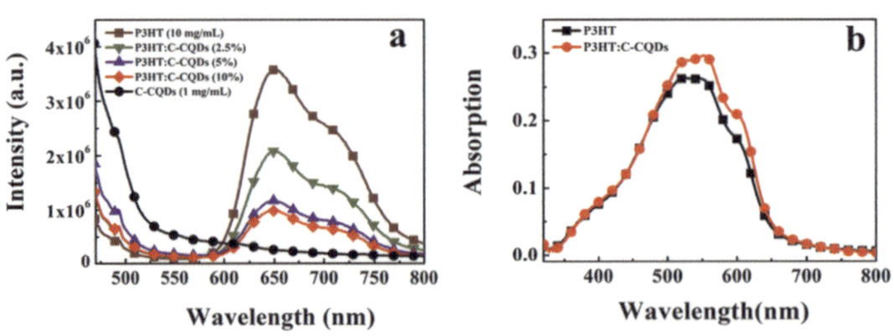

Figure 6.
(a) PL spectra of P3HT (10 mg ml^{-1}), C-CQDs (1 mg ml^{-1}) and P3HT:C-CQDs (P3HT: 10 mg ml^{-1}, C-CQDs content: 0, 2.5, 5, and 10 wt% vs. P3HT) composite films at an excitation of 450 nm. (b) Absorption spectra of P3HT and P3HT:C-CQDs (1: 5%) in the solid state.

are currently very few studies that concentrate on CQDs as electron donors, leaving plenty of possibility for additional research to fully realise its potential.

2.6 Dopants

Doping is one of the popular methods used to modify the optoelectronic properties of the base material. According to the reports, CQDs may soon be viable alternative eco-friendly dopant materials for commercial solar cells.

Cui et al. synthesized CQDs (size = 3.5 nm) by CVD in an Ar atmosphere using C_2H_2 as a carbon source. Their group fabricated the several OSCs with P3HT (donor), $PC_{61}BM$ (acceptor) and CQDs (dopant with different concentrations 0, 0.025, 0.05, 0.075 and 0.1 wt%). They noticed when CDQ's (dopant) climbed from 0 to 0.075 wt %, the PCE efficiency increased from 3.3% to 3.67% (11% enhancement) while the PCE reduced to 3.55% with a further increase in doping concentration to 0.1 wt%.

The improved PCE of CQDs based solar cell was attributed to (i) the absorption and scattering by well-dispersed CQDs in the active layer, which might improve J_{SC}, and (ii) reduced series resistance of the device that improved J_{SC} and the fill factor (FF) [33].

Shejale et al. synthesised CQDs (size = 2–6 nm) from citric acid and urea by microwave-assisted pyrolysis method. Their group fabricated DSSCs and discovered that after doping the active layer with manufactured N-CQDs (co-active layer) under 1 solar irradiation, the cell's PCE increased to 8.75%. The incorporation of the N-CQDs in the photoactive layer had (i) a synergistic effect on the absorbance and (ii) reduced the recombination between the photo-anode and electrolyte. Their group also employed the N-CQDs as sensitizers and co-sensitizers in DSSCs and consequently, enhanced performance was observed in both situations. However, the highest improvement was recorded when N-CQDs were doped in the active layer (co-active layer). This improvement was attributed to (i) many dye anchoring sites, (ii) strongly conducting photo-anode, (iii) rapid charge carrier transfer, and (iv) intrinsic light-emitting photo-fluorescent characteristics of N-CQDs in mesoporous titania [35].

Table 1 below provides a comparative study of the impacts of CQDs on the performances of various solar cell types.

3. Application of CQDs in WLEDs

WLEDs exhibit interesting characteristics like (1) wider emission spectrum (400–760 nm) (2) high luminous efficiency, (3) high luminance, and (4) energy conservation which make WLEDs a strong candidate for future solid-state lighting sources [8, 37].

There are mainly three methods to realize the WLEDs as shown in **Figure** 7: (1) First, WLED is obtained by encapsulating tricolor chips emitting red, green, and blue (RGB) light. These RBG lights combine to generate white light. However, as service time increases, these tricolor chips show distinct aging properties that can result in different brightness decay and optical stabilities, which restricts the use of such WLEDs. (2) Second, WLEDs are realized by combining a blue LED chip (which emits blue light) and yellow phosphor (which emits yellow light upon excitation by a blue LED chip). The mixture of blue and yellow light produces white emission from one chip. But these WLEDs exhibit (i) low CRI because of less red component, and (ii) the blue LED chip and the yellow phosphor show

Synthesis technique	Type of solar cell#	Roll of CQDs	Solar cell structure (with CQD)	A % PCE (without CQD)	B % PCE (with CQD)	% ΔPCE (*B*–*A*)	Ref.
Dehydration [chemical oxidation]	NCTiSC	Sensitizer	FTO/TiO_2/CQD/electrolyte/ Pt-FTO	0.03%	0.13%	0.10%	[6]
Hydrothermal	SiNwSC	EDS layer	Si/SiNWs/CQDs/PVA	10.85%	10.96%	0.11%	[14]
CVD	OSC	ETL	ITO/PEDOT:PSS/active layer/ ETL (CQDs)/Al (Active layer = P3HT:PC61BM)	1.75%	3.11%	1.36%	[24]
			ITO/PEDOT:PSS/active layer/ ETL (CQDs)/Al (Active layer = PTB7:PC61BM)	6.28%	6.85%	0.57%	
			ITO/PEDOT:PSS/active layer/ ETL (CQDs)/Al (Active layer = PTB7-TH:PC71BM)	7.59%	8.23%	0.64%	
Hydrothermal	PSC	HTM	FTO/d-TiO_2/mp-TiO_2/MAPI/ HTM/Au	0.71%	3.00%	2.29%	
Pyrolysis	QDSC	Sensitizer	FTO/TiO_2/N-CQD/gel electrolyte/Pt-FTO	—	0.79%	NA	[20]
Alkali-assisted ultrasonic chemical method	PSC	ETL	ITO/CQDs/TiO_2/perovskite/ spiro-OMeTAD/Au	15.15%	18.89%	3.75%	[25]
CVD	OSC	Electron acceptor	ITO/ZnO/P3HT:C-CQDs/ MoO_3/Al	0.08%	0.29%	0.21%	[33]
CVD	OSC	Dopant	ITO/ZnO/P3HT:C-CQDs: PC61BM/MoO_3/Al	3.30%	3.67%	0.37%	[33]
Hydrothermal	DSSC	EDS layer	FTO/TiO_2/CQDs/electrolyte/Pt-FTO	7.30%	8.20%	0.90%	[15]
Microwave irradiation	i-PSC	ETL	FTO/PEDOT:PSS/$MAPbI_3$/ PCBM:CQDs/BCP/Ag	16.01%	18.10%	2.09%	[26]
Solvothermal	NCTiSC	Sensitizer	FTO/TiO_2/mp-TiO_2/CQD/ electrolyte/Pt-FTO	—	0.56%	NA	[19]
Hydrothermal	DSSC	Co-sensitizer	FTO/TiO_2/N719 dye/CQDs/ electrolyte/Pt–Ni-FTO	8.50%	9.15%	0.65%	[22]
Hydrothermal	c-SiSC	EDS layer	Not mentioned	17.39%	18.40%	5.80%	[16]
Microwave irradiation	i-OSC	ETL	ITO/PEI-CQDs/PTB7-Th: PC71BM/MoO_3/Ag	8.55%	9.47%	0.92%	[27]
Solvothermal	OSC	HTL	ITO/PEDOT:PSS-CQDs/PTB7: PC71BM/TiO_x/Al	7.26%	8.57%	1.31%	[31]
Pyrolysis	S-TFSC	Absorber layer	ITO/CdS/CCTS:CQD/Al	0.30%	7.00%	6.70%	[36]

Measured %PCE is Under one standard sun (AM 1.5).
 # NCTiSC: nanocrystalline TiO_2 solar cell, SiNwSC: silicon nanowire solar cells, QDSC: quantum dot solar cell, OSC: Organic solar cell, PSC: perovskite solar cell, DSSC: Dye sensitized Solar cell, i-PSC: inverted perovskite solar cell, c-SiSC: crystalline silicon solar cell, i-OSC: inverted organic solar cell, S-TFSC: Superstrate Thin Film Solar Cell.

Table 1.
Effect of CQDs on the performance of solar cells.

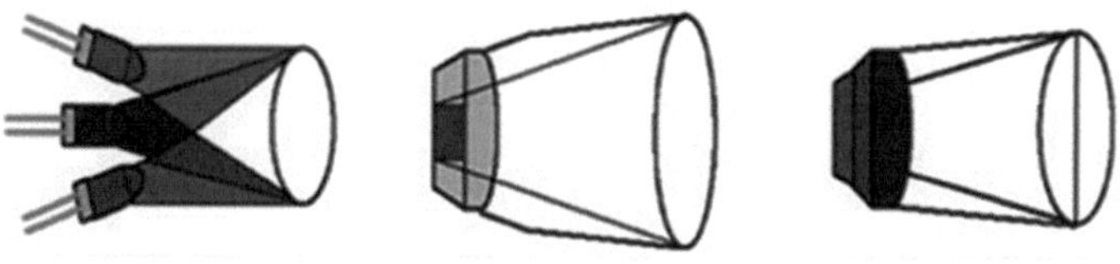

Figure 7.
Diagrammatic representation for three methods to realize the WLEDs.

different changing trends in luminous efficiency over time, resulting in the color change of the WLED.

(3) Third, WLED is obtained by combining near-ultraviolet LED chip (emits UV light) and a fluorescent material (which can effectively produce fluorescence under excitation of near ultraviolet light). This fluorescent material consists of the combination of red/green/blue phosphors (RGB tricolor fluorescent material) or/and the single white-light fluorescent material which upon excitation by UV light generates the white light. Warm WLEDs have been successfully reported using this method [38].

WLED phosphor materials generally include rare earth phosphors, SQDs, and CQDs. Since their research and preparation techniques are rather well-established, rare earth fluorescent elements are frequently employed in WLEDs. However, their extensive application is constrained by the high cost of their raw ingredients and the difficulty in controlling their emission spectrum. Traditional SQDs have poisonous components (Cd, Pb, Hg, etc.), which significantly restricts their practical use. On the other hand; CQDs are inexpensive, environmentally friendly, and easily synthesizable which makes CQDs a promising alternative fluorescent material for WLED application [1, 8, 38]. In addition, CQDs exhibit wide emission characteristics [Vis-NIR, full width half maxima (FWHM) > 80 nm] in contrast to commercial rare-earth phosphors and traditional SQDs.

It results from the robust electron-phonon interaction of the CQD and the greatly dispersed particle size which makes CQDs a perfect material for WLEDs [8, 38].

CQDs are used in three ways for WLEDs: (1) Combination of different CQDs, (2) CQDs with different phosphors, and (3) single CQDs. In WLEDs based on (1) and (2) methods CQDs or/and phosphor materials are excited which consequently generates the white light. However, in both these methods, it is difficult to tune the mixture ratio of multiple phosphors during fabrication device and the distinct light stability of phosphors causes the color shifting of the WLED. On the other hand, WLEDs based on single CQDs consist of CQDs which generate white light by adjusting RGB spectral composition during the phosphor synthesis. This, single SCWE-CQDs based WLEDs have attracted much attention owing to their advantages such as high color quality white emission, easy preparation, no color fading with time, and long lifetime [8, 10, 11, 38, 39].

In order to further imitate the sunlight to achieve high color quality white emission researchers are trying to broaden the emission spectra of CQDs to achieve SCWE emission. As a result, in the realm of current research on fluorescent material for WLEDs, WLEDs based on a single CQD phosphor are emerging.

On the basis of the source of excitation (optical or electrical) WLEDs are divided into two categories which are discussed below:

3.1 Phosphor-converted WLEDs

One of the methods uses CQDs as light-converting phosphors that produce white light when optically pumped by blue or UV LED chips, which are utilised as primary light sources. Sometimes this kind of WLED is referred to as phosphor-converted WLEDs (pc-WLEDs). CQDs-based WLEDs were first prepared simply by coating yellow CQDs on blue GaN chips. These WLEDs exhibited CCT > 5000 K (cold WLEDs), inferior CRI, and excessive blue light which was due to the lack of a sufficient red component. The strong blue light from these chips is harmful to human retina and therefore, warm WLEDs (CCT < 4000 K) containing enough red emission are becoming increasingly important [8, 9].

Wang et al. have developed R-CQDs (QY up to 53%) as the red phosphors and using this in combination with our blue emissive CQD (B-CQD) and green emissive CQD (G-CQD) phosphors, we have realized a UV-pumped CQD phosphor-based warm WLED with the commission internationale de l'elcairage (CIE) coordinates, CCT and CRI of (0.3924, 0.3912), 3875 K and 97 (**Figure 8**). The luminous efficiency of the optimized warm WLED was as high as 31.3 lm W^{-1}, which is similar to that of SQD and rare earth phosphor-based WLEDs [40].

Zhu et al. fabricated WLEDs by combining white-emitting CQDs dispersed in epoxy resin and a UV-chip (370 nm), with the CIE coordinate (0.277, 0.362) which approached the coordinate (0.33, 0.33) of pure white light [41].

Meng et al. coated solid ultra-BWCQDs on the surface of the UV-LED chip (400 nm) to produce warm WLEDs. The WLEDs exhibited full width at half maximum over 200 nm throughout the entire visible light window, with a sufficient red component with CIE coordinates of (0.42, 0.38) and a high CRI value of 91 [10]. In comparison to the commercial WLED lamp (CRI $\approx$ 85), the warm WLED lamp is more capable of showing the true colors of the fruits. **Figure 9** shows (a) photograph of the warm WLED lamp, (b) EL spectrum of the solid ultra-BWCQDs, (c) Fruit colors under the commercial warm WLED lamp, and (d) fruit color under the solid ultra-BWCQD warm WLED lamp.

The CCT and CRI values showed a very slight change with increase of drive current from 20 to 120 mA. These results show that the warm white light generated from the warm WLED lamp has a high color chromatic stability against an increase in drive current [10].

Recently, Li. et.al. prepared environmentally friendly white fluorescent carbon dots (CDs)/ZnO QDs and employed them for making WLEDs. Prepared WLED exhibited fluorescence emission spectra ranging from 425 nm to 750 nm, CIE chromaticity coordinate of (0.30, 0.34) (as shown in **Figure 10**), and a color temperature of 7093 K [42].

3.2 Electroluminescent WLEDs

Yuan et al. synthesised multicolor bandgap fluorescent carbon quantum dots (MCBF-CQDs) from blue to red with quantum yield up to 75% by solvothermal method.

Using MCBF-CQDs as an active emission layer, their group fabricated monochrome electroluminescent LEDs (mc-ELLEDs) for the first time. The maximum luminance of blue LEDs reaches 136 cd m^{-2}, which is the best performance for CQD-based mc-ELLEDs. MCBF-CQDs-based energy level diagram of mc-ELLEDs, emission spectrum of MCBF CQDs and mc-ELLEDs are shown in **Figure 11a** and **b**.

Their group also fabricated WLED using green bandgap fluorescent-CQDs (G-BF-CQD) blended *N*-vinyl carbazole (PVK) as an emissive layer. Prepared WLED showed a broad EL emission spectrum with two emission peaks centered at about

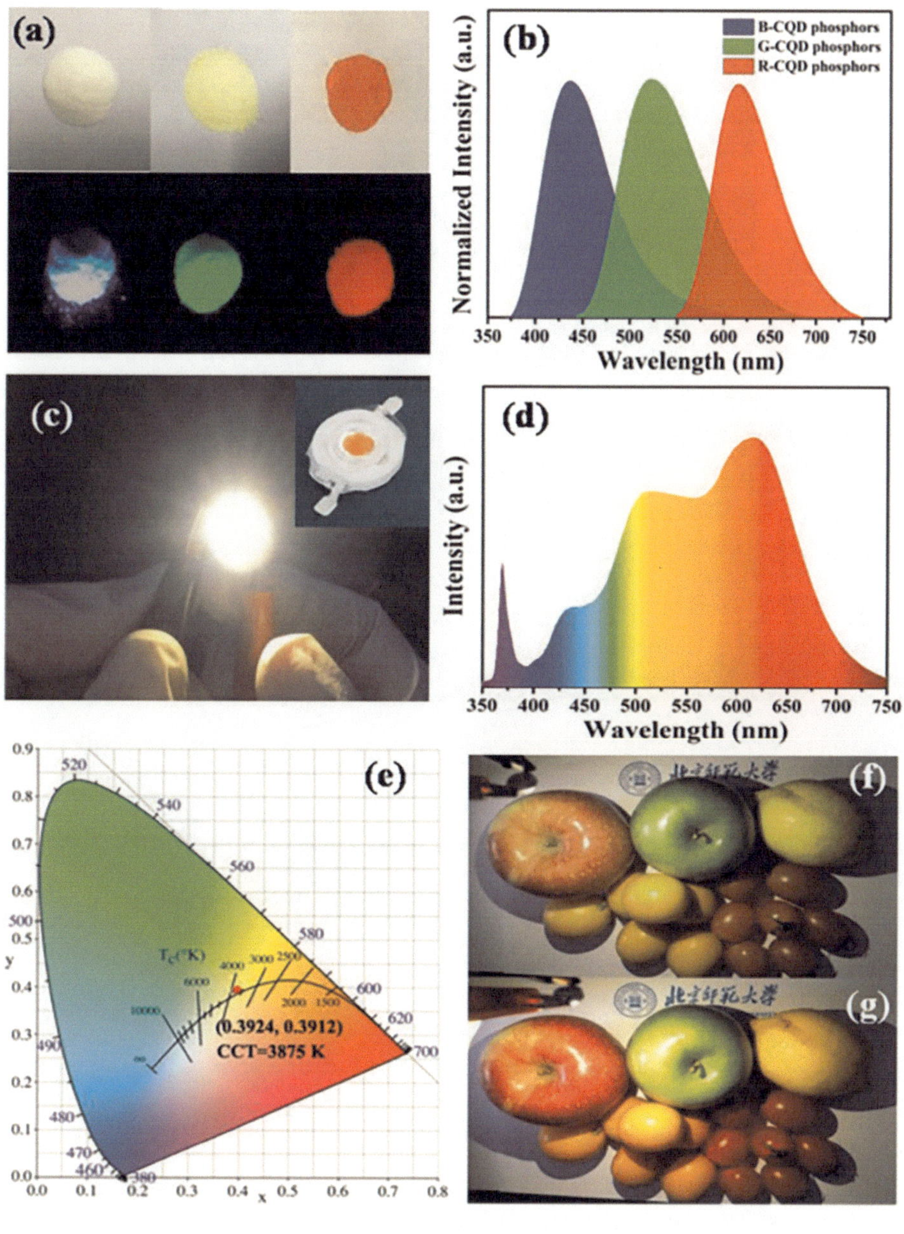

Figure 8.
(a) Image of Blue, Green and Red Phosphor CQDs under sunlight (above) and UV light (below), (b) their corresponding spectrum emission, (c) Photograph of warm WLED lamp (inset) and the operating warm WLED lamp (d) Electro luminance Spectrum, (e) CIE color coordinate of the warm WLED lamp. Fruit color under (f) commercial WLED lamp and (g) CQD warm WLED lamp.

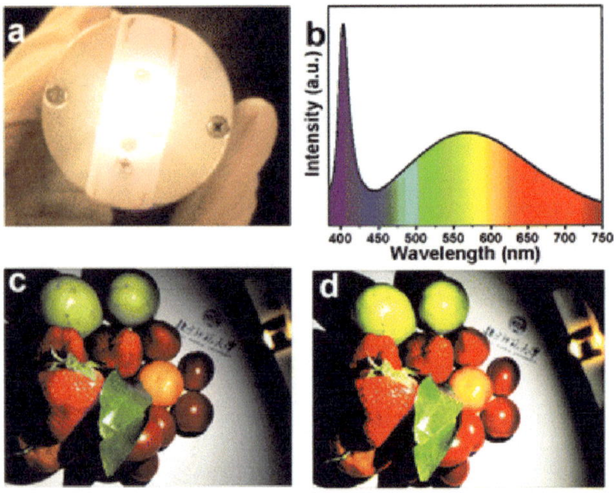

Figure 9.
(a) Photograph of the warm WLED lamp, (b) EL spectrum of the solid ultra-BWCQDs, (c) Fruit colors under the commercial warm WLED lamp, and (d) fruit color under the solid ultra-BWCQD warm WLED lamp.

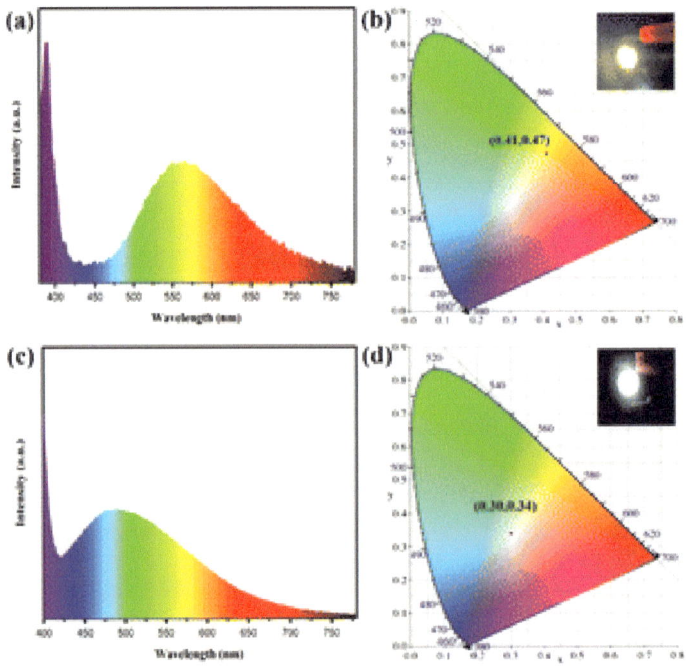

Figure 10.
CIE chromaticity coordinate.

410 and 517 nm (**Figure 11c**), which are associated with the emission of PVK and G-BF-CQDs respectively with CIE coordinates at (0.30, 0.33) as shown in **Figure 11d**. The maximum luminance (L_{max}) and η_c reached about 2050 cd m^{-2} and 1.1 cd A^{-1} respectively with low turn ON voltage (V_{on}) of about 3.9 V which is comparable to QDs–LEDs [43].

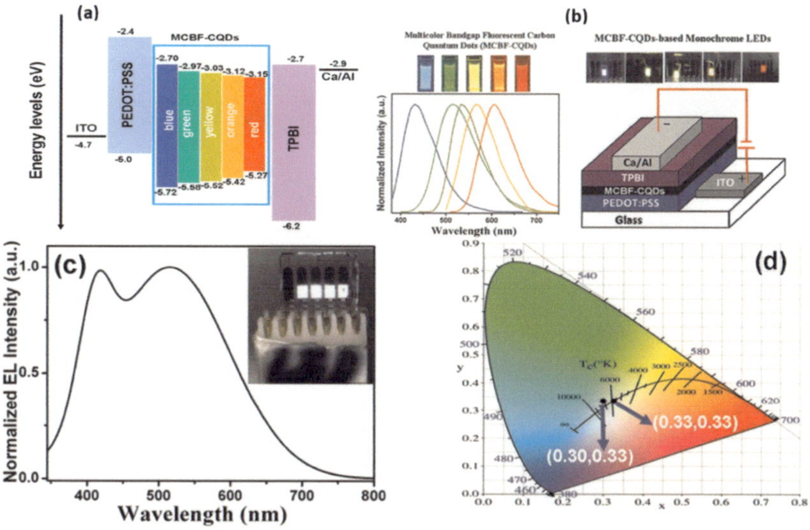

Figure 11.
(a) MCBF-CQDs-based energy level diagram of monochrome electroluminescent LEDs, (b) Emission spectrum of MCBF-CQDs from blue to red along with the monochrome electroluminescent LEDs, (c) broad EL emission spectrum with two emission peaks centered at about 410 and 517 nm, (d) two emission peaks which are associated with the emission of PVK and G-BF-CQDs respectively with CIE coordinates at (0.30, 0.33).

However, owing to the lack of red component, these cold WLEDs are not the best choice for indoor lighting. Combined with the blue emission stemming from PVK and the broad red emission from the R-CQDs-NMe_2, -NEt_2, and -NPr_2, the corresponding CIE coordinate and CCT warm WLEDs-1, -2, and -3 are (0.379, 0.314)/3365 K, (0.383, 0.311)/3168 K, and (0.388, 0.309)/2987 K (WLEDs-3), respectively [44]. Taking advantage of the high QYs which were up to 86% in ethanol, the L_{max} and ηc_{max} of the LEDs reach as high as about 5248–5909 cd m^{-2}, 3.65–3.85 cd A^{-1}. It was the first time to realize high-performance CQDs based electroluminescent warm-WLEDs which was attributed to the CQD's high-efficiency bandgap red emission and good solubility. Additionally, portions of the SCWE-CQDs were utilised as the active emission layers in the electroluminescent WLEDs [45]. However, the efficiency of these types of WLEDs was decreased due to the surface defect-state emission nature, the weaker luminescence, and the poorer carrier transfer performance.

Luo et al. synthesized white emission graphene quantum dots (GQDs) using graphite (precursor) via microwave-assisted hydrothermal method, and thus fabricated a solution-processed WLED, which shows CIE (0.24, 0.25) at the applied voltage of 14 V (**Figure 12a**) near to that of pure white light (0.33, 0.33). The EQE of this WLED was 0.24%, 0.21%, 0.20%, and 0.19% for 11, 12, 13, and 14 V applied voltages, respectively (**Figure 12b**) [46].

Paulo-Mirasol et al. reported white CDs based WLEDs at voltage close to 5 V with a CIE of (0.284, 0.291) as well, while the L_{max} is only 21 cd A^{-1} [45].

Recently, Rao et al. obtained extremely photoluminescent SCWE-CQDs by carefully regulating the dilution ratios between the pure red carbon quantum dots (RCQDs) solution and *N,N*-dimethylformamide (DMF) with relative PLQY around 53%.

They were able to produce white light with CIE (0.328, 0.336), which is close to pure white light (0.33, 0.33), and corresponding CRI and CCT values of 84 and 5710 K, respectively [47].

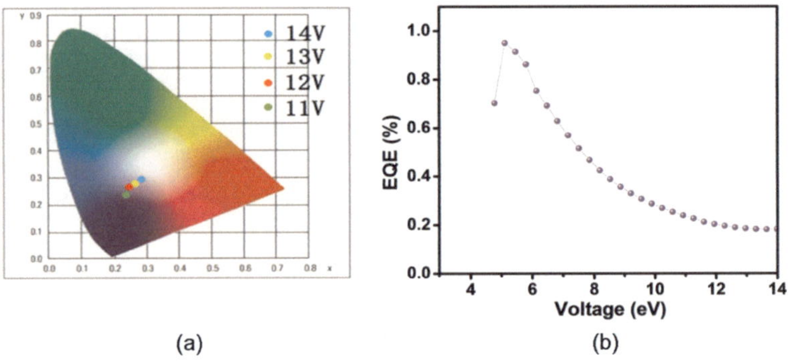

Figure 12.
(a) CIE coordinate at varied applied voltage of where CIE closer to that of pure white light (0.33, 0.33) at 14 V, (b) The EQE of this WLED was 0.24%, 0.21%, 0.20%, and 0.19% for 11, 12, 13, and 14 V applied voltages respectively.

4. Conclusion

CQDs have remarkable potential for PV use due to their large surface area, high conductivity, and rapid charge transfer. CQDs have been employed in a variety of solar cells as sensitizers and co-sensitizers, energy downshifting layers (EDSL), ETL, and hole transport materials (HTM), and their inclusion has enhanced the overall efficiency of solar cells. In addition, solar panels perform better when CQDs are included in the encapsulation layer.

Although warm WLEDs with a high color quality can now be realised using R-CQDs with long-wavelength and highly efficient emission, doing so is still challenging. Hardly reported SCWE-CQDs due to their surface defect-state emission character usually displayed faint emissions. Because of this, bandgap SCWE-CQDs with high QYs and solubility are urgently needed to enhance the luminescence and carrier transfer performance of SCWE-CQD-based electroluminescent WLEDs. Furthermore, highly effective electroluminescent WLEDs require efficient CQDs with triplet-excited-state involved features, such as room-temperature phosphorescence and thermally induced delayed fluorescence.

Author details

Pawan Kumar[1], Shweta Dua[2], Balaram Pani[3] and Geeta Bhatt[4*]

1 Department of Electronic Science, University of Delhi, India

2 Bhaskaracharya College of Applied Sciences, University of Delhi, India

3 University of Delhi, India

4 Non-Collegiate Women Education Board, University of Delhi, India

*Address all correspondence to: geeta.bhatt@bcas.du.ac.in

References

[1] Kumar P, Dua S, Kaur R, Kumar M, Bhatt G. A review on advancements in carbon quantum dots and their application in photovoltaics. RSC Advances. 2022;**12**(8):4714-4759. DOI: 10.1039/D1RA08452F

[2] Xu X, Ray R, Gu Y, Ploehn HJ, Gearheart L, Raker K, et al. Electrophoretic analysis and purification of fluorescent single-walled carbon nanotube fragments. Journal of the American Chemical Society. 2004; **126**(40):12736-12737. DOI: 10.1021/ja040082h

[3] Wang Y, Hu A. Carbon quantum dots: synthesis, properties and applications. Journal of Materials Chemistry C. 2014; **2**(34):6921-6939. DOI: 10.1039/C4TC00988F

[4] Kumar P, Bhatt G, Kaur R, Dua S, Kapoor A. Synthesis and modulation of the optical properties of carbon quantum dots using microwave radiation. Fullerenes, Nanotubes and Carbon Nanostructures. 2020;**28**(9):724-731. DOI: 10.1080/1536383X.2020.1752679

[5] Izatt RM, Izatt SR, Bruening RL, Izatt NE, Moyer BA. Challenges to achievement of metal sustainability in our high-tech society. Chemical Society Reviews. 2014;**43**(8):2451-2475. DOI: 10.1039/C3CS60440C

[6] Mirtchev P, Henderson EJ, Soheilnia N, Yip CM, Ozin GA. Solution phase synthesis of carbon quantum dots as sensitizers for nanocrystalline TiO_2 solar cells. Journal of Materials Chemistry. 2012;**22**(4):1265-1269. DOI: 10.1039/C1JM14112K

[7] Baker SN, Baker GA. Luminescent carbon nanodots: emergent nanolights. Angewandte Chemie International Edition. 2010;**49**(38):6726-6744. DOI: 10.1002/anie.200906623

[8] He P, Shi Y, Meng T, Yuan T, Li Y, Li X, et al. Recent advances in white light-emitting diodes of carbon quantum dots. Nanoscale. 2020;**12**(8):4826-4832. DOI: 10.1039/C9NR10958G

[9] Yuan F, Li S, Fan Z, Meng X, Fan L, Yang S. Shining carbon dots: synthesis and biomedical and optoelectronic applications. Nano Today. 2016;**11**(5): 565-586. DOI: 10.1016/j.nantod.2016.08.006

[10] Meng T, Yuan T, Li X, Li Y, Fan L, Yang S. Ultrabroad-band, red sufficient, solid white emission from carbon quantum dot aggregation for single component warm white light emitting diodes with a 91 high color rendering index. Chemical Communications. 2019; **55**(46):6531-6534. DOI: 10.1039/C9CC01794A

[11] Yuan T, Yuan F, Li X, Li Y, Fan L, Yang S. Fluorescence–phosphorescence dual emissive carbon nitride quantum dots show 25% white emission efficiency enabling single-component WLEDs. Chemical Science. 2019;**10**(42): 9801-9806. DOI: 10.1039/C9SC03492G

[12] Sharma K, Sharma V, Sharma SS. Dye-sensitized solar cells: fundamentals and current status. Nanoscale Research Letters. 2018;**13**(1):1-46. DOI: 10.1186/s11671-018-2760-6

[13] Molaei MJ. The optical properties and solar energy conversion applications of carbon quantum dots: a review. Solar Energy. 2020;**196**:549-566. DOI: 10.1016/j.solener.2019.12.036

[14] Han X, Zhong S, Pan W, Shen W. A simple strategy for synthesizing highly

luminescent carbon nanodots and application as effective down-shifting layers. Nanotechnology. 2015;**26**(6): 065402. DOI: 10.1088/0957-4484/26/6/065402

[15] Riaz R, Ali M, Maiyalagan T, Anjum AS, Lee S, Ko MJ, et al. Dye-sensitized solar cell (DSSC) coated with energy down shift layer of nitrogen-doped carbon quantum dots (N-CQDs) for enhanced current density and stability. Applied Surface Science. 2019; **483**:425-431. DOI: 10.1016/j.apsusc.2019.03.236

[16] Ali M, Riaz R, Bae S, Lee HS, Jeong SH, Ko MJ. Layer-by-Layer self-assembly of hollow nitrogen-doped carbon quantum dots on cationized textured crystalline silicon solar cells for an efficient energy down-shift. ACS Applied Materials & Interfaces. 2020; **12**(9):10369-10381. DOI: 10.1021/acsami.9b21087

[17] Lv T, Tang Y, Fan H, Liu S, Zeng S, Liu W. Carbon quantum dots anchored on the anti-reflection silica layer as solid luminescence down-shifting materials in solar panel encapsulation. Solar Energy Materials and Solar Cells. 2022;**235**: 111450. DOI: 10.1016/j.solmat.2021.111450

[18] Essner JB, Baker GA. The emerging roles of carbon dots in solar photovoltaics: a critical review. Environmental Science: Nano. 2017; **4**(6):1216-1263. DOI: 10.1039/C7EN00179G

[19] Gao N, Huang L, Li T, Song J, Hu H, Liu Y, et al. Application of carbon dots in dye-sensitized solar cells: a review. Journal of Applied Polymer Science. 2020;**137**(10): 48443. DOI: 10.1002/app.48443

[20] Wang H, Sun P, Cong S, Wu J, Gao L, Wang Y, et al. Nitrogen-doped carbon dots for "green" quantum dot solar cells. Nanoscale Research Letters. 2016;**11**(1):1-6. DOI: 10.1186/s11671-016-1231-1

[21] Mistry B, Machhi HK, Vithalani RS, Patel DS, Modi CK, Prajapati M, et al. Harnessing the N-dopant ratio in carbon quantum dots for enhancing the power conversion efficiency of solar cells. Sustainable Energy & Fuels. 2019;**3**(11): 3182-3190. DOI: 10.1039/C9SE00338J

[22] Yang Q, Yang W, Zhang Y, Ge W, Yang X, Yang P. Precise surface state control of carbon quantum dots to enhance charge extraction for solar cells. Nanomaterials. 2020;**10**(3):460. DOI: 10.3390/nano10030460

[23] Singh T, Singh J, Miyasaka T. Role of metal oxide electron-transport layer modification on the stability of high performing perovskite solar cells. ChemSusChem. 2016;**9**(18):2559-2566. DOI: 10.1002/cssc.201601004

[24] Yan L, Yang Y, Ma CQ, Liu X, Wang H, Xu B. Synthesis of carbon quantum dots by chemical vapor deposition approach for use in polymer solar cell as the electrode buffer layer. Carbon. 2016;**109**:598-607. DOI: 10.1016/j.carbon.2016.08.058

[25] Li H, Shi W, Huang W, Yao EP, Han J, Chen Z, et al. Carbon quantum dots/TiO_x electron transport layer boosts efficiency of planar heterojunction perovskite solar cells to 19%. Nano Letters. 2017;**17**(4):2328-2335. DOI: 10.1021/acs.nanolett.6b05177

[26] Zhu X, Sun J, Yuan S, Li N, Qiu Z, Jia J, et al. Efficient and stable planar perovskite solar cells with carbon quantum dots-doped PCBM electron transport layer. New Journal of Chemistry. 2019;**43**(18):7130-7135. DOI: 10.1039/C8NJ06146G

[27] Park S, Lee H, Park SW, Kim TE, Park SH, Jung YK, et al. Improved exciton dissociation efficiency by a carbon-quantum-dot doped work function modifying layer in polymer solar cells. Current Applied Physics. 2021;**21**:140-146. DOI: 10.1016/j.cap.2020.10.019

[28] Pitchaiya S, Natarajan M, Santhanam A, Asokan V, Yuvapragasam A, Ramakrishnan VM, et al. A review on the classification of organic/inorganic/carbonaceous hole transporting materials for perovskite solar cell application. Arabian Journal of Chemistry. 2020;**13**(1): 2526-2557. DOI: 10.1016/j.arabjc.2018.06.006

[29] Vercelli B. The role of carbon quantum dots in organic photovoltaics: a short overview. Coatings. 2021;**11**(2): 232. DOI: 10.3390/coatings11020232

[30] Paulo S, Stoica G, Cambarau W, Martinez-Ferrero E, Palomares E. Carbon quantum dots as new hole transport material for perovskite solar cells. Synthetic Metals. 2016;**222**:17-22. DOI: 10.1016/j.synthmet.2016.04.025

[31] Nguyen DN, Roh SH, Kim DH, Lee JY, Wang DH, Kim JK. Molecular manipulation of PEDOT: PSS for efficient hole transport by incorporation of N-doped carbon quantum dots. Dyes and Pigments. 2021;**194**:109610. DOI: 10.1016/j.dyepig.2021.109610

[32] Feng X, Zhao Y, Yan L, Zhang Y, He Y, Yang Y, et al. Low-temperature hydrothermal synthesis of green luminescent carbon quantum dots (CQD), and optical properties of blends of the CQD with poly (3-hexylthiophene). Journal of Electronic Materials. 2015;**44**(10): 3436-3443. DOI: 10.1007/s11664-015-3893-3

[33] Cui B, Yan L, Gu H, Yang Y, Liu X, Ma CQ, et al. Fluorescent carbon quantum dots synthesized by chemical vapor deposition: an alternative candidate for electron acceptor in polymer solar cells. Optical Materials. 2018;**75**:166-173. DOI: 10.1016/j.optmat.2017.10.010

[34] Privitera A, Righetto M, Mosconi D, Lorandi F, Isse AA, Moretto A, et al. Boosting carbon quantum dots/fullerene electron transfer via surface group engineering. Physical Chemistry Chemical Physics. 2016;**18**(45): 31286-31295. DOI: 10.1039/C6CP05981C

[35] Shejale KP, Jaiswal A, Kumar A, Saxena S, Shukla S. Nitrogen doped carbon quantum dots as Co-active materials for highly efficient dye sensitized solar cells. Carbon. 2021;**183**: 169-175. DOI: 10.1016/j.carbon.2021.06.090

[36] Sivagami D, Priyadarshini BG. Role of carbon quantum dot for enhanced performance of photo-absorption in Cu 2 CoSnS 4 superstrate solar cell device. Materials Advances. 2022;**3**(5):2405-2416. DOI: 10.1039/D1MA01117K

[37] Kumari R, Sahu SK. A new insights into multicolor emissive carbon dots using Trachelospermumjasminoides leaves for the application of WLEDs. Colloids and Surfaces A: Physicochemical and Engineering Aspects. 2022;**647**:128959. DOI: 10.1016/j.colsurfa.2022.128959

[38] Cui B, Feng XT, Zhang F, Wang YL, Liu XG, Yang YZ, et al. The use of carbon quantum dots as fluorescent materials in white LEDs. New Carbon Materials. 2017;**32**(5):385-401. DOI: 10.1016/S1872-5805(17)60130-6

[39] Zheng J, Wang Y, Zhang F, Yang Y, Liu X, Guo K, et al. Microwave-assisted hydrothermal synthesis of solid-state carbon dots with intensive emission for white light-emitting devices. Journal of Materials Chemistry C. 2017;**5**(32):8105-8111. DOI: 10.1039/C7TC01701D

[40] Wang Z, Yuan F, Li X, Li Y, Zhong H, Fan L, et al. 53% efficient red emissive carbon quantum dots for high color rendering and stable warm white-light-emitting diodes. Advanced Materials. 2017;**29**(37):1702910. DOI: 10.1002/adma.201702910

[41] Zhu J, Bai X, Chen X, Shao H, Zhai Y, Pan G, et al. Spectrally tunable solid state fluorescence and room-temperature phosphorescence of carbon dots synthesized via seeded growth method. Advanced Optical Materials. 2019;**7**(9):1801599. DOI: 10.1002/adom.201801599

[42] Li W, Wu M, Jiang H, Yang L, Liu C, Gong X. Carbon dots/ZnO quantum dots composite-based white phosphors for white light-emitting diodes. Chemical Communications. 2011;**58**(12):1910-1913. DOI: 10.1039/D1CC06180A

[43] Yuan, F., Wang, Z., Li, X., Li, Y., Tan, Z.A., Fan, L. and Yang, S. Bright multicolor bandgap fluorescent carbon quantum dots for electroluminescent light-emitting diodes. Advanced Materials, 2017, 29(3), p. 1604436. DOI: 10.1002/adma.201604436

[44] Jia H, Wang Z, Yuan T, Yuan F, Li X, Li Y, et al. Electroluminescent warm white light-emitting diodes based on passivation enabled bright red bandgap emission carbon quantum dots. Advanced Science. 2019;**6**(13):1900397. DOI: 10.1002/advs.201900397

[45] Paulo-Mirasol S, Martínez-Ferrero E, Palomares E. Direct white light emission from carbon nanodots (C-dots) in solution processed light emitting diodes. Nanoscale. 2019;**11**(23):11315-11321. DOI: 10.1039/C9NR02268F

[46] Luo Z, Qi G, Chen K, Zou M, Yuwen L, Zhang X, et al. Microwave-assisted preparation of white fluorescent graphene quantum dots as a novel phosphor for enhanced white-light-emitting diodes. Advanced Functional Materials. 2016;**26**(16):2739-2744. DOI: 10.1002/adfm.201505044

[47] Rao L, Zhang Q, Wen M, Mao Z, Wei H, Chang HJ, et al. Solvent regulation synthesis of single-component white emission carbon quantum dots for white light-emitting diodes. Nanotechnology Reviews. 2021;**10**(1):465-477. DOI: 10.1515/ntrev-2021-0036

Chapter 4

Quantum Dots Sensitized Solar Cell

Vikram P. Bhalekar

Abstract

In the modern and automated twenty-first century, with technological advancements, the scientific society has gifted a new alternative clean energy source (dye/quantum dot sensitized solar cells) to mankind as one of the applications of nanomaterials. Nowadays, the world is using a tremendous amount of fossil fuel for energy creation, the solar energy by using nanomaterials in the form of solar cells is a perfect alternative. In the present chapter, the emphasis has been given on the different techniques used by the researcher for synthesis of nanoparticles. The synthesis of quantum dots by simple cost-effective technique is covered with respect to PbS quantum dots. The working of quantum dot sensitized solar cells is also explained with its basic components. The narrow-band-gap semiconducting materials, which are suitable for solar cell application, are also listed in this context.

Keywords: solar cells, quantum dots, PbS, nanomaterials, UV, HRTEM, SAED

1. Introduction

The large existing need of the global energy is accomplished by burning fossil fuels. However, the inherent problems associated with the use of fossil fuel are their limited availability and the environmental issues. This forces mankind to look for new, more sustainable long-term energy solutions to provide the future energy resource. As the major renewable energy source, i.e., solar energy has the potential to become an essential component of future global energy production. The solar cell converts sunlight directly into electricity. In recent year, the concept of utilizing nanomaterial-based architectures in light energy conversion devices has emerged as an alternative to single-crystalline-based photovoltaic devices.

Besides the silicon solar cells, quantum dots sensitized solar cell (44%) and dye sensitized solar cells (33%) are two promising alternatives due to their cost-effective concepts for solar to electric energy conversion. The configuration of QDSSC consists of Photoanode (TiO_2, ZnO, ZrO_2, SnO_2, etc., nanoparticle thin films), quantum dots (PbS, PbSe, CdS, CdSe, etc.), electrolyte (polysulfide, cobalt complex, etc.), and counter electrode (MoO_3, CuS, CuSe, Pt, carbon black, etc.). Upon the illumination of light, the QDs generate the exciton (electron-hole pairs). Consequently, electron is injected into the photoanode to generate photocurrent, the hole is transported to the counter through redox couple toward electrode (Cathode). The overall power conversion efficiency (PCE) of a QDSSC is governed by the light harvesting efficiency, quantum yield for charge injection, and charge collection efficiency.

IntechOpen

The chapter focuses on different approaches of preparation of nanomaterials. Synthesis of Quantum Dots (PbS) by simple cost-effective route. The structure of Quantum Dot Sensitized Solar Cell is given with requirement of basic components. The use of narrow-band-gap semiconductor, the materials suitable for the solar cell are also discussed in this topic.

2. Methods of preparation of nanomaterials

Numerous methods of nanomaterials synthesis have been developed. Generally, the two approaches top-down and bottom-up are used.

a. **Top-down approach:** In this approach, the etching and ball milling techniques are generally used to crush down large structure into small one, till the desired size is achieved.

b. **Bottom-up approach:** This approach uses the atomic or molecular precursors to form nanostructure; the nanoparticles are formed atom by atom or layer by layer. This approach is more widespread and generally used to synthesis of nanomaterials having different morphologies and properties. Hence, bottom-up approach has given a stress for further important discussion.

The method of synthesis of nanoparticle is given in **Figure 1.**
Synthesis of nanoparticles:
There are four methods used for the synthesis of nanoparticles:

1. Physical

2. Chemical

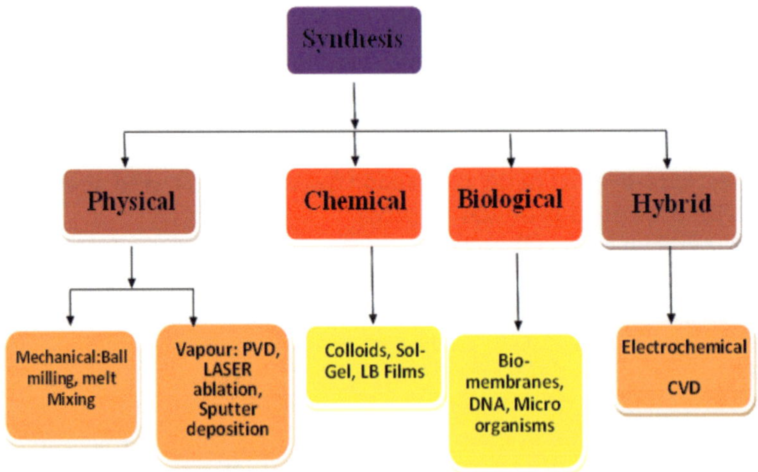

Figure 1.
Methods of synthesis of nanomaterials.

3. Biological

4. Hybrid

2.1 Physical methods

It is broadly divided into two types, namely mechanical and vapor-based methods [1–3].

2.1.1 Mechanical methods

In this process, containers are used to make nanoparticles, hard balls of tungsten carbide, zirconia, or steel are plunged in the containers along with the materials whose nanoparticles are to be made. So this high-energy ball milling is a type of mechanical method.

2.1.2 Vapor-based methods

The nanostructures are formed by evaporating different material on various types of substrates. It is divided into **a**. Physical vapor deposition (PVD) and **b.** Chemical vapor deposition (CVD).

2.1.2.1 Physical vapor deposition (PVD) method

It is general term used to describe any of the variety of methods to deposit thin films by the condensation of vaporized form of the desired films material on to various working surfaces (e.g., onto semiconductor wafers). PVD is used in the manufacturing of semiconductor devices.

2.1.2.1.1 LASER ablation

LASER ablation is the process of removing material from a solid (or occasionally liquid) surface by using pulse of LASER beam. At low LASER flux, material is heated by the absorbed energy and evaporates or sublimates, and at high flux, the material is converted to plasma. Single-wall carbon nanotubes are synthesized by LASER ablation method.

2.1.2.1.2 Biological method or green synthesis

Synthesis of nanomaterial using biological ingredient, which is divided by using microorganism such as fungi, yeasts, and bacteria, plants extracts or enzymes and use of templates such as DNA, membrane, viruses, diatoms.

2.1.2.1.3 Hybrid method

Hybrid method utilized for the preparation of nanoparticles involves the process such as electromechanical, chemical vapor deposition, particle arresting in glass or zeolites or polymers micro-emulsion.

2.1.2.2 Chemical vapor deposition

There are techniques depending on the source of excitation and the conditions of deposition, which include Atmospheric-Pressure CVD (APCVD), Low-Pressure CVD (LPCVD), Ultrahigh-Vacuum CVD (UVCVD), Metal Organic CVD (MOCVD), and Hot Filament CVD (HFCVD). In this method, when a substrate is exposed to one or two volatile precursors, they chemically react in vapor phase or decompose on its surface under thermal, plasma, or laser excitation and form nanostructures of high quality.

2.2 Chemical methods

These are less expensive, simple, and easily scalable techniques and used to synthesize nanostructures of different shape, morphologies, and particle sizes.

Various chemical methods are discussed as follows:

2.2.1 Colloidal synthesis

Colloids are the class (few nm sizes) of material where two or more phases exist with a size less than a micron. Particles are generally suspended in some host matrix. The colloidal particles are stabilized by columbic repulsion or stearic hindrance against aggregation.

Colloidal synthesis is carried out generally in a three-neck flask and to avoid processes such as oxidation of products; the reaction is carried out in inert atmosphere.

2.2.2 Chemical bath deposition

The high-quality metal chalcogenide nanocrystalline thin films (metallic or non-metallic substrate) are fabricated by using this technique. This requires the metal and chalcogen ions. In other way, deposition of the film is possible if the ionic product should exceed the solubility product.

Thin film deposition proceeds as nucleation on the surface of the substrate, i.e., formation of stable second phase by the combination of a minimum number of ions or molecules in contact with a solution. The adsorbed cations/anions deposited on the substrate act as nucleation centers, and subsequent deposition takes place through adsorption of more ionic species present in the solution that results in the homogeneous and uniform film. The film deposition takes place ion by ion or cluster by cluster. The former results into uniform, adherent thin film, whereas latter thick, powdery, and diffusely reflecting films. The pH is also one of the deciding factors, at higher pH values, solubility product exceeds ionic product, i.e., unfavorable condition for deposition. Also as the temperature increases, metal ion gets dissociated from the complexing agent and film deposits through nucleation. Moreover, concentration, type of complexing agent, stirring rate of the solution also decide the quality of films [4, 5].

Doctor blade method:

Doctor blade method is very simple technique for coating of substrate, in this method precursor, solution is dropped/placed on the substrate. The film is produced by moving the blade over the substrate. Drying and annealing steps are used to form a solid film on the substrate. Various steps involved in doctor blade method are illustrated in **Figure 2**.

In doctor blade technique, well-mixed slurry of nanoparticles along with other constituents such as binder, dispersants, etc., are used. The slurry is spread over on a

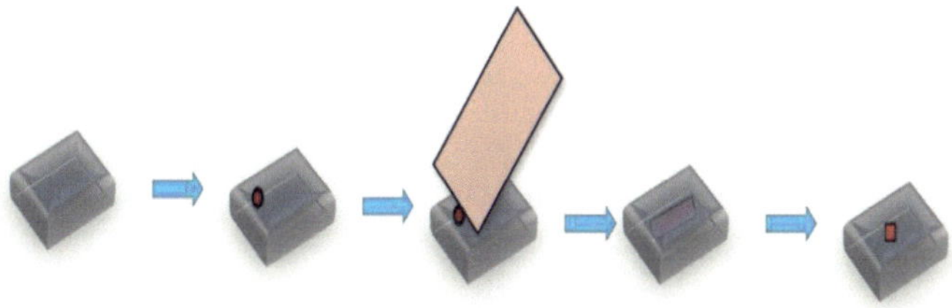

Figure 2.
Doctor blade technique for the fabrication of photoanode.

substrate and thin layer is produced, thickness of films is depending on the distance between blade and substrate [6, 7]. The advantages of Doctor Blade method are

- Formation of large area film
- No wastage of material
- Uniform film deposition

2.2.3 Successive ionic layer and adsorption reaction (SILAR) method

SILAR is also known as modified chemical bath deposition technique. It has advantages such as

1. It is the simplest method for deposition of any element in any proportion.
2. SILAR method does not require very good quality substrate, target, and vacuum like closed vapor deposition method.
3. Thickness and size of film are easily controlled by changing deposition cycles.
4. Deposition is carried out at room temperature.

It is mostly based on the phenomenon, i.e., adsorption and reaction of the ionic species. In which, the substrate is dipped in the separate solution for the fixed interval of time. One SILAR cycle is completed when substrate is dipped in both the precursors once successively. In order to avoid the precipitation, after every withdrawal of the films, it is followed by rinsing the substrate by alcohol or water.

A large number of metal chalcogenides (PbS, CdS, CdSe) [8] and metal oxides are prepared using this method. It is the simplest technique with which QD size and coverage can be controlled. The schematic of SILAR [9] deposition is given in **Figure 3**.

2.2.4 Electrodeposition

The binder-free and inexpensive technique used to produce nanoparticle with controlled morphology, size, and composition. In this technique, varying operating parameters of deposition such as concentration of solution, operating voltage, deposition potential and electric current flow, the thickness and its crystallographic orientation can be controlled.

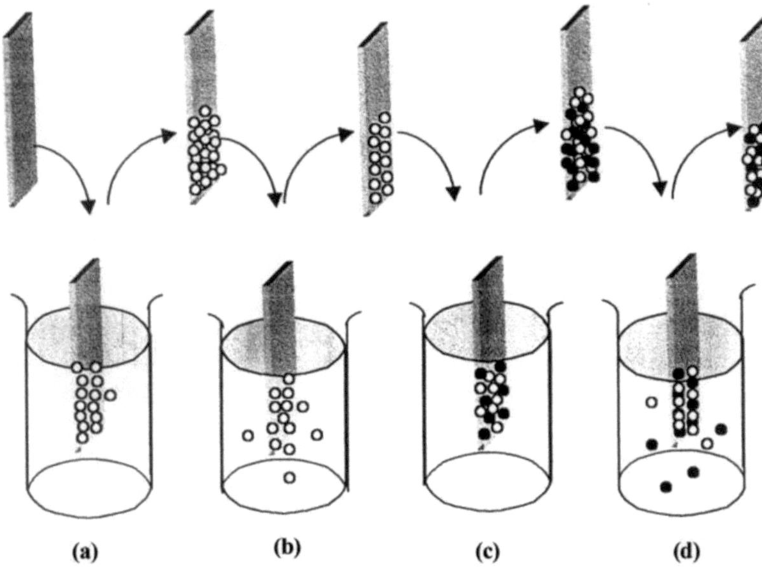

Figure 3.
Schematic of SILAR depositions. Source: Ref. [9].

The films produced by this technique are uniform, adherent. As per the reports, the synthesis of nanoparticles, nanowires, metal oxides, and metal nanoparticles is possible using electrodeposition [10].

2.2.5 Sol-gel method

This technique generally has hydrolysis of precursors, condensation followed by polycondensation to form particles, gelation, and drying process. In this technique, sol is generated and converted into a viscous gel, and then it gets converted into a solid material.

This technique is commonly used to synthesize ceramic or metal oxides [11].

2.2.6 Langmuir-Blodgett method (L-B)

L-B method is used to deposit the molecular monolayers and multilayers. It is suitable to deposit various organic materials salts and fatty acids [12]. This technique transfers organic layer at air liquid interface on to solid substrate in which amphiphilic long-chain molecule like oleic epoxide. Although the layers of L-B films are ordered, there is only Vander Waals interaction between these layers. Thus, even large number of layer present in the film preserves its two dimensional properties.

2.2.7 Spray pyrolysis

This is very simple and cost-effective deposition technique to deposit thin and thick films of metal oxides, metal sulfides [13, 14], etc. The dense, porous, multilayer films can be prepared by this technique.

The spray pyrolysis unit consists of atomizer, which sprays a metal salt solution onto a heated substrate; the droplets sprayed on undergo a thermal decomposition.

There are also other factors such as precursor solution, temperature controller, and substrate heater influencing the decomposition.

The synthesis of PbS quantum dot by simple facile technique without capping agent and clumsy vacuum techniques is discussed here.

3. Synthesis of lead sulfide quantum dots

QDs possess size-dependent and discrete electronic energy spectra due to quantum confinement effect [15, 16]. Quantum dots such as CdS, CdSe, InP, PbS, PbSe, etc., are synthesized by researchers using a methods mentioned above and have a wide range of applications. Among metal chalcogenides, lead chalcogenides, especially PbS and PbSe QDs have been interesting nanostructures due to their characteristic property to display Multiple Exciton Generation, where a single photon can yield three excitons; hence, it is useful in highly efficient photovoltaic conversion. Also, PbS quantum dot sensitized solar cell gave a very high photocurrent. In view of these, the synthesis and application of PbS have assumed great importance. PbS is a IV–VI semiconductor with Bohr excitonic radius of 18 nm. It has a bulk band gap of 0.41 eV that can be tuned up to 1.5 eV at the QD level, and hence, it shows a strong quantum confinement effect. PbS has its applicability in sensors, photography, IR detector (due to absorption near IR region), solar absorber, etc.

To synthesize QDs, hydrothermal, sono-chemical, micro-emulsion, and organometallic techniques have been developed by the researchers. The organometallic method gives a better size distribution of PbS nanocrystals, but the formation involves hazardous and unstable chemicals such as $(TMS)_2S$, trioctylphosphine, etc.; therefore, it is significant to find a simple route. Previously PbS QDs resulted in either polycrystalline or single crystalline with the help of a coordinating agent, but Bhalekar and Pathan synthesized PbS quantum dots by sono-chemical technique using precursors such as lead nitrate $Pb(NO_3)_2$ and sodium sulfide Na_2S in aqueous media, without any harmful element.

The experimental procedure involved in the synthesis is as follows:

About 0.01 M of $Pb(NO_3)_2$ and Na_2S (very well dispersed) solutions were prepared in double distilled water (DDW) at room temperature separately. These solutions are then added to 400 ml DDW drop-wise and the color variation is observed. The solutions are sonicated further using probe sonicator (ENUP-250A). The procedural steps [15] are schematically shown as in **Figure 4**. The solutions with color variation are named from A to D. Sonication is an important because it produces cavitation in the medium, which is equivalent to pressures of few hundreds of atmospheres. The hot spots in the aqueous solution are due to ultrasonication, and it de-agglomerates/ slows down the rate of agglomeration of the particles in the medium.

The sonication creates the conditions such as high temperature and pressure, which is not possible by other techniques. The rate of reaction increases with the rate of agglomeration. Y. Dong et al. have used sodium sulfide along with oleic acid as coordinating agent for controlling growth, stabilizing the resulting colloidal dispersion, and electronically passivating the semiconductor surface. Bhalekar and Pathan [15] controlled the growth only by large aqueous bath and sonication. Thus, growth is not rapid to form bulk; hence, it results in well-defined PbS QDs. The drop-wise addition of precursors changes the color of the solution from colorless to faint yellow [15] as in **Figure 4** and with further addition changes it from yellow to faint brown and at last to brown. A rapid nucleation occurs when there is rapid injection, which turned the solution immediate black color, in this condition, it is difficult to achieve narrow size distribution.

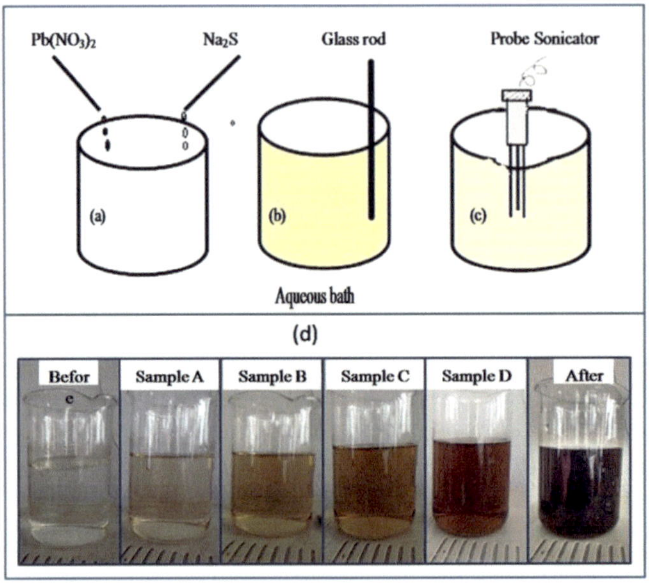

Figure 4.
Synthesis procedure. (a) Addition of precursors in the aqueous bath. (b) Mixing of the solution using glass rod. (c) Ultrasonication of the solution using probe sonicator. (d) Actual photograph of synthesized quantum dots. Source: Ref. [15].

The reaction in the bath is, as initially $Pb(NO_3)_2$ and Na_2S produces Pb^{2+}, $(NO_3)^-$, Na^+ and S^{2-} free ions respectively in aqueous bath and in a large bath it yield as,

$$Pb^{2+} + 2(NO_3)^- + 2Na^+ + S^{2-} \xrightarrow[aq.]{} PbS + 2NaNO_3$$

The solubility of the $Pb(NO_3)_2$ in aqueous solution can be given as,

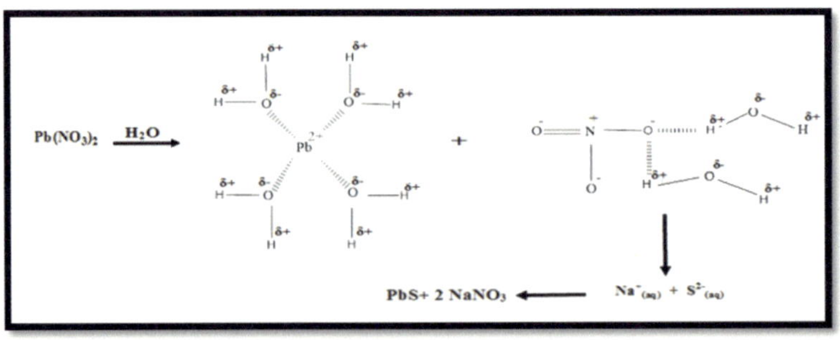

The reaction involved in the aqueous bath proceeds as per the Hard and Soft Acids and Bases (HSAB) Theory. In which Na^+ is a hard acid and $(NO_3)^-$ is a hard base and hence they combine. Pb^{2+} is a border line acid and S^{2-} is a soft base and due to prominent interaction PbS is formed.

Optical Properties of Lead Sulfide Quantum Dots are as absorption is observed in QDs due electron-hole pair generation persuaded by absorption of photons. The optical absorption spectra [15] in **Figure 5** reveals that, initially when the color of the solution is faint yellow, the peak observed at 1155 nm in near infrared region known as

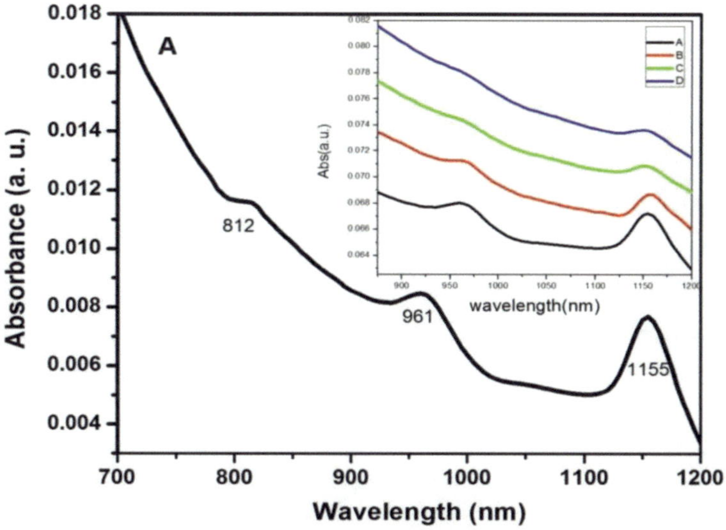

Figure 5.
Absorption spectra for synthesized lead sulfide quantum dots. Source: Ref. [15].

first excitonic peak indicating the formation of PbS QDs. The quantum confinement effect is also evident from the large blue shift in absorption spectra (shown in the inset). In the main spectrum, the second and third excitonic peaks are at 965 nm and 812 nm respectively. The average radius and effective band gap energy of PbS QDs are 2.55 nm and 1.073 eV respectively. The peak positions for sample B, C are same as that of sample "A" except sample "D" in which the second and third excited peaks are not visible due to large particle size. Also, from the sample "A" onward, sharpness of peaks decreases with the color variation. This specifies a broader particle size distribution.

TEM images [15] of synthesized lead sulfide quantum dots as in **Figure 6** are taken at 500 nm, 200 nm, 50 nm, and 20 nm, and it confirms the formation of QDs from [15] HRTEM image **Figure 7a** and selected area electron diffraction (SAED) pattern **Figure 7b**. The regular circular particles as observed are in between 3 and 8 nm and the average mean radius of PbS QD was 6 nm as evident from particle size distribution curve [15], i.e., histogram as in **Figure 8**. Since the inter-planar spacing (0.347 nm) observed here is in conformity with the standard PbS data. **Figure 7b**, i.e., SAED pattern of the sample "A" confirms the formation of predominantly single nanocrystalline structure. This diffraction fringes of QDs matches with the cubic phase of PbS and labeled rings have been identified to the (220), (222), (400), (440), (511) planes [JCPDS no.: 05-0592].

3.1 Nanostructured solar cells: the QD class

Quantum dot sensitized solar cells are excitonic solar cells, the basic idea behind the emergence of third-generation photovoltaics is to design solar cells with efficiency that exceeds the limit proposed by Shockley and Quiesser. QDs with their unique characteristics are widely used to improve the efficiencies of QDSSC. They are the structures having properties such as size-dependent optical band gap, high molar extinction coefficients, high intrinsic dipole moments, which giving rise to good charge separations and also the multiple charge carriers are created by using a single photon, in order to build a stable solar cell.

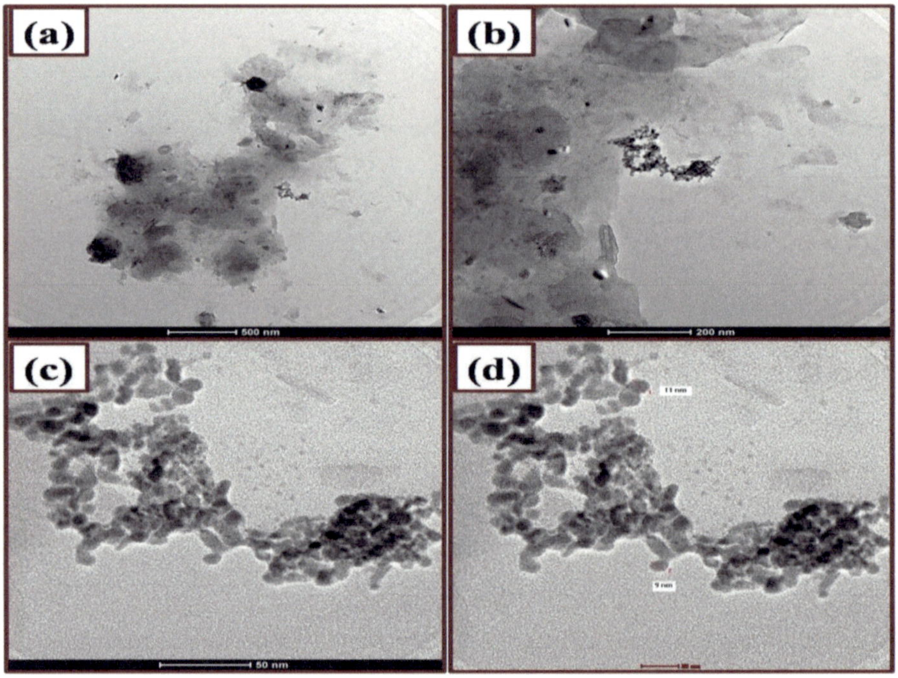

Figure 6.
Transmission electron microscopic images of lead sulfide quantum dots at various magnifications. Source: Ref. [15].

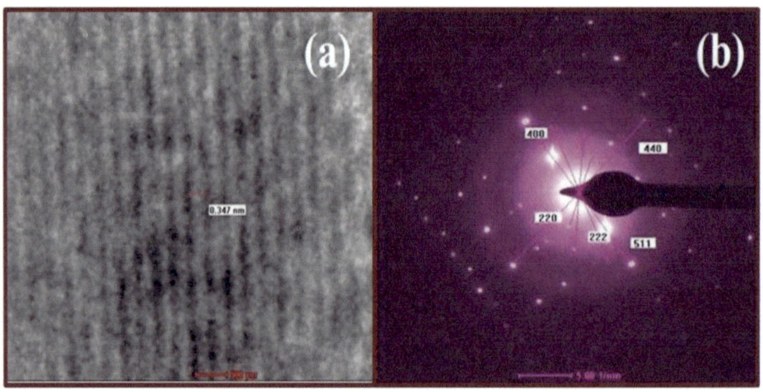

Figure 7.
(a) HRTEM image and (b) SAED pattern of lead sulfide quantum dots. Source: Ref. [15].

3.2 Quantum dot sensitized solar cells: an excitonic class

The cost-effectiveness and simple method of fabrication make the Dye Sensitized Solar Cell (DSSC) the popular customer among scientists. The maximum thermodynamic photoconversion efficiency is 31%, which is smaller than the Shockley Quiesser limit. The light harvesting capabilities and band of absorption are disadvantages of DSSC. Solar spectrum may not be completely utilized by the dye due to the limited absorption band. Dye absorption duration reduces the performance of the solar cell due to aggregation. When the dyes are replaced by zero-dimensional structures

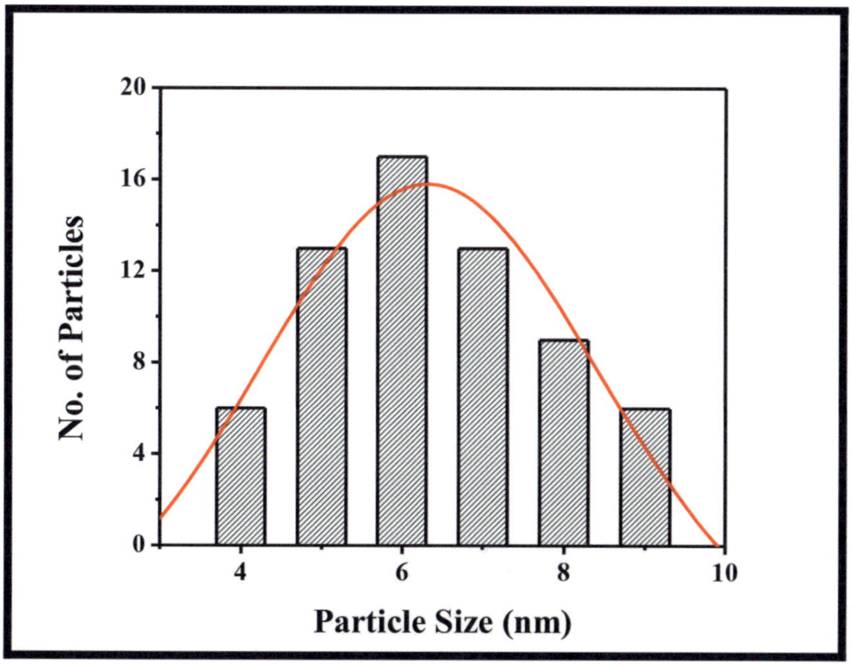

Figure 8.
Particle size distribution of lead sulfide quantum dots. Source: Ref. [15].

(Quantum dots), the thermodynamic limits will get altered. For QDSSC, 44% [17] is the maximum projected thermodynamic efficiency.

3.3 Basic architecture of QDSSC

QDSSC [16] consists of three essential components, namely photoanode, electrolyte containing redox couple, and counter electrode. **Figure 9** shows the schematic of QDSSC including the electron transfer processes.

3.4 Working of QDSSC

When the light is incident on the solar cell through a transparent conducting oxide, photons are absorbed to generate excitons. As the band positions between the Conduction Band (CB) of the Photoelectrode and the CB of QD is greater than the binding energy of the exciton, then it gets separated at the Metal Oxide (MO)/QD interface into electrons and holes. The electron moves into the CB of MO to reach the transparent Conducting Oxide (TCO) surface before it reaches the band. In the other way, the electrolyte reduces the reductant species (Re), i.e., hole created in the QD. The Re turns into Oxidant species (Ox) after losing electron to the hole, which diffuses toward the Counter Electrode (CE) to receive an electron coming from the external circuit. With the hole reduction, the QD is ready to absorb another photon for the creation of an exciton.

Thus, a photovoltaic effect converting light into electricity does useful work in the external load. The constant photoconversion process is continual till the light is incident on the active area of the solar cell. The photovoltaic processes discussed above are summarized as,

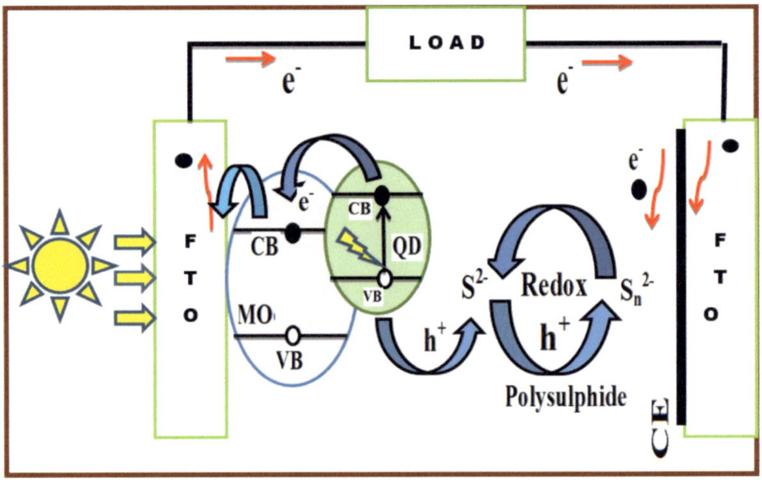

Figure 9.
Schematic of quantum dot sensitized solar cells. Source: Ref. [16].

$$QD + hv \rightarrow QD^{*} \quad (\text{formation of exciton } \textit{due} \text{ to absorption of photon}) \tag{1}$$

$$QD^{*} + MO \rightarrow QD^{+} + MO^{-} \quad (\text{Electron tranfer from } QD \text{ to } MO) \tag{2}$$

$$QD^{+} + \text{Re} \rightarrow QD + Ox \quad (\text{charged } QD \text{ getting reduced to neutral } QD) \tag{3}$$

$$Ox + e^{-} \rightarrow \text{Re} \quad (\text{Reduction of oxidant frome - supplied } \textit{by CE}) \tag{4}$$

Components of QDSSC:

Photoelectrode: Electronic structure of metal oxide and its effect on electron injection.

The wide-band-gap semiconductors are needed for transmitting maximum solar spectrum (visible and infrared region) for effective power conversion efficiency. The conduction band of Photoanode should be more positive with respect to the conduction band of the sensitizer. That creates a band counterbalance between the QD-MO and has electron injection from QD to MO.

In addition, electrical band gap, the electronic mobility, and electronic structure are taken into consideration. For smooth passage of the carriers through the metal oxide semiconductors SnO2 and ZnO are preferred materials as photoelectrodes. In spite of that, TiO2 has appeared as a model photoanode semiconducting system, which demonstrated the best of the efficiencies in both DSSC and QDSSC.

The electron transfer from sensitizer to MO in both DSSC and QDSSC is directed by Marcus theory [18, 19]. According to this theory, the rate of electron injection is directly proportional to Density of States (DOS) in the conduction band of metal oxide. But, the DOS depends on the effective mass of the electron (me*). me* of CB electrons in titania is around 5–10 me (me is the mass of the electron) and about 0.3 me in ZnO and SnO_2 [20]. TiO_2 has two orders of magnitude densities of vacant states

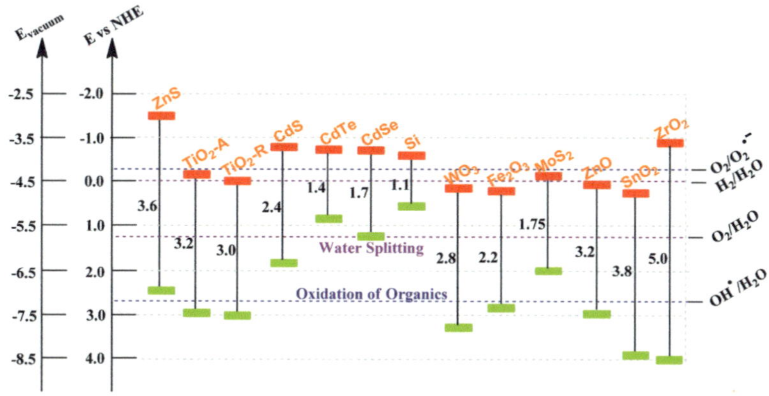

Figure 10.
Band positions of metal oxides with respect to vacuum. Source: Ref. [21].

greater than in ZnO and SnO_2. Some of the MO and their band positions with respect to vacuum level are shown in **Figure 10** [21].

The effective area offered by the photoelectrode or photoanode surface is many times greater than its geometrical area. This assists the wider coverage of sensitizer for better light harvesting capability in the electrode. Various studies have proven the influence of particle size of photoanode on electron transport properties such as charge transport resistance (Rt) and charge recombination resistance Rr. Improved charge transport was observed in particles with greater sizes leading to longer diffusion length and minimal collisions at grain boundaries.

(*The grain boundaries defect (the interface between two grains) reduces the electronic transport through a material)

With a porosity of 60%, the calculated electron diffusion length in a mesoporous titania photo electrode is 15–20 μm. The use of one-dimensional nanostructure improves electron transport properties such structures includes nano tubes, nano rods, nano wires, and nano fibers. Some of the metal oxides are listed below:

3.4.1 Titanium dioxide (TiO_2)

Titanium dioxide is in general n-type a wide direct band gap semiconductor. It is inexpensive and nontoxic. This metal oxide has three morphologies, namely Anatase, Rutile, and Brookite having band gap 3.2 eV, 3.05 eV, and 3.26 eV, respectively.

Rutile is the most stable morphology of the titania, it is used in device application when light scattering is important. The Fermi level of anatase is 100 mV, which is higher than rutile; due to this anatase is attaining higher open-circuit voltages when used as electrode in solar cell applications.

In addition, anatase has greater surface area and hence large amount of sensitizer loads on it, rather than rutile, which results in to greater photocurrents when used in DSSC/QDSSC.

3.4.2 Zinc oxide (ZnO)

Zinc Oxide (ZnO) is wide-band-gap semiconducting material II–VI having direct band gap 3.2 eV, and bond energy is 60 meV at room temperature. ZnO is frequently used as the domain of solar energy conversion due to its stability against photo corrosion and photochemical property and alternative material to TiO2 owing huge surface area and high catalytic activity. It has a great benefit to be applied in a catalytic reaction process. Zinc oxide crystallizes in three main forms, hexagonal wurtzite, cubic zinc blende, and rocksalt. The wurtzite structure is most stable at environmental conditions, and thus, it is most common. ZnO is used in gas sensing, photocatalyst, solar cell, ultraviolet light emitting materials, field effect transistors, and transparent conductors.

3.4.3 Zirconium dioxide ZrO_2

The white crystalline oxide ZrO_2 adopts monoclinic crystal structure at room temperature and transitions to tetragonal and cubic at higher temperatures. The band gap of ZrO_2 depends on the phase (cubic, tetragonal, monoclinic, or amorphous) and synthesis method, due to trap states, its band gap decreases up to 2.8–3.7 eV.

ZrO_2 is used as a photoelectrode for the fabrication of solar cell, due to its high refractive index, wide band gap, low absorption, and dispersion in the visible and near infrared spectral region. Along with these materials, tin oxide, cerium oxide are also used for fabrication of photoanodes.

3.4.4 Photoelectrode sensitization with QDs

The photoelectrode (mesoporous nanocrystalline) is the means of electron transport in solar cells, after the dissociation of exciton at the MO/QD interface. But it stands unworthy unless it is sensitized by the QDs and the photovoltaic process cannot be initiated. The QDs need to absorb maximum part of solar spectrum to create better exciton density at the interface. In order to have better charge transportation, the mere absorption is not sufficient for power conversion, but this demands proper band positioning between MO and QD, i.e., band offset should be greater than the exciton binding energy. Different types of QDs are involved in QDSSC; these includes CdS, CdSe, CdTe, ZnSe, PbS, PbSe, Bi2S3, Bi2Se3, Sb2S3, Sb2Se3, and InAs ternary compounds such as $CuInS_2$, $CuInSe_2$.

The band position, for example, lead chalcogenide QD system is shown in **Figure 11** with favorable band edges with respect to MO [16].

Multiple exciton generation: Under the illumination condition, a high-energy photon excites the electron from the valence band (VB) and jumps to conduction band (CB), and it generates one electron-hole pair, i.e., exciton.

As the energy of light (Photon) is twice the band gap of QD, the exciton has a very high kinetic energy, which can be used to excite another electron hopping from the VB to CB by impact ionization. Hence, one photon creates two excitons permitting the internal quantum efficiency to drive beyond 100%. Indeed, more excitons are produced, if the energy of the photon is sufficiently high. Multiple exciton generation is observed in PbS PbSe and CdSe/ZnS QDs. This property of QDs is useful to improve the efficiency of QDSSC and light-emitting devices. In PbS-based QDSSC, it is observed that an incident photon to current conversion efficiency (IPCE) is over 100%, which leads to the Multiple Exciton Generation.

3.4.5 Surface passivation of quantum dots: Influence on QDSSC performance

QDs have defect states, due to high surface to volume ratio. This results in low photocurrents obtained in QDSSC. Passivation of these surface or defect states using organic or inorganic passivating agents, which include ZnS, ZnSe, Cu-ZnS, and

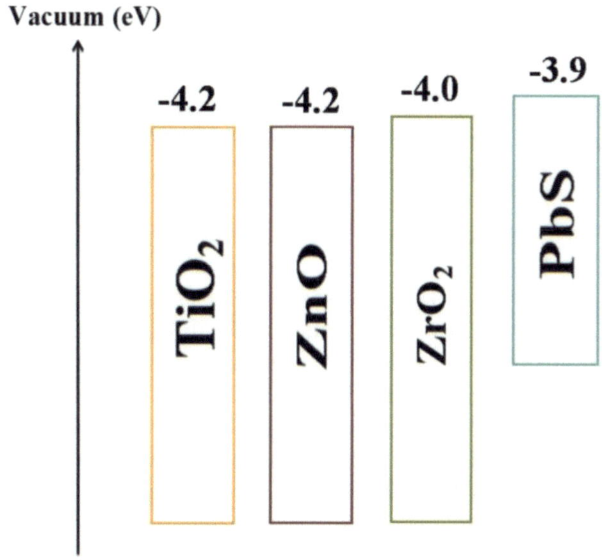

Figure 11.
Energy band positions of TiO2, ZnO, ZrO2 with respect PbS. Source: Ref. [16].

halides, is seen to have immensely enhanced the performance of QDSSC. The original trapped states on QDs are reduced by such treatments and also reduce the recombination of charge carriers. ZnS is a very often used as surface passivator in QDSSC. It is observed that ZnS coating of MO besides QD has further improved the cell efficiency by improving the electrode/electrolyte interface [22]. Combined treatment of ZnS/SiO_2 on photoelectrode sensitized with $CdSe_xTe_{1-x}$ yielded a relatively very high efficiency in QDSSC, of around 8.2% [23]. Thus, a surface treatment of QDs as well as photoanodes improves the performance of QDSSC.

3.4.6 Electrolyte

Electrolyte (redox) consists of species of which is an electron donor called reductant and an electron acceptor called oxidant. Dye/Quantum dot sensitized solar cell shows a photovoltaic action due to the circulation of charge carriers through an electrolyte. Polyiodide is the most favorable electrolyte with respect to DSSC, due to its electron transfer kinetics, which justifies their high photoconversion efficiencies. Though, the I-/I3- (Iodide/polyiodide) electrolyte displays amiable charge transfer kinetics due to its better hole evacuation capability, it is corrosive in case of QDSSC where the photocurrent degrades continuously.

Sulfide/polysulfide (S^{2-} /Sn^{2-}) is used as the electrolyte in QDSSC due to its stability with the semiconductor QDs and solar cell performance. The other redox couples such as Fe^{3+}/Fe^{2+}, Co^{2+}/Co^{3+} and $Fe(CN)_6{}^{3-}$/$Fe(CN)_6{}^{4-}$ [24–27] are employed. Polysulfide (PS) has a high redox potential, which supports in faster and better hole evacuation, but it reduces the open-circuit voltage and has poor fill factor (FF).

The stability of QDSSC is a question as long as liquid electrolytes are employed. Electrolytes volatilization, permeation of oxygen and water vapor from the atmosphere lead to quick degradation of the device performance. Rather than all disadvantages of the PS as an electrolyte, it has reported the best of efficiencies for QDSSC so far.

3.4.7 Counter electrode (CE)

The collection of the electrons from the external circuit and the reduction of the oxidant species in the electrolyte are done with the help of a component known as counter electrode. This completes the circuit and with the electrolyte, it makes the device ready for a constant photovoltaic action. Thus, a counter electrode reduces the oxidant species.

Pt and Au are the most popular CE in DSSC, and they are highly efficient and compatible with Iodide/Polyiodide redox species [28–31]. In QDSSC, Pt results in a poor device performance due to the chemical activity, which reduces chemisorption of sulfide species. The charge transfer at the CE/Electrolyte is also affected due to poisoning effect of electrolyte on the surface of CE leading to reduction in FF.

With their very good electronic mobility and corrosion resistance for electrolyte species, carbonaceous materials such as graphene are considered as CE with Graphene/CoS, graphene/PbS, Carbon nanofiber/CuS, Carbon black/PbS, multiwall carbon nanotubes (MWCNT)/Cu2ZnSnSe(CZTSe), etc.

4. Conclusion

The chapter focused on different types of nanomaterial synthesis techniques such as physical, chemical, etc. In the latter part, it covers the synthesis of PbS quantum dots, by hazardous-chemical-free simple technique using ultrasonication and without capping agent. This technique produces quantum dots of 6 nm size, and it is confirmed from UV spectra, HRTEM images, and SAED pattern. In order to understand the structure of solar cells, the QDSSC is discussed in detail with its essential components and materials used in the architecture.

Author details

Vikram P. Bhalekar
Department of Physics, Shirdi Sai Rural Institutes, ASC College Rahata, (SPPU, Pune), India

*Address all correspondence to: bhalekarvp@gmail.com

References

[1] Kulkarni SK. Handbook, Nanotechnology: Principles and Practices. New Delhi, India: Springer; 2015

[2] Martín-Palma RJ, Lakhtakia A. Nanotechnology: A Crash Course. Bellingham: SPIE; 2010

[3] Wong MS, Stucky GD. The facile synthesis of nanocrystalline semiconductor quantum dots. In: MRS Proceedings. England: Cambridge University Press; 2001

[4] Hodes G. Handbook, Chemical Solution Deposition of Semiconductor Films. Florida, USA: CRC Press; 2002

[5] Gorer S, Hodes G. Quantum size effects in the study of chemical solution deposition mechanisms of semiconductor films. The Journal of Physical Chemistry. 1994;**98**(20):5338-5346

[6] Chopra KL. Handbook, Thin Film Phenomena. New York: McGraw-Hill; 1969. p. 61

[7] Aegerter MA, Mennig M. Sol-Gel Technologies for Glass Producer and Users. New York: Springer Science, Business Media LLC

[8] Pathan H. M. amd C. D. Lokhande (2003) —Deposition of metal chalcogenide thin films by successive ionic layer adsorption and reaction (SILAR) method bull. Materials Science 27: 85-111

[9] Pathan HM, Lokhande CD. Deposition of metal chalcogenide thin films by successive ionic layer adsorption and reaction (SILAR) method. Bulletin of Materials Science. 2004;**27**(2):85-111

[10] Mohanty US. Electrodeposition: A versatile and inexpensive tool for the synthesis of nanoparticles, nanorods, nanowires, and nanoclusters of metals. Journal of Applied Electrochemistry. 2011;**41**(3):257-270

[11] Lakshmi BB, Patrissi CJ, Martin CR. Sol-gel template synthesis of semiconductor oxide micro-and nanostructures. Chemistry of Materials. 1997;**9**(11):2544-2550

[12] Ulman A. An Introduction to Ultrathin Organic Films: From Langmuir--Blodgett to Self-Assembly. USA: Academic Press; 2013

[13] Nalwa HS, editor. Handbook of Nanostructured Materials and Nanotechnology, Five-Volume Set. USA: Academic Press; 1999

[14] Okuyama K, Lenggoro IW, Tagami N, Tamaki S, Tohge N. Preparation of ZnS and CdS fine particles with different particle sizes by a spray-pyrolysis method. Journal of Materials Science. 1997;**32**(5):1229-1237

[15] Bhalekar VP. Strong quantum confined lead sulphide quantum dots using ionic reaction and their properties. ES Materials & Manufacturing. 2019;**3**:52-56

[16] Bhalekar VP. Studies on chemical synthesis of PbS quantum dots for third generation solar cell. 2018

[17] Hanna MC, Nozik AJ. Solar conversion efficiency of photovoltaic and photoelectrolysis cells with carrier multiplication absorbers. Journal of Applied Physics. 2006;**100**(7):074510

[18] Marcus RA. On the theory of oxidation-reduction reactions involving electron transfer. The Journal of Chemical Physics. 1956;**24**(5):966-978

[19] Tvrdy K, Frantsuzov PA, Kamat PV. Photoinduced electron transfer from semiconductor quantum dots to metal oxide nanoparticles. Proceedings of the National Academy of Sciences. 2011;**108**(1):29-34

[20] Tiwana P, Docampo P, Johnston MB, Snaith HJ, Herz LM. Electron mobility and injection dynamics in mesoporous ZnO, SnO_2, and TiO_2 films used in dye-sensitized solar cells. ACS Nano. 2011;**5**(6):5158-5166

[21] Wei W, Jiang C, VellaisamyA L. Roy Recent progress in magnetic iron oxide–semiconductor composite nanomaterials as promising photocatalysts. Nanoscale. 2015;**7**:38-58

[22] Lee Y-S, Gopi VM, et al. High performance of TiO2/CdS quantum dotsensitized solar cells with a Cu–ZnS passivation layer. New Journal of Chemistry. 2017;**41**:1914-1917

[23] Guijarro N, Campiña JM, Shen Q, Toyoda T, Lana-Villarreal T, Gómez R. Uncovering the role of the ZnS treatment in the performance of quantum dot sensitized solar cells. Physical Chemistry Chemical Physics. 2011;**13**(25):12024-12032

[24] Tachibana Y, Akiyama HY, Ohtsuka Y, Torimoto T, Kuwabata S. CdS quantum dots sensitized TiO2 sandwich type photoelectrochemical solar cells. Chemistry Letters. 2007;**36**(1):88-89

[25] Tachibana Y, Umekita K, Otsuka Y, Kuwabata S. Performance improvement of CdS quantum dots sensitized TiO_2 solar cells by introducing a dense TiO_2 blocking layer. Journal of Physics D: Applied Physics. 2008;**41**(10):102002

[26] Lee HJ, Chen P, Moon SJ, Sauvage F, Sivula K, Bessho T, et al. Regenerative PbS and CdS quantum dot sensitized solar cells with a cobalt complex as hole mediator. Langmuir. 2009;**25**(13):7602-7608

[27] Lee HJ, Yum JH, Leventis HC, Zakeeruddin SM, Haque SA, Chen P, et al. CdSe quantum dot-sensitized solar cells exceeding efficiency 1% at full-sun intensity. The Journal of Physical Chemistry C. 2008;**112**(30):11600-11608

[28] Calogero G, Calandra P, Irrera A, Sinopoli A, Citro I, Di Marco G. A new type of transparent and low cost counter-electrode based on platinum nanoparticles for dye-sensitized solar cells. Energy & Environmental Science. 2011;**4**(5):1838-1844

[29] Thomas S, Deepak TG, Anjusree GS, Arun TA, Nair SV, Nair AS. A review on counter electrode materials in dye-sensitized solar cells. Journal of Materials Chemistry A. 2014;**2**(13):4474-4490

[30] Saito Y, Fukuri N, Senadeera R, Kitamura T, Wada Y, Yanagida S. Solid state dye sensitized solar cells using in situ polymerized PEDOTs as hole conductor. Electrochemistry Communications. 2004;**6**(1):71-74

[31] Duan Y, Tang Q, Liu J, He B, Yu L. Transparent metal selenide alloy counter electrodes for high-efficiency bifacial dye-sensitized solar cells. Angewandte Chemie International Edition. 2014;**53**(52):14569-14574

Chapter 5

Solar Energy

Murtadha Jasim Hasan AlTimimi

Abstract

Solar energy schemes based on photovoltaic cells have attracted extensive interest in recent years due to their capabilities of clear and seemingly limitless generated energy. This chapter proposed a simple, cost effective and efficient system for solar photovoltaic applications. Solar energy is considered as an fastest growing renewable energy resource like wind energy for electricity generation. Solar energy is a free, clean abundant sun energy considered an inexhaustible source for electricity generation. The solar photovoltaic system is characterized by variable output power due to its operation dependency on solar irradiance and cell temperature. To maximize the energy generation potential of solar PV, a research effort is focused on solar cell manufacturing technology to increase its generation efficiency and explore advancements in power electronic devices for small and large scale deployment. The main aim is to find the unknown parameters of the nonlinear current–voltage (*I-V*) equations by adjusting the *I-V* curve at three remarkable points when the circuit is open (open circuit), when the circuit is delivering maximum power, and when it is short circuited. The data of these three studies are mentioned by all commercial PV module makers in their datasheets and with all these parameters of the adjusted *I-V* equation.

Keywords: PV panel, MPPT, current–voltage, voltage-power

1. Introduction

Since the Industrial Revolution, fossil fuels have become widespread in the world. For energy stored in fossil fuels, this process requires combustion and therefore cannot prevent emissions and particles in the atmosphere. In addition, the transport and extraction of fossil fuels cause pollution with serious consequences such as air pollution, water deterioration, soil degradation and global warming. Currently, fossil fuels dominate global energy consumption [1].

With global energy shortages and pollution problems, the protection of energy and the environment becomes the main problem for People. The development and application of clean sources of renewable energy such as solar, wind, fuel, tidal and geothermal, etc., are increasingly important. Among them, solar energy will dominate due to its availability and reliability. How As predicted by [2], by the end of this century the sun will provide up to 64% of the total energy, as shown in **Figure 1**.

IntechOpen

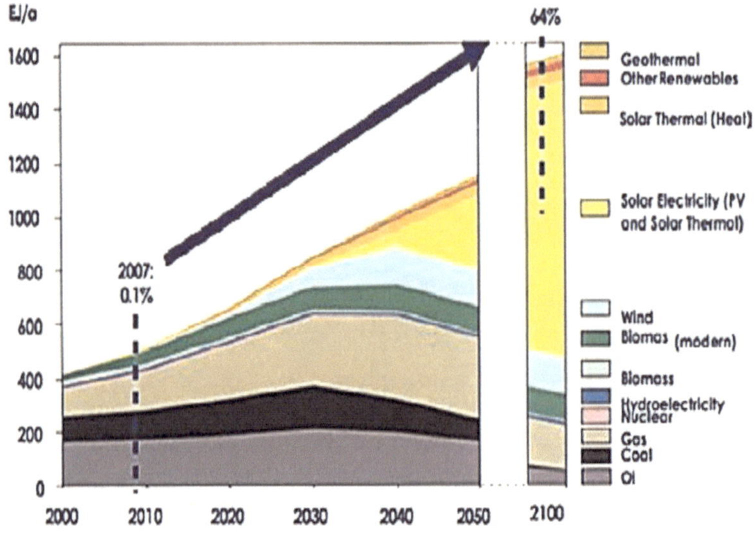

Figure 1.
Global energy used by a source in the 21st century [2].

1.1 Overview of photovoltaic cells

The generation of photovoltaic energy (PV) has become one of the most important ways to use solar energy. And the regenerative energy system based on decentralized generation (DG) is generally connected to the network through inverters or electronic power inverters. Therefore, the development of a photovoltaic inverter system is important to reduce energy and environmental problems.

The first discovery of the effect of photovoltaic energy was made in 1839 by Edmund Becquerel, who demonstrated the production of a device that generates electricity when exposed to sunlight. More than a century later, in 1954, D. Chapin, C.S. Fuller and G. Pearson [2] manufactured the first silicon solar cell at Bell's Laboratories.

1.1.1 Type of photovoltaic panels

Although there are different types of panels, the most common types of solar panels can be classified into the following categories [3]:

1.1.1.1 Polycrystalline silicon panels

This type shows that the solar cells in monocrystalline panels are slices cut from pure crystal silicon bars. Where the paintings work best in bright weather with sunlight directly on them. Absorption of sunlight radiation is very high, due to the presence of uniform black color. This type of solar panel is very effective. It is made from the silicon mass obtained from the fusion of pure silicon parts. Efficiency ranges between 15 and 18% and its disadvantages are difficult to manufacture and high cost.

1.1.1.2 Monocrystalline silicon panels

Also known as polycrystalline plates, these multi-crystalline panels consist of silicon screws, encapsulated to form blocks and create a cell made of pure crystal. This type is not as efficient as this type of monocrystalline silicon (single crystal) tapes. Efficiency of between 12 and 15%. It is difficult to manufacture and costs high but without monocrystalline.

1.1.1.3 Amorphous silicon panels

The Amorphous Silicon Panel is also known as a thin film board. This plate is made of thin layers of silicon on a layer of glass or metal. They are manufactured by attaching thin layers of silicon to a layer of glass or metal. They also have less than 10% efficiency. Its advantage is easy to manufacture and at a low cost. This is the technique used in this work.

1.1.2 Development of a PV cell

The solar cell is essentially a semiconductor of two materials known as Type P and Type N semiconductor material. Most frequently Highly purified silicon is a semiconductor material. Through the targeted addition of impurities such as boron and phosphorus, the various types of P and N are produced [4]. Photovoltaic (PV) cells are converted into electrical currents by incident radiation in the solar spectrum. Photovoltaic cells are most commonly made of silicone and are available in two variants, crystalline, and thin film type, as detailed in **Table 1** [5].

In real applications, the light absorbed by a solar cell will be a combination of direct solar radiation and diffused light reflected from the surrounding surfaces. Solar cells are usually coated with anti-reflection material to absorb as much radiation as possible [5–7]. The PV cells may be arranged in a series arrangement to form a panel, and the panels may be connected in parallel row configurations to form arrangements as in **Figure 2**. When cells or plates are connected in series, they must have the same amperage to produce an additive voltage output, and similarly, the plates must have the same nominal voltage when Connected in parallel to produce larger currents.

1.2 Characteristics of photovoltaic cell

The photovoltaic cell consists of a p-n junction in a semiconductor layer in a slice or slice. In general, the PV cell can be achieved by connecting the parallel DC source to the diode. In addition, the model contains its resistor in parallel and series. A parallel or parallel resistor (Rsh) indicates the relationship between a resistor and a resistor parallel to a working device. The resistor chain switches the current flow through semiconductor materials, metal sheets, communications and power buses. These Umayyad patterns can be combined with the Resistance Series (Rs) [8–10]. The equivalent circuit of the PV cell model is shown in **Figure 3**.

The characteristic equation *I-V* of a PV cell is given as [11]:

$$I = I_{\text{ph}} - I_{\text{D}} - (V + \text{IR}_{\text{s}})/R_{\text{sh}} \tag{1}$$

Where I: is the output current of the cell (A).

I_{ph}: is the current generated by the light or photocurrent (A).

Crystalline silicon (waver based)		Thin film
Monocrystalline	**Polycrystalline**	
Made of a big Silicon crystal.	Manufacturing Process of using Multiple silicon Crystals.	Made with Silicon (for example Amorph Silicon (a-Si), copper Indian and Gallium Selenide (CIGS), or Cadmium Telluride (CdTe).
More efficient and Expensive.	A little less efficient than is monocrystalline Panels, but also less expensive.	They are cheaper and they are not dripping from shading.
Used in applications where the installation is Square foot is Comprehends.	Most frequently, Solar panels are seen.	Together for the construction of integrated photovoltaic (BIPV) applications because of his many application options.

Table 1.
Crystalline (wafer-based) and photovoltaic thin-film cell.

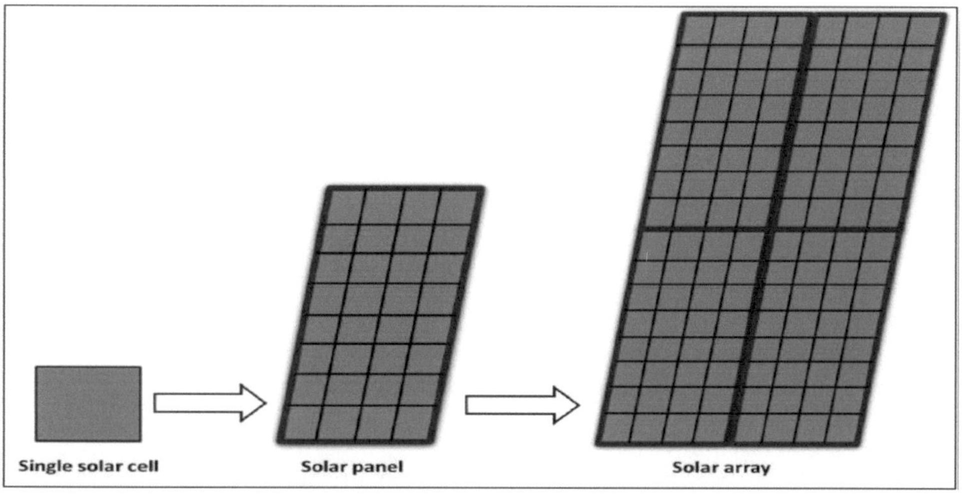

Figure 2.
Solar array implementation.

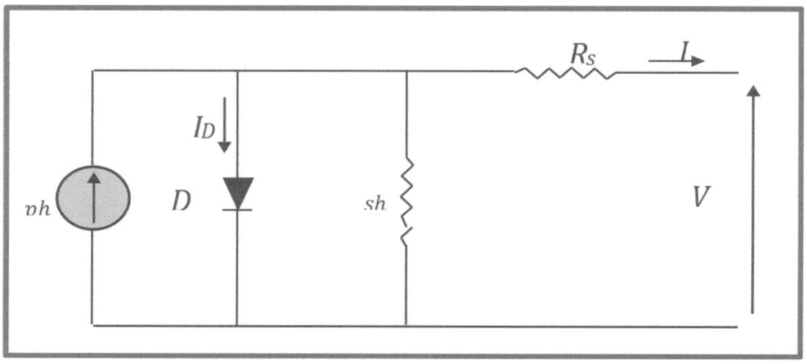

Figure 3.
The equivalent circuit of a PV cell.

ID: is the current of the diode (A).
V: is the output voltage of the cell (V).
R_{sh}: is the leakage resistance (Ω).
R_s: is the series resistance (Ω).
The diode current ID is given by the Shockley diode equation [11]:

$$I_D = I_o \left(e^{((q\mathrm{Vd})/\mathrm{KTA})} - 1\right) \tag{2}$$

Where I_o: is the reverse saturation current of the diode (A).
q: is the charge of the electron (q = 1.6 × 10(−19) Coulomb).
VD: is the voltage of the diode (V).
K: is the Boltzmann constant (K = 1.38 × 10(−23) J/K).
T: is the temperature of the cell (K).
A: It's the ideality of the constant-current diode, it depends on the PV technology and it takes the value between one and two.

Replacing I_D of the Eq. (1) by the Eq. (2) gives the I-V relation of the PV cell:

$$I_D = I_{ph} - I_o\left(e^{\left(\frac{qVd}{KTA}\right)} - 1\right) - \frac{V + IR_s}{R_{sh}} \tag{3}$$

The photocurrent (I_{ph}) depends mainly on the solar radiation and the working temperature of the cell, which is given as [11, 12]:

$$I_{ph} = \frac{S}{S_R}(I_{SC} + K_I(T - T_r)) \tag{4}$$

Where ISC: is the cell's short-circuit current at 25°C and 1 kW/m^2.
K_I: is the short-circuit current/temperature coefficient.
T_r: is the cell reference temperature (K).
S: is the solar insolation in W/m^2.
S_r: is the solar reference insolation which is normally (1000 W/m^2).
The inverse saturation current of the diode (I_o) depends on the ambient temperature. The inverse saturation current of the diode at the set point temperature (I_o,n) can be determined as follows by setting the no-load condition (without output current) as follows:
Using the Eqs. (4) & (3), let I = 0 and T = 25°C and solve for I_o = I_o,n, I_{ph} = Isc and R_{sh} = ∞

$$0 = I_{sc} - I_{o,n}\left(e^{\frac{qV_d}{KTA}} - 1\right)$$

$$I_{sc} = I_{o,n}\left(e^{\frac{qV_d}{KTA}} - 1\right)$$

$$\therefore I_{o,n} = \frac{I_{sc}}{\left(e^{\frac{qV_d}{KTA}} - 1\right)} \tag{5}$$

Where Io,n is the nominal diode saturation current at the nominal condition (25°C and 1 kW/m^2).
The reverse saturation current of the diode (I_o) at any temperature (T) can be calculated as [11, 12]:

$$I_o = I_{o,n}\left(\frac{T}{T_r}\right)^3 \exp\left[\frac{qE_g}{KA}\left(\frac{1}{T_r} - \frac{1}{T}\right)\right] \tag{6}$$

Where E_g is the band-gap energy of the semiconductor used in the cell and equal (1.12 electron-volt) for the silicon.

1.3 PV panel model

Since the energy generated by a PV cell is very low (**Figure 4**). Therefore, to generate enough energy, the cells should be assembled in a parallel-serial configuration of a module. As mentioned above, the photovoltaic panel is a set of photovoltaic modules that are connected in parallel and in series to produce a required current and voltage and therefore current. The equivalent circuit for a PV array formed in the row N_S and in the parallel cells N_P is shown in **Figure 4** [11].

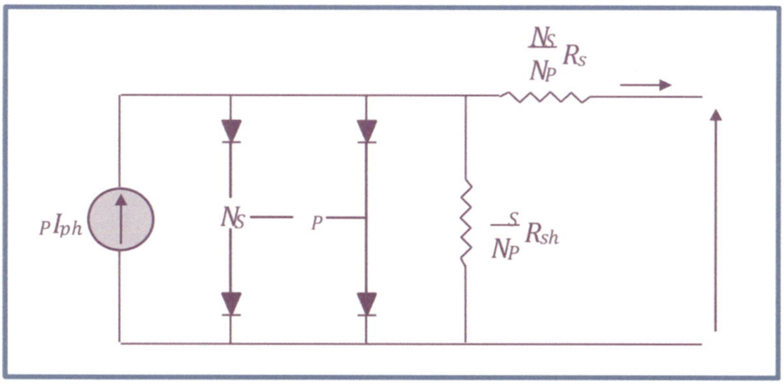

Figure 4.
Equivalent circuit model of the generalized PV panel.

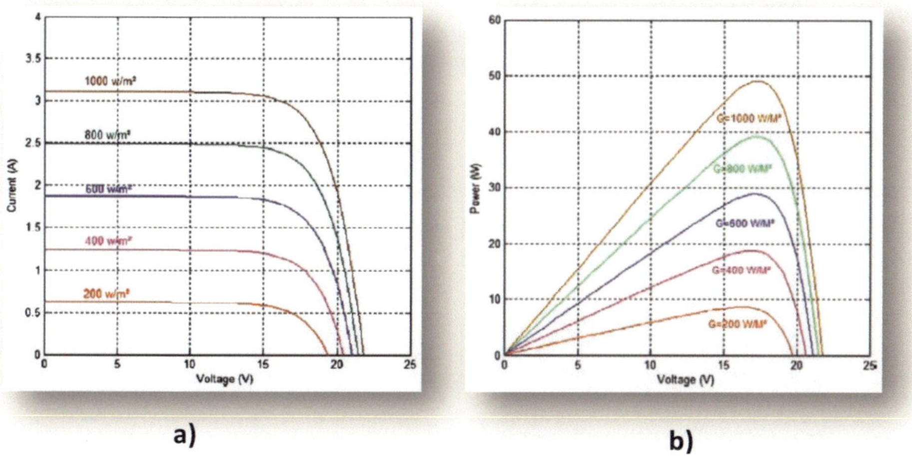

Figure 5.
Characteristics of a photovoltaic panel for different values of solar irradiance (S) at constant temperature (25°C) [13] (a) I-V *characteristics (b)* P-V *characteristics.*

The terminal equation for the current and voltage of the panel becomes as follows [13]:

$$I = N_P I_{ph} - N_P I_O \left(\exp\left[\frac{q}{KTA} \left(\frac{V}{N_S} + \frac{IR_s}{N_P} \right] - 1 \right) - \frac{\frac{N_P}{N_s} V + IR_s}{R_{sh}} \qquad (7)$$

Where NP: is the number of cells connected in parallel.

NS: is the number of cells connected in series.

Therefore, the *I-V* characteristics, as illustrated in Eq. (7), are nonlinear and varied by varying the radiation intensity and temperature. **Figures 5** and **6** illustrates these characteristics of PV panel under different conditions of radiation intensity and temperature [13].

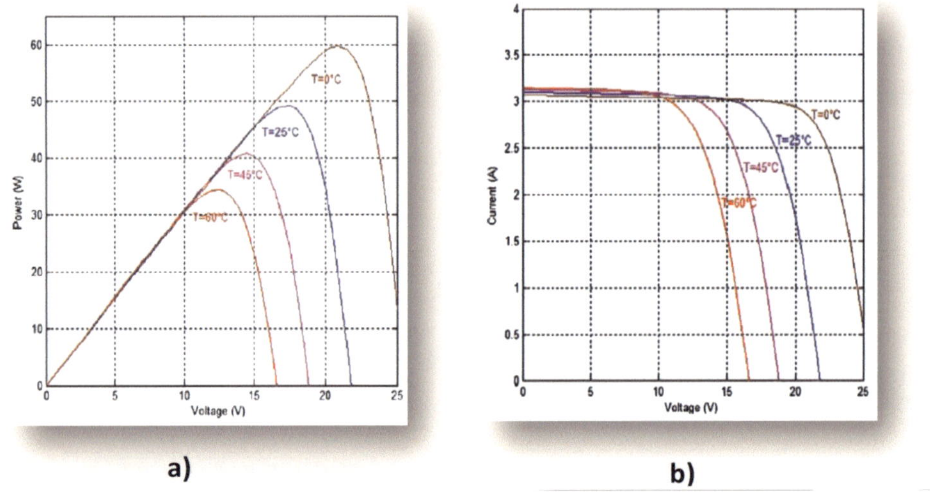

Figure 6.
Characteristics of a photovoltaic panel for different values of temperature (C) at a constant solar irradiance (1000 W/m^2) [13] (a) I-V *characteristics (b)* P-V *characteristics.*

Author details

Murtadha Jasim Hasan AlTimimi
Department of Electrical Power Technologies Engineering, Technical College of Electrical Engineering, Middle Technical University, Iraq

*Address all correspondence to: mu.ti.2291@gmail.com; murtadha.jh1990@mtu.edu.iq

References

[1] Zhang Q. Optimization and Design of Photovoltaic Microinverter, [PhD thesis]. Toronto: University of Central Florida; 2013

[2] Zhao Z. High Efficiency Single-Stage Grid-Tied PV Inverter for Renewable Energy System, [PhD thesis]. Virginia: University of Virginia; 2012

[3] Rodrígueza JA, Riverola FF. Applying rough sets for the identification of significant variables in photovoltaic energy production with isolated systems. Jurnal Teknologi. 2013;**63**(3):9-16

[4] Jasim Hasan M, Abbas FM. Design and implementation of two-stage 3-level inverter grid connected for photovoltaic applications. Journal of Engineering and Applied Sciences. 2019;**6430–6443**: 1816-949x

[5] McEvoy A, Markvart T, Castaner L. Practical Handbook of Photovoltaics. 2nd ed., Elsevier Ltd; 2012

[6] Massawe HB. Grid Connected Photovoltaic Systems with Smart Grid Functionality. 2013

[7] Martins DC, Demonti R. Photovoltaic energy processing for utility connected system. In: The 27th Annual Conference of the IEEE Industrial Electronics Society. Vol. 2. Denver, CO; 2001. pp. 1292-1296

[8] Nanakos AC, Tatakis EC, Papanikolaou NP. A weighted efficiencyoriented design methodology of flyback inverter for AC photovoltaic modules. IEEE Transactions on Power Electronics. 2012;**27**(7)3221-3233

[9] Chen S-M, Liang T-J, Yang L-S, Chen J-F. A safety enhanced, high step-up DC–DC converter for AC references 100 photovoltaic module application. IEEE Transactions on Power Electronics. 2012; **27**(4):1809-1817

[10] Jiang S, Cao D, Li Y, Peng FZ. Grid-connected boost halfbridge photovoltaic microinverter system using repetitive references 98 current control and maximum power point tracking. IEEE Transactions on Power Electronics. 2012; **27**(11):4711-4722

[11] Hersch P, Zweibel K. Basic Photovoltaic Principles and Methods. Technical Information Office, U.S. Department of Energy; 1982

[12] Jasim Hasan M, Abbas FM. Single-phase, H-bridge 3-level inverter of wide range input voltage for grid connected solar photovoltaic applications. International Journal of Computer Applications. 2018;**181**(26)

[13] Jasim Hasan M. Modeling and simulation of 1kw single-phase grid-tied inverter for solar photovoltaic system. IOP Conference Series: Materials Science and Engineering. 2020;**881**(1):1-16

Section 2

Quantum Dots in Bio and Food Technology

Chapter 6

Application of Quantum Dots in Bio-Sensing, Bio-Imaging, Drug Delivery, Anti-Bacterial Activity, Photo-Thermal, Photo-Dynamic Therapy, and Optoelectronic Devices

Karunanithi Rajamanickam

Abstract

Quantum dots (QDs) are of prevalent scientific and technological consideration because of their tunable size and thus frequency change (band-gap energy) in the NIR optical region. QDs have exceptional properties such as optical, physiochemical, electrical, and capacity to be bound to biomolecules. These selective size-dependent attributes of QDs assist them with having versatile applications in optoelectronic and biomedical fields. Their capacity to emit light at various frequencies because of an outer stimulus makes quantum dots perfect for use in imaging, diagnostics, tests for individual particles, and medication transportation frameworks. Ongoing advances in quantum dot design incorporate the potential for these nanocrystals to become therapeutic agents to restore numerous disease conditions themselves via bioconjugation with antibodies or medications. In this chapter, a few advances in the field of biomedical applications, such as bio-sensing, bio-imaging, drug loading capacity, targeted drug delivery, anti-stacking limit hostile to bacterial activity, photo-thermal treatment, photodynamic treatment, and optical properties for biomedical applications are presented, further to a short conversation on difficulties; for example, the biodistribution and harmful toxic effects of quantum dots is also discussed.

Keywords: quantum dots, bio-sensing, bio-imaging, drug-delivery, anti-bacterial activity, photo-thermal therapy, photodynamic therapy, optoelectronic devices, functional QDs

1. Introduction

Nanotechnology is a part of applied science and innovation that deals with the control of matter on a nuclear and sub-atomic scale, ranging from 1 to 100 nm. This is a profoundly interdisciplinary field that advantages from the endeavors and progressions of many disciplines, including applied physical science, materials science, point of interaction and colloid science, supramolecular science, compound designing, mechanical designing, organic designing, and electrical designing [1–5]. Nanotechnology has prospered since the introduction of group science and the development of the scanning tunneling microscope in the 1980s, with the capacity to gauge and imagine novel peculiarities, as well as control and production materials and gadgets with nanostructures of 100 nm or more modest. Semiconductor nanocrystals, otherwise called quantum dots, are a recently arising nanomaterial that has provoked the curiosity of numerous scientists. QDs are ordinarily synthesized from periodic table elements in groups III–V and II–VI [6–8]. Nonetheless, ongoing improvements in nanotechnology and biomedical utilization of nanomaterials have driven us to expand the meaning of QDs to incorporate more extensive classifications of nanoparticles (like carbon, silicon, and gold) which show the quantum-confinement occurrence related to a sensational alteration in electrons' way of behaving at the limit of the Bohr radius. Because of their tiny size (contrasted with most cell structures or biomolecules), they can communicate with biological structures on the nuclear scale. However, a few scientists use "traditional" semiconductors III–V and II–VI nanocrystals, these are as yet the significant short utilized in research and biomedical applications for a few goal reasons. Firstly, they have remarkable physical characteristics that recognize them from different kinds of QDs; second, they have been effectively used in medical procedures, ultimately, proceeded with use by industry requirements the broad assessment of dangers related with exposure. Electrons are depicted by quantum confinement effects in terms of their energy levels, likely wells, valence band, conduction band, and electron energy band [9]. At the point when the molecule's size is too little to possibly be similar to the frequency of the electron, the quantum confinement effect is noticed. Clearly, the confinement of an electron and a hole in nanocrystals is profoundly subject to material properties, explicitly the Bohr radius a_B. These impacts rely upon material properties, [10] special to greater nanocrystals, and are Electrons in bulk dielectric materials can be portrayed by energy bands or electron energy levels (bigger than 10 nm). There are different energy levels or bands for electrons. These energy levels are suggested to as continuous since the energy difference is tiny in mass materials. The main part of electrons oscillates in the valence band under a prohibited energy level, known as the band gap, as they balance out at different energy levels. There are no electron states in this energy range. Fewer energy levels that are over the restricted hole make up the conduction band. The Bohr span is equivalent to 2.34 nm [11]. Materials' electrical and optical qualities contrast emphatically with mass materials when they are at the nanoscale. As the material gets more uncertain until it comes to the nanoscale, the confining aspect normally gets more unassertive. The properties, in any case, are presently at the quanta level and consequently discrete as [12] opposed to arriving at the midpoint of by mass and accordingly constant. All in all, the energy range becomes discrete, estimated as quanta, as opposed to consistent, as in mass materials. The bandgap, or the small and limited space between energy levels, results thus. Discrete energy levels are the situation that is alluded to as quantum confinement. Electrons in mass dielectric materials can be described by energy groups or electron energy levels (bigger than 10 nm). There are different energy levels or groups for electrons. These energy levels are indicated as persistent since the energy distinction is little in bulk materials. The majority of electrons waver in valence groups under a forbidden energy level, known as the band gap, as they balance out at different energy levels. There are no electron states in this energy range.

Fewer energy levels that are over the restricted gap make up the conduction band. Moreover, disconnected islands of electrons that can shape at the designed connection point of two unique semiconducting materials make up quantum confinement distinctiveness. Normally, electrons are contained in plate molded regions called quantum dots. As was at that point referenced, electron control in these frameworks significantly changes how they collaborate with electromagnetic radiation. The bandgap limits are modified by the expansion or evacuation of a couple of iotas on the grounds that the electron energy levels of quantum spots are discrete instead of constant. Changes in the quantum dots surface shape additionally influence the bandgap energy as a result of the speck's little size and the impacts of quantum confinement [13, 14].

Quantum dots, which have excitons bound in every one of the three spatial dimensions, offer attributes somewhere between those of discrete particles and bulk semiconductors. Quantum dots can have their structure and molecule size modified subsequent to being excited by a reasonable laser pulse by exploiting the notable impacts of the quantum size confinement effect. When contrasted with quantum dots, fluorophores, (for example, rhodamine 6G and fluorescein) have various disadvantages in biosensing applications [15–17]. High yields, broad absorption, restricted size-dependent emission spectra, and incredible photo quenching obstruction are a couple of advantages of QDs' extraordinary optical elements [18]. The quantum confinement phenomenon gives an upsurge to the dimension-tunable absorbance and luminescence bands of QDs. At the point when the size of the semiconductor meets the material's Exciton Bohr Radius (EBR), the discrete electron band-gap energy levels are denoted as being under quantum confinement. On the off chance that the size of the QDs diminishes, the absorption and photoluminescence will go through a hyperchromatic shift and the energy of the band-gap expands [19]. Subsequently, QDs of similar material composition yet different sizes can emit fluorescence at a pool of frequencies. With innumerable applications, this optoelectronic nature of QDs gives a higher multiplexing potential. These QDs can possibly carry out numerous roles subsequent to being altered with explicit affinity ligands (antibody, peptide, aptamers, etc.). Subsequently, they might have the option to fulfill the rules for the ideal theranostic framework, including yet not restricted to the characteristics mentioned beneath.

Targeting specific cell types, diminished cytotoxicity, effective intracellular trafficking, conquering intracellular boundaries, responding to external or internal stimuli, delivering curative agents, and bearing a symptomatic specialist (optical or MRI) all add to a medication's capacity to treat a patient progressively. Thus, QDs can be utilized for drug delivery, biosensing, and bio-imaging. Creating QDs that are not poisonous to cells could have an enormous clinical effect. Given the adaptability of QDs, research is in the works to foster tests that can follow continuous cell processes, exhibiting selectivity and particularity towards cells and the capacity to beat excretory issues, for example, non-poisonousness. The examination into QD/drug nanoparticle plans will show huge development in different medical services-related regions, for example, diagnostics and therapeutics for conditions like malignant growth and heart and immune system problems/illnesses. In this chapter, the role of QDs in various medical applications like bio-sensing, bio-imaging, drug delivery applications, anti-bacterial activity, photo-thermal, photodynamic therapy, and optoelectronic devices were discussed in detail. Furthermore, their specific targeting abilities were also discussed when they are formulated with multifunctional moieties, they offer a versatile platform that is suitable for theranostics.

2. Quantum dots in medical applications

Researchers are excited up for new nano-theranostic stages including quantum dots in bio-detecting, as they can detect, picture, and treat simultaneously. Their benefit over old advancements lies in the control of conductive properties through crystal size. However, quantum dots enjoy many benefits, they have huge constraints. For instance, they are just comprised of a few molecules, and they are difficult to eliminate from the body. Researchers are attempting to take care of these issues so quantum dots can be utilized in additional clinical applications. The remarkable optical properties of QDs have prompted their utilization in various fields, including imaging, following, diagnostics, drug delivery, tissue designing, malignant growth treatment, multicolor optical coding, and single molecule probes. For example, quality treatment can fix infections brought about by hereditary anomalies, like malignant growth or diabetes. QDs can stop quality action, and furthermore track down application in RNA advancements. For instance, in situ hybridization (ISH) can utilize QDs to identify mRNA particles, and RNA mediation might join siRNAs with CdSSe/ZnS quantum dot—polyethylenimine (QDs-PEI) to really target qualities more successfully [20]. Various benefits of utilizing QDs are investigated by different research groups because of their adaptable properties that are featured in **Table 1**.

Radiotherapy (RT) is a typical malignant growth treatment technique, with a few incidental effects. Researchers have fostered nanozymes by doping Mn (II) and Silver Selenide quantum dots particles (Ag2Se QDs) transmitting in the second near-infrared window. These nanozymes are attached with cancer targeting arginine-glycine-aspartate (RGD) tripeptides and polyethylene glycol and built into in vivo nanoformulations for NIR-II imaging-directed radiation treatment of malignancy [21]. Tumor particularity and NIR-II radiating abilities of the nanoprobes empower exact confinement, giving exact RT. The ultra-stability of these nanoprobes in the living body likewise improves RT adequacy through persistent creation of oxygen and help from hypoxia of cancers. Nanoprobe-intervened RT supported by real-time, enhanced clarity imaging advances against growth invulnerability and altogether hinders cancers or fixes them totally. Another investigation discovered that Mercaptopropionic corrosive (MPA)-coated QDs were exceptionally

The advantages of using QDs
Large Stoke's shift and sharp emission spectra
Novel optical and electronics properties—quantum confinement/electron and photons in nanostructure
High resistant to degradation
Easy to conjugate with variety of biomolecules
Many-fold brightness compared to other organic dyes
Easy moldability
High scattering and thus better contrast with electron microscope

Table 1.
The advantages of using QDs in various biomedical applications.

biocompatible and enact the lysosomal pathway, which clears cell garbage. The decreased ROS could assist the cells with adapting to nanomaterial-initiated pressure, clearing the way for malignant growth treatment. Scientists fostered a new pseudo-homogeneous vector for gene delivery that uses the cationic CQDs got from chitosan to make a non-viral gene exchange framework [22]. This new vector shows better uptake in cells. Another sort of ZnO QDs nanoplatforms to deliver genes that assist with treating Parkinson's illness has been grown as of late. Glutathione changed ZnO QDs composites stacked with quality and NGF to safeguard the brain and reverse the impacts of neurodegenerative issues, which are normal in Parkinson's patients. QDs were utilized to address the reasons for later-stage nearsightedness (myopia). Carboxylated CuInS/ZnS QDs (ZCIS QDs) were added to intraocular lenses by a facial initiation-immersion technique to warm up the lenses and keep the cells from sticking to the lens surface. This technique could assist with regarding cataract development as well as posterior capsule opacification (PCO) [21].

3. Bio-sensing application

The utility of semiconductor QDs in biosensing applications has been growing, as they offer several advantages over other types of sensors. All the more significantly, it is not difficult to present nucleic acid enhancement methodologies or potentially nanomaterials to work on the conjugation of aptamer-based detecting structures. In this manner, the composite of QDs attached with aptamers acquires added open doors bioanalytical methods. QD-based fluorescent nanoformulations are generally utilized for bio-sensing for DNA and protein discoveries [23]. Heavy metal cytotoxicity was minimized or eliminated by developing metal-free quantum dots. QDs synthesized using silicon quantum dots (Si QDs), graphene quantum dots (GQDs), carbon dots (C-dots), and near-infrared (NIR) QDs (silver selenide, silver sulfide, (Ag_2Se and Ag_2S) are the alternate possibilities. Apart from this, metal nanoclusters have exhibited noticeable benefits as fluorescence probes for optical bio-sensing and imaging procedures in current biomedical research [24]. QD-based biosensors are molecular networks made of organic receptors and are proficient substitutes for conventional sensors [25]. These biosensors have gained significant attention, because of their function in bridging the gap between pure organic receptors and inorganic materials [25]. The physical properties of photo-sensors can be altered by the environment of surface-coated and biological ligand-attached quantum dots. This includes heat, ions, and pH of solution as well as confirms the use of QDs in bio-sensor development. The chemical sensors are receptor molecules that show selective responses to particular ions (cations/anions) or neutral groups. For the selection and accounting of the compounds present in biological environment, development of chemosensors is significant, particularly those specific ions that are possibly dangerous to surroundings and individuals.

3.1 Core-ligand interaction

Connecting biomolecules like ligands, antibodies, peptides, or nucleic acids with nanoparticles have drawn broad interest for researchers in the biosensing region as it offers practical bionanomaterials for targeting and drug delivery applications [26]. Semiconductor QDs are broadly utilized in the biosensing region due to their one-of-a-kind property, for example, confined and symmetric emission with colors that are adjustable, considerable quantum yield, appropriate stability, and very much directed shape and size [27]. Among these ligands, aptamers show a few advantages containing lesser aspects, great synthetic stability, and straightforward cycle in combination with high cluster to-bunch homogeneity and more flexibility. Further to this, it is not difficult to present nucleic acid intensification strategies and nanomaterials that can offer significantly better sensitivity. Subsequently, the mix of semiconductor quantum dots and aptamers gets added possibilities in bio examination. In this way, nanoformulation offers a few applications in different sign transducing components, including optical, electrochemical, and electro-produced discharge of light because of synthetic response (chemiluminescence) approaches. These two unique parts connect at this basic biomolecular-materials point of interaction and offer further developed action and promising qualities. The interactions are represented by different factors, for example, Förster reverberation energy move (FRET), the presence of electrostatic and other useful appealing powers (energy electron move) between the biomolecule and the QD surface.

A comprehension of these basic collaborations at this point of interaction can yield a bunch of methodologies that will allow the reasonable plan of ensuing researcher high-movement bionanocomposites and theranostic nanoformulations [28]. Four significant methodologies can be utilized to adjust aptamers onto ODs. (i) Self-gathering among DNA and QDs. (ii) Biospecific collaborations, e.g., biotin-avidin (or streptavidin) cooperation. (iii) Covalent interactions. (iv) Nucleic acid hybridization [29]. The blend of Quantum dots and aptamers will give different identification stages, including optical, electrochemical, and electrochemical luminescence (**Figure 1**).

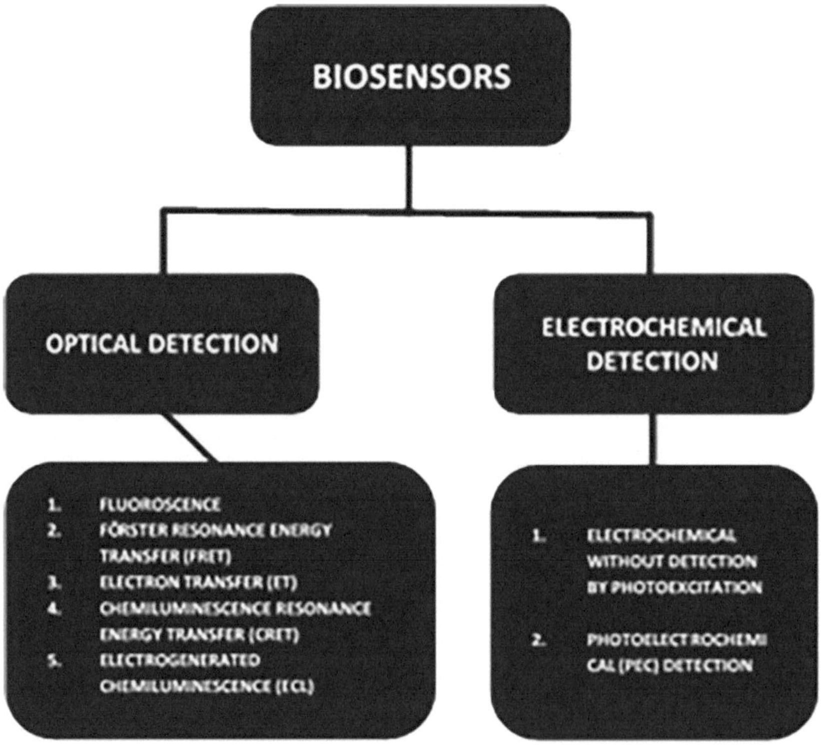

Figure 1.
QD based various biomolecule detection platforms [26].

This takes into consideration the identification of an assortment of analytes like proteins, little particles, and cancer cells [30]. Advanced signal level intensification systems incorporate utilizing nanoscaled materials as devices for ultrasensitive bioanalysis. High-level sign enhancement procedures take into account ultrasensitive bioassays. These procedures likewise consider exceptionally unambiguous biosensing-recognizing follow measures of a wide assortment of analytes in clinical, ecological, or modern applications. The quick advancement in the utilization of other utilitarian materials for detecting and the clarification of their novel photosensitive qualities proposes that by conjugating the aptamer-QD with graphitic-carbon nanoformulation, novel detecting and identification stages might be planned. In what way to integrate these cross breeds into cells and complete in vivo detecting might be a test from here on out.

4. Bio-imaging application

Water-soluble quantum dots (QDs) can be conjugated with various biological molecules like peptides, proteins, aptamers, drugs, and antibodies. The benefits of QD-based biomolecule conjugation help to deal with several immunohistochemistry, labeling and imaging the single-molecule and tracks that are well explored in numerous studies [31–33]. The exclusive wavelength tunable properties of QDs offer wide prospects for designing systems for numerous analyses by multi-colored imaging for the concurrent recognition of numerous targets. Linking drugs with QDs or their combination into QD-based drug delivery molecules marks it a potential candidate for monitoring the drug release and undertaking image-guided therapy. In light of the versatile nature of their photosensitive properties, QDs emanating in the NIR region have turned into an interesting device for intense tissue single photon and multi-photon *in-vivo* imaging. Conventional green fluorescent proteins and fluorescent biological dyes have several restrictions such as spectral crossing, low signal intensity, photo-bleaching when compared to QDs which have substantial benefits in chemosensors and biosensor development [34]. In recent times, QDs are demonstrated to have a significant candidate for fluorescent probes and tags in several organic procedures, ranging from molecular-level histopathological studies to whole-body diagnosis [35, 36]. While a small number of studies have tried the utility of QDs for *in-vitro* or *in-vivo* diagnostic imaging studies, the important restriction on possible harmfulness of group II–IV QDs (like cadmium telluride and cadmium selenide) have hampered their practical application in science and medication. The leakage of the ions belonging to the heavy metal present inside the core of these quantum dots by physiochemical reactions may contribute to the harmful toxic effect, which raises several questions about the utility and biocompatibility of these QDs [37, 38]. Furthermore, the long-term cumulative effects of these ions' augmentation and prolonged periods within the body may produce unnecessary damage to tissues and organs. At the same time, the quick clearance by the reticuloendothelial system (RES) or getting into the liver and spleen can result in noise in the image and thus the quality may get affected. The application of quantum dots as contrast agents for in vivo imaging has been an area of high prospect since they were initially studied, and they have recently been granted repeated attention. The optical imaging system even at present utilizes these conventional fluorescent molecules for animal imaging. The NIR wavelengths (~730 nm) are significant, as they have reduced light scattering and low tissue absorption also relatively easy to detect with apparatuses [39]. Different types of quantum dots emit

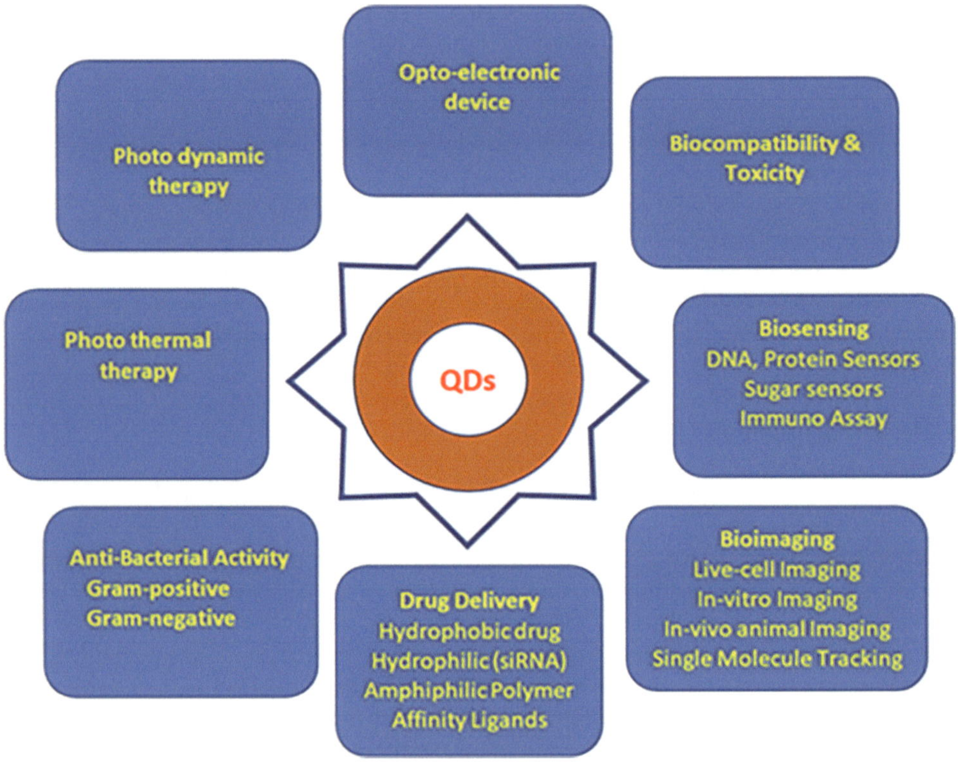

Figure 2.
QDs plausible medical utility and cautiousness.

different wavelengths—one type emits red light (730 nm). QD bioconjugates are widely reported in several studies for their application in in-vitro imaging and diagnostics, like immuno-fluorescent labeling of tissues and cells, intracellular delivery of QDs, tracking of single-molecule in living cells, and in-vivo tracking by fluorescence labeling drug localization and biodistribution [29]. Hence, quantum dots hold great potential for molecular and cellular imaging both in vitro and in vivo conditions. In recent times, researchers have used QDs for imaging stem cells in mice embryo. This unlocks the opportunity to use quantum dots for imaging stem cell and stem cell therapy tracking in a fast and accurate manner (**Figure 2**). Stem cell therapy is the most common way of infusing stem cells into a living being with the expectation that they will separate and supplant harmed tissue or develop new organs.

5. Drug delivery application

Targeted drug delivery using QDs has appeared to have potential applications in recent times. Since the enhanced efficacy of existing drugs and new developments in therapeutics are made possible by adopting various functionalized QDs for this purpose. Several preliminary and drug trials have demonstrated the potential application of this QD based on theranostic systems at the same time also to achieve reduced drug harmfulness, better quality in bio-compatibility [40] and bio-availability [41], enhanced circulation times [42], precise drug release [43] and targeting [44]. Nevertheless, translation of QDs vehicles from bench side to bedside requires a detailed understanding of the impact of these QDs in organic systems when their utility towards in vivo conditions. Recent development in the properties of QDs includes optimized brightness, diminished hydrodynamic dimensions, chemically inert to the environment, and functional groups attached to ligands that are site-specific. Recent studies have shown that Si QDs and fluorescent nanodiamonds are biocompatible, which positions them as excellent diagnostic imaging candidates. Song et al. [45]. Several preliminary and drug-trials have demonstrated the potential application of this QD based on theranostic systems at the same time also to achieve reduced drug harmfulness, better quality in bio-compatibility [40] and bio-availability [41], enhanced circulation times [42], precise drug release [43] and targeting [44]. Nevertheless, translation of QDs vehicles from bench side to bed side requires an in-depth understanding of the impact of these QDs in biological systems when their utility towards in vivo conditions [46]. Recent development in the properties of QDs includes optimized brightness, diminished hydrodynamic dimensions, chemically inert to the environment, and functional groups attached to ligands that are site-specific [47]. Recent studies have shown that Si QDs and fluorescent nanodiamonds are biocompatible, which positions them as excellent diagnostic imaging candidates. Song et al. [48] developed nanoparticles with dual modes of contrast-enhanced CT (CECT) and fluorescence imaging, made of a combination of silica coated-gold nanoformulation/quantum dots (Au-SiO_2-QDs). Several research groups are developing nanoparticle-based agents that can diagnose and treat a person through one product.

Quantum dots (QDs) can go about as the primary nanocarrier or be important for additional perplexing structures. Paclitaxel (PTX), cancer-battling drug, is frequently bundled with nanostructured lipid transporters intended to have a theranostic way to deal with malignant growth treatment [49]. It was found to have exemplification adequacy of ~80%. PTX could be supported and delivered in 12 h in mixture of silica

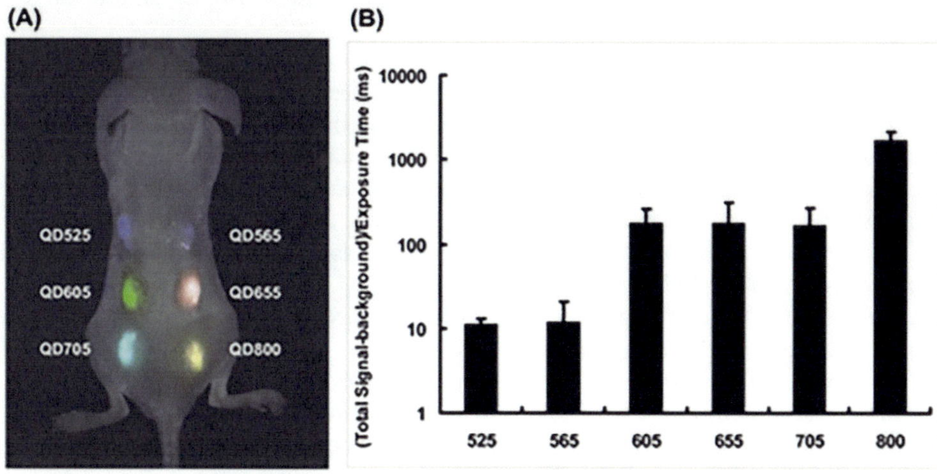

Figure 3.
Quantitative multiplex imaging capability in live animals using QD. (A) 1 × 106 ES cells labeled with QD 525, 565, 605, 655 and 705 were subcutaneously injected on the back of nude mice right after labeling and the image was taken with a single excitation light source right after injection. The quantification of fluorescent signal intensity defined as total signal-background/exposure time in millisecond was shown in (B). (image reused under unrestricted reuse permission) [34].

nanocapsules that were stacked with ZnSe:Mn/ZnS. This core-shell nanoformulation with the anticancer medication PTX. Another study by Cai et al. used Doxorubicin (DOX) on pH-responsive Zinc Oxide quantum dots [50, 51]. In another study, very small QDs were synthesized by the group with a size of approximately 3 nanometer attached with a functional group of poly-ethylene-glycol (PEG) and hyaluronic acid to target glycoprotein CD44 that are overexpressed in malignant growth cells with DOX as the drug model [51]. Under acidic intracellular environment, this medication (DOX) will be delivered. In another study, Yang et al. worked on quercetin (QE) attached with cadmium selenide-zinc sulphide QDs as anticancer and antibacterial nanoformulation and exhibited that QE-attached CdSe-ZnS were more attractive contrary to drug-nontoxic *Escherichia coli* and *Bacillus subtilis* and subsequently quercetin attached cadmium selenide nanoformulation without ZnS [52]. The anti-cancer activity measure was centered around the multiplication and relocation of human gastric epithelial cell lines (BGC-823), which displayed an expansion in cellular toxicity of two-to six-times contrasted with crude Quercetin-CdSe QDs (**Figure 3**) [53].

6. Anti-bacterial activity

Antibiotics are the main drugs that are used for fighting bacterial contamination and have a significant responsibility in the maintenance of society's well-being. Bacterial infections have risen to a public health crisis, with widespread resistance to antibiotics. Graphene or Carbon quantum dots may be used as a new form of antibiotic and could even be applied to biomedical research [54–56]. An increasing threat to the public from bacteria has raised concerns about the spread of drug resistance. A method to limit this development and fight bacterial infections has been found in the use of quantum dots [57]. Widespread bacterial infections—particularly of the hospital-acquired kind—and the spread of antibiotic-resistance plague the medical world. Carbon quantum dots, also known as CQDs, have been investigated for a variety of uses—including as an antibacterial agent.

Bacterial diseases and the spread of antibiotic obstruction address a developing danger to general wellbeing, and the blend of CQDs with antibodies could be a promising sort of antibacterial treatment. The spread of antibacterial-safe microbes is a developing danger to general wellbeing. Carbon quantum dots have turned into a promising antibacterial option since they are non-poisonous to people, they have no known obstruction mechanism. All Gram-positive, and Gram-negative bacteria were found to be inhibited by Carbon QDs [58–62]. Examination of the antibacterial component of positively charged carbon QDs (PC-CQDs) demonstrated that little estimated PC-CQDs functionalized with –NH2 and –NH prompted solid adherence manner on bacterial cell layers [63–65].

Additionally, the passage of PC-CQDs caused conformational changes in the gene and generation of receptive oxygen species in microscopic organisms. Safety assessment represented that PC-CQDs did not set off noticeable drug confrontation or hemolysis. Moreover, PC-CQDs really advanced injury recuperating in rats tainted with blended *Staphylococcus aureus* and *Escherichia coli* by repressing bacterial development while advancing the arrival of inflammatory cytokines and development factors fundamental for tissue repair. A few examinations have revealed the expected antibacterial action of these QDs for genuine injury mending applications in complex bacterial contaminations and, surprisingly, safe microorganisms caused diseases

[63, 66–69]. Recent research showed the successful application of Quaternized carbon quantum dots (CQDs) which have excellent broad-spectrum antibacterial activity. These Q-CQDs contain electron-withdrawing hydroxyl group (−OH), and electron donating methyl (−OCH3), in addition to tri-methylamin (−N+(CH3)3) groups. Q-CQDs functionalized with −N+(CH3)3 displayed a strong attachment to the membrane of the bacterial cell. Hence it damages the bacterial cell by crossing the membrane and entering the cell. The production of reactive oxygen species (ROS) due to the permeation of Q-CQDs into these bacteria resulted and the efflux of the RNA and DNA in the cytoplasm, which ultimately kills the bacteria. Q-CQDs were also shown to have anti-bacterial activity in treating *S. aureus*, *E. coli*, and mixed *S. aureus*-co-*E. coli* diseased injuries by killing bacteria and stimulating septic wound therapeutic. Further functional groups can be attached to the surface of these QDs and chemical modifications can be done by effective surface engineering strategy to widen their utility to meet specific targeting and guiding needs in numerous disease treatments [70].

7. Photo thermal therapy

Recent years have seen a lot of interest in photo-thermal therapy (PTT), a negligibly invasive and possibly effective procedure [71, 72]. It is based on the generation of heat for the thermal melting of malignant tumors by activating photosensitizing chemicals by pulsed laser irradiation at near-infrared wavelengths. The main advantages of photothermal therapy over traditional radiotherapy or chemotherapy include the capacity to penetrate deep tissue layers and the little impact of unselected cell death on the healthy adjacent tissue. Large optical frequency absorption cross-sections, low toxicity, simplicity in functionalization, biocompatibility, and is highly soluble in organic solutions are all desirable characteristics of a photosensitizer. Recent years have seen the usage of QDs by researchers to accomplish PTT in malignant cells [73]. In vitro research on the PTT application of CQDs for malignancy treatment was reported by Zhang et al. [74]. When exposed to NIR light, PEGylated (stealth) Carbon-QDs were found to have an abundant capability for photo-thermal treatment with no harmful toxicity beside breast cancer cell lines (MCF-7). Advanced studies have succeeded by the combination of photo-thermal and chemotherapy by linking these CQDs silicon dioxide (SiO_2) nanoformulation loaded with doxorubicin (DOX) in the treatment of cancer cells [75]. This study's in vivo tests showed that tumor growth in mouse models might be efficiently controlled without causing cancer to return. Doxorubicin (DOX) administration using nanoplatforms in conjunction with PTT and chemotherapy has been investigated in a number of studies in cancer cell lines.

In addition to DOX, thermo-acoustic nanoplatforms (CdTs) were created that targeted cancer cells that contained lysosomes while also rapidly raising temperature in response to laser irradiation [76]. For targeted gene therapy to malignant cells for suppressing cancer both in vitro and in vivo, therapeutic genes were directly attached to CdTs via electrostatic interactions. Additionally, it was able to simultaneously use CdTs for photo-thermal ablation of cancer cells and photo-acoustic imaging based on the recognition of ultrasonic pulses [77, 78]. In order to create an effective system for image-guided positron treatment, it is crucial to strike a balance between radiative degeneration (fluorescence emission), and non-radiative degeneration (dissipated as heat). Intramolecular rotation restriction prevents non-radiative decay and makes it easier for brilliant emission to occur in the aggregation stage. On the contrary, the PTT effect got elevated as the result of heat released by the aggregation-induced emission (AIE) molecules.

Likewise, in a biological environment, more intramolecular communication in donor-acceptor (D-A) based coplanar NIR particles has obstructed all thermal intensity discharge pathways. Subsequently, controlling atomic movement in the collected state is one method for adjusting both heat intensity generation and emission for broadened accuracy in the analysis and exhaustion of cancer growths. In the making of novel photothermal treatment, researchers proved that boron quantum dots have shown potential for the treatment of growths [78]. These unique compounds exhibit great photoacoustic imaging performance, and are biocompatible, and are thought to be non-toxic. The safety of BQDs was established in vitro, and in vivo research demonstrated their potent photothermal conversion effect and photothermal ablative capability. It is safe to assume that these innovative formulations could provide a solid platform for upcoming cancer therapy research and development.

8. Photo dynamic therapy

Photodynamic treatment is an arising and promising helpful methodology for fighting a loathsome illness like malignant growth [79, 80]. Photodynamic treatment is a treatment methodology that joins the photophysical and photochemical cycles to achieve natural impact [81]. Photodynamic treatment utilizes natural photosensitizers that, when presented to light, produce singlet oxygen. QDs have broadly been utilized for this reason, and specialists have created water-dissolvable nanocomposites in view of CdSe/ZnS QDs and hydrophobic tetraphenylporphyrin particles passivated by chitosan. These nanocomposites showed a 45% typical productivity in creating singlet oxygen due to the intracomplex Förster reverberation energy move with TPP.

It incorporates the actual course of photochemically responding to an energized photosensitizer with either cell substrates or sub-atomic oxygen, which at last outcomes in the passing of disease cells. The photosensitizer has two electrons with contradicting spins in a low-energy sub-atomic orbital in its ground state. One of these electrons is excited to higher energy atomic orbitals without changing spins when light is ingested [82]. Because of the photodrug's concise lifetime in this state, which is known as the singlet excited state, it cannot take part in responses with cell substrates (going from nano to picoseconds).

The excited photosensitizer can either go through fluorescence, which delivers light energy, to get back to its ground state, or non-radiative decay, which delivers heat energy through inward conversion (IC). The improvement of photosensitizers from the original to the ongoing third era, delivery systems, the development or suppression of immunity, combinational treatment, and other basic components of photodynamic treatment should be in every way completely talked about. The utilization of late made quantum dots is growing in photodynamic treatment (PDT). The primary benefits of utilizing these QDs are that they beat the downsides of conventional PDT compounds through having great photostability, high quantum yield, areas of strength for and as a result of their huge change dipole moment [83, 84].

Moreover, the photophysical qualities and fluid solvency of the core can be custom fitted to address specific issues by shifting their size and content, which offers a huge surface region for linkage to biomolecules like peptides and antibodies. Be that as it may, the results of QDs utilized freely in photodynamic treatment (PDT) were disappointing. They can notwithstanding, participate in fluorescence reverberation energy move (FRET) by filling in as energy benefactors, which various gatherings effectively utilized for PDT. The FRET interaction is shown by the contributor particles' extinguished fluorescence and the acceptor atoms' expanded fluorescence. QDs alone are not effective at creating singlet oxygen, however, when joined with normal dyes, the combination produces singlet oxygen with a high quantum yield. Their closeness, estimated in nanometers, is the principal essential for FRET from QD to PS. They can remain joined by noncovalent complexes or covalent bonds. The problem of QD instability in biological media has been addressed by a variety of surface modification techniques. Commonly, amphiphilic molecules are used to encase QDs. The hydrophobic cavity of the QD is encircled by the hydrophobic end of the molecule, which permits the QD to disperse in solution. This is a widely used encapsulation method because it addresses the issues of luminescence quantum yields and colloidal stability [85].

The size of the particle will determine whether or not effective QDs can be developed for diagnostic and therapeutic uses. A QD can pass through biological barriers like the alveolo-capillary, blood-brain barrier, gastric, and barriers in the dermis as well as the renal filtration barrier if its core material has a minimum diameter of 8–10 nm (which is typically increased with surface features) [86]. Before commercializing QDs for use in humans, we must conduct additional research on their potential toxicity, which will take time overall [87]. According to a recent study, the creation of noble carbon dots from curcumin and folic acid improved the effectiveness of photodynamic therapy in treating cancer cells. For nucleus-targeting PDT, conjugated carbon dots (CCDs), a novel two-photon active photosensitizer was proved to release lethal reactive oxygen species (ROS). Through a pathway mediated by folate-receptor, a combination of carbon dots, curcumin, and folic acid (CDcf) was discovered to interact with malignant cells, leading to significant localization within the nucleus. PDT efficiency eventually increased as a result of the increased reactive oxygen species (ROS) generation in the nucleus brought on by two-photon excitation. As a result, by directly attacking cancer cells' DNA, cancer cells were eliminated more successfully. The development of multi-functional dual-photon active nanoformulation on a solo stage for improved PDT in oral malignancy diagnosis and therapy was made possible by CDs' inherent ROS generation and nucleus-targeting capacity. In order to aim and distribute PDT proxies accurately and efficiently, CDcf has developed a new approach. Furthermore, the technique can be easily expanded for better enactment in the nanoprobes with added malignant cell lines using targeted therapy and diagnosis [88].

9. Quantum dots in optoelectronic devices

Optoelectronic systems are very interested in quantum dots because of their unique characteristics [89, 90]. A new generation of semiconductor devices known as quantum dots has been the focus of the nanotechnology sector. Due to their ability to provide special properties, quantum dots, also known as "artificial atoms," are now being used in a variety of high-performance devices. They have improved the performance of solar cells and lasers, for example. The adequacy and usefulness of these promising nanomaterials for cutting-edge optoelectronic gadgets in both modern and biomedical fields have been further developed through various endeavors made around the world [91]. Lasers, photo-detectors (photodiode), amplifiers, and solar cells are a few examples of optoelectronic devices. Researchers have developed quantum dot devices over the past two decades that outperform earlier devices based on quantum dots in terms of performance. Self-assembled nanoparticles have drawn a lot of attention for many years. A semiconductor can develop specific unique properties by reducing its size to the nanoscale. Nanometer-sized semiconductors can have their electrical and optical properties altered by manipulating their morphology. Similar to electrons, semiconductor quantum dots (QDs) are an intriguing nanoscale structure of confined carriers in all dimensions [92]. Zero-dimensional semiconductor nanostructures are more tunable and delicate to outer boundaries than customary mass semiconductors; furthermore, in light of the fact that QDs are zero-dimensional, their energy levels can be dense to an assortment of delta functions. The key advantages of low-layered semiconductors have produced a lot of interest. Early research established the wavelength tunability and threshold reduction capabilities of quantum-organized semiconductor lasers, which were subsequently theoretically supported. Since there was not a practical method for making QDs for several years after these groundbreaking studies, efforts to develop QDs and use them in devices were limited to experimental research. Quickly following the effective production of self-assembled quantum dot laser diodes (QD based 77 K operational diodes) were created [93, 94]. A worldwide drive was begun to further develop quantum dots development control after an exhibit of self-collected quantum dots and the main quantum dot laser. Indium arsenide/indium phosphide (InAs/InP) and gallium arsenide (In(Ga)As/GaAs) [92] are two instances of mismatched grids that definitely stand out from researchers [95]. Heteroepitaxial development modes in thin films are regularly made sense by utilizing thermodynamic justifications. Further developed QD self-gathering got prompt prizes. Lasers, as well as quantum dab enhancers, quantum dot solar cells (QDSCs), quantum dot infrared photodetectors (QDIPs), and quantum dot super-luminescent diodes (QDSLDs), have all been accounted for [96]. Between subband assimilation in QDs in the last part of the 1990s prompted the advancement of mid-frequency and long-frequency infra-red photo-detectors. Self-coordinated quantum dots optoelectronic gadgets have progressed essentially since these nanostructures were first created [97].

In terms of efficiency, quantum dot lasers are now on par with quantum well lasers. Inter-sublevel instruments have improved efficiency and unique properties, and they can be used as light sources and detectors. Future systems are predicted to heavily rely on quantum dot optoelectronic sensors. Additionally, high-performance QDIPs are made possible by active research in QDs [98]. The fact that QDIPs can operate at higher temperatures and have lower dark currents gives them a fundamental advantage over QWIPs. Efficiency of quantum-well IR photo-detectors (QWIPs) has been surpassed by QD-IR photo-detectors (QDIPs) [99]. Moreover, there are as yet various issues, including deficient quantum execution and the requirement for further developed QDIP engineering and creation to completely use their true capacity in third-age infrared detecting [100]. In the impending years, QDIPs with effectiveness tantamount to present state-of-the-art advancements like QWIPs and HgCdTe photodetector might be utilized [101]. Quantum wells are less developed than quantum dots (QDs) to the extent that optoelectronic systems plan on account of different head imperfections in QDs' slow retention and the very thermal relationship between the intermediate and conduction band [102]. To accomplish high transformation productivity far in excess of the record worth of GaAs single-intersection solar cells, extra improvements in gadget physical science, plan, development, and portrayal of QDSCs would be required [103, 104]. QD-based biosensors are widely used in science, engineering, and technology to monitor many facets of contemporary life. To advance research in these fields, it is necessary to comprehend the various types of sensors, how to use them for different applications, as well as how to optimize and utilize them. Quantum dots (QDs), which are zero-dimensional semiconductors, exhibit optical gain, laser technology, strong light absorption, and intense narrow-band radiation over the visible and infrared bands. Imaging, solar energy deriving, display units, and signal transportation can all benefit from these qualities. One of the modest light-emitting technologies is the use of LECs (light-emitting electrochemical cells). Numerous LECs have been confirmed by means of a variety of biological materials, including fluorescent polymers. The understanding of LECs using materials of inorganic form, predominantly materials of low-dimensional is interesting to relate to light-emitting diodes/instruments, that have been made possible by recent advances in this QD device technologies. Recent developments in two-dimensional and low-dimensional materials, like chiral light-emitting devices and materials based on quantum dots, offer significant functionalities that hold promise for new optoelectronic applications based on emergent materials [105].

10. Challenges and future perspectives

Even though QDs have an extensive choice of applications, including in vivo biomedical imaging and detection, they also have the potential to be harmful to both human health and the environment. Typically, QDs can impair organism function and cause metabolic disability or death through at least three distinct pathways. The QDs' composition comes first and is of utmost importance [106]. Toxic ions could be released from QDs and poison cells during internal corrosion. As the nanoparticles have a high surface-to-volume ratio, they are more prospective than bulk material to undergo partial decomposition and release ions [107–109]. Regardless of composition, QDs' small size creates another potential drawback because particles can adhere to cell membranes or be ingested and retained inside cells, impairing organ function. It was discovered that many QDs have toxic effects, with earlier studies attributing the cause to the presence of hazardous metals. Commercial quantum dots frequently contain the toxic heavy metal core CDs cadmium, lead, mercury, and arsenic [110, 111]. According to reports, the Cd-core QDs were indeed cytotoxic, particularly when surface oxidation from air or UV exposure caused reduced Cd to form on the particle surface and the release of free $Cd2+$ ions. The cells were damaged after being cultured with CdSe QDs for seven days, showing diffused nuclei and ill-defined cell boundaries [112, 113]. Additionally, it was demonstrated that cadmium telluride (CdTe) QDs had negative effects on cellular functions, with the smaller ones emitting green light is more harmful than the larger ones emitting red light [114]. In imaging, clinical applications, and basic biomedical research, QDs hold great promise. To comprehend the potential of these new generation materials, a review of QD properties and general perception is conducted. The toxicity of these nanoparticles is one significant barrier, but it is not yet fully understood. Before commercializing QDs for use in humans, we must conduct additional research on their potential toxicity, which will take time overall [105, 115–118].

11. Summary and conclusion

Numerous implications for celllabelling, biomedical imaging, diagnostics, and drug delivery are provided by quantum dots. They have benefits over traditional fluorescent dyes and green fluorescence proteins, including a size similar to antibodies that enables combined applications using these recognition molecules, narrow-emission and broad-excitation spectra, high intense photons, and anti-quenching ability, wavelength tunability making them suitable for multi-wavelength applications, and high brightness and photostability. Quantum dots might not be able to replace all of the other fluorophores currently employed in labeling and imaging, though. In addition to their excellent photophysical characteristics, quantum dots face some difficulties, such as restricted in vivo applications due to complex surface chemistry. Quantum dots are frequently used in conjunction with other kinds of fluorophores to great effect. They may be advantageous in some applications but disadvantageous in others. QDs must be small enough to penetrate biological barriers if they are to be used in routine clinical settings for diagnostic and therapeutic purposes. Additionally, they must have low toxicities. The theranostic applications of QDs-based biosensors are covered in detail in this chapter. Although there are opposing views on QDs' bio-compatibility and toxicity, they have made significant strides in the field of research thus far. The most recent article was published a few years ago, and now scientists are starting to realize how QDs can be used for therapeutic, diagnostic, and sensing purposes. Despite all of this, creating smart formulations using QDs still presents a number of difficulties.

Author details

Karunanithi Rajamanickam
Faculty of Allied Health Sciences, Chettinad Academy of Research and Education (CARE), Tamilnadu, India

*Address all correspondence to: rkarunanithi@care.edu.in

References

[1] Chaturvedi S, Dave PN. Nanotechnology: History and Future. 21st Century Nanoscience–A Handbook: Public Policy, Education, and Global Trends. 2020;**10**:4-1

[2] Marzo JL, Jornet JM, Pierobon M. Nanonetworks in biomedical applications. Current Drug Targets. 2019;**20**(8):800-807

[3] Hornyak GL, Moore JJ, Tibbals HF, Dutta J. Fundamentals of Nanotechnology. US: CRC Press; 2018

[4] Khanna V. Nanosensors. US: CRC Press; 2016

[5] Hatzikraniotis E, Kyratsi T. Materials Science: Trends, Material Properties and Educational Perspectives. In: Psillos D, Kariotoglou P, editors. Iterative Design of Teaching-Learning Sequences. Dordrecht: Springer; 2016

[6] Xu S, Cui J, Wang L. Recent developments of low-toxicity NIR II quantum dots for sensing and bioimaging. TrAC Trends in Analytical Chemistry. 2016;**80**:149-155

[7] Viana OS, Ribeiro MS, Fontes A, Santos BS. Quantum Dots in Photodynamic Therapy. In: Batinić-Haberle I, Rebouças J, Spasojević I, editors. Redox-Active Therapeutics. Oxidative Stress in Applied Basic Research and Clinical Practice. Cham: Springer; 2016

[8] McHugh KJ, Jing L, Behrens AM, Jayawardena S, Tang W, Gao M, et al. Biocompatible semiconductor quantum dots as cancer imaging agents. Advanced Materials. 2018;**30**(18):1706356

[9] Grushevskaya H, Krylov G, Kruchinin S, Vlahovic B, Bellucci S. Electronic properties and quasi-zero-energy states of graphene quantum dots. Physical Review B. 2021;**103**:235102

[10] Pandey P. Role of nanotechnology in electronics: A review of recent developments and patents. Recent Patents on Nanotechnology. 2022;**16**(1):45-66

[11] Leigh WB. Devices for Optoelectronics (1st ed.). US: CRC Press; 2021

[12] Kambhampati P. Nanoparticles, nanocrystals, and quantum dots: What are the implications of size in colloidal nanoscale materials? Journal of Physical Chemistry Letters. 2021;**12**(20):4769-4779

[13] Ghasemi H, Mozaffari MH. Synthesis and optoelectronic properties of CdSe quantum dots. 18 May 2021. pp. 1-4

[14] Duan L, Hu L, Guan X, Lin CH, Chu D, Huang S, et al. Quantum dots for photovoltaics: A tale of two materials. Advanced Energy Materials. 2021;**11**(20):2100354

[15] Slavica B. Applicability of Quantum Dots in Biomedical Science. Djezzar B, editor. Ionizing Radiation Effects and Applications. IntechOpen;2017. DOI: 10.5772/ intechopen.71428

[16] Bera D, Qian L, Tseng T-K, Holloway PH. Quantum dots and their multimodal applications: A review. Materials. 2010;**3**(4):2260-2345

[17] Bakalova R, Zhelev Z, Gadjeva V. Quantum dots versus organic fluorophores in fluorescent deep-tissue imaging—Merits and demerits. General Physiology and Biophysics. 2008;**27**(4):231-242

[18] Lee SF, Osborne MA. Brightening, blinking, bluing and bleaching in the life of a quantum dot: Friend or foe? ChemPhysChem. 2009;**10**(13):2174-2191

[19] Sukhanova A, Devy J, Venteo L, Kaplan H, Artemyev M, Oleinikov V, et al. Biocompatible fluorescent nanocrystals for immunolabeling of membrane proteins and cells. Analytical Biochemistry. 2004;**324**(1):60-67

[20] Lin G, Chen T, Zou J, Wang Y, Wang X, Li J, et al. Quantum Dots-siRNA Nanoplexes for Gene Silencing in Central Nervous System Tumor Cells. Frontiers in Pharmacology. 4 Apr 2017;**8**:182. DOI: 10.3389/fphar.2017.00182. PMID: 28420995; PMCID: PMC5378761

[21] Wang M, Li H, Huang B, Chen S, Cui R, Sun ZJ, et al. An Ultra-Stable, Oxygen-Supply Nanoprobe Emitting in Near-Infrared-II Window to Guide and Enhance Radiotherapy by Promoting Anti-Tumor Immunity. Advanced Healthcare Materials. Jun 2021;**10**(12):e2100090. DOI: 10.1002/adhm.202100090. Epub 2021 Apr 22. PMID: 33885213

[22] Rezaei A, Hashemi E. A pseudohomogeneous nanocarrier based on carbon quantum dots decorated with arginine as an efficient gene delivery vehicle. Scientific Reports. 2021;**11**(1):13790

[23] Zhang Y, Wang T-H. Quantum dot enabled molecular sensing and diagnostics. Theranostics. 2012;**2**(7):631-654

[24] Xiao Y, Wu Z, Yao Q, Xie J. Luminescent metal nanoclusters: Biosensing strategies and bioimaging applications. Aggregate. 2021;**2**(1):114-132

[25] Naresh V, Lee N. A review on biosensors and recent development of nanostructured materials-enabled biosensors. Sensors. 2021;**21**(4):1109

[26] Wang J, Han S, Ke D, Wang R. Semiconductor quantum dots surface modification for potential cancer diagnostic and therapeutic applications. Journal of Nanomaterials. 2012;**2012**:129041

[27] Wen L, Qiu L, Wu Y, Hu X, Zhang X. Aptamer-modified semiconductor quantum dots for biosensing applications. Sensors, Basel. 2017;**17**(8):1-14. OI: 10.3390/s17081736. PMID: 28788080; PMCID: PMC5579848

[28] Nienhaus K, Wang H, Nienhaus GU. Nanoparticles for biomedical applications: Exploring and exploiting molecular interactions at the nano-bio interface. Materials Today Advances. 2020;**5**:100036

[29] Wen L, Qiu L, Wu Y, Hu X, Zhang X. Aptamer-Modified Semiconductor Quantum Dots for Biosensing Applications. Sensors (Basel). 2017 Jul 28;**17**(8):1-14. DOI: 10.3390/s17081736. PMID: 28788080; PMCID: PMC5579848

[30] Freeman R, Girsh J, Willner I. Nucleic acid/quantum dots (QDs) hybrid systems for optical and photoelectrochemical sensing. ACS Applied Materials & Interfaces. 2013;**5**(8):2815-2834

[31] Manzoor O, Soleja N, Mohsin MJ. Nanoscale gizmos—The novel fluorescent probes for monitoring protein activity. Biochemical Engineering Journal. 2018;**133**:83-95

[32] Gopalan D, Pandey A, Alex AT, Kalthur G, Pandey S, Udupa N, et al. Nanoconstructs as a versatile tool for detection and diagnosis of Alzheimer biomarkers. Nanotechnology. 2021;**32**(14):142002

[33] Younis MR, He G, Qu J, Lin J, Huang P, Xia XH. Inorganic nanomaterials with intrinsic singlet oxygen generation for photodynamic

therapy. Advanced Science. 2021;**8**(21):2102587

[34] Konkar A, Lu S, Madhukar A, Hughes SM, Alivisatos AP. Semiconductor nanocrystal quantum dots on single crystal semiconductor substrates: High resolution transmission electron microscopy. Nano Letters. 2005;**5**(5):969-973

[35] Xing Y, Chaudry Q, Shen C, Kong KY, Zhau HE, Chung LW, et al. Bioconjugated quantum dots for multiplexed and quantitative immunohistochemistry. Nature Protocols. 2007;**2**(5):1152-1165

[36] Smith AM, Dave S, Nie S, True L, Gao X. Multicolor quantum dots for molecular diagnostics of cancer. Expert Review of Molecular Diagnostics. 2006;**6**(2):231-244

[37] Derfus AM, Chan WCW, Bhatia SN. Probing the cytotoxicity of semiconductor quantum dots. Nano Letters. 2004;**4**(1):11-18

[38] Guo G, Liu W, Liang J, Xu H, He Z, Yang X. Preparation and characterization of novel CdSe quantum dots modified with poly (d, l-lactide) nanoparticles. Materials Letters. 2006;**60**(21):2565-2568

[39] Hong G, Antaris AL, Dai H. Near-infrared fluorophores for biomedical imaging. Nature Biomedical Engineering. 2017;**1**(1):0010

[40] Yang ST, Wang X, Wang H, Lu F, Luo PG, Cao L, et al. Carbon dots as nontoxic and high-performance fluorescence imaging agents. The Journal of Physical Chemistry C, Nanomaterials and Interfaces. 2009;**113**(42):18110-18114

[41] Jackson B, Bugge D, Ranville J, Chen C. Bioavailability, toxicity, and bioaccumulation of quantum dot nanoparticles to the amphipod Leptocheirus plumulosus. Environmental Science & Technology. 2012;**46**:5550-5556

[42] Clift MJD, Stone V. Quantum dots: An insight and perspective of their biological interaction and how this relates to their relevance for clinical use. Theranostics. 2012;**2**(7):668-680

[43] Jain NS, Somanna P, Patil BA. Application of quantum dots in drug delivery. Nanoscience & Nanotechnology-Asia. 2022;**12**(1):16-31

[44] Probst CE, Zrazhevskiy P, Bagalkot V, Gao X. Quantum dots as a platform for nanoparticle drug delivery vehicle design. Advanced Drug Delivery Reviews. 2013;**65**(5):703-718

[45] Matea CT, Mocan T, Tabaran F, Pop T, Mosteanu O, Puia C, et al. Quantum dots in imaging, drug delivery and sensor applications. International Journal of Nanomedicine. 2017;**12**:5421-5431

[46] Gidwani B, Sahu V, Shukla SS, Pandey R, Joshi V, Jain VK, et al. Quantum dots: Prospectives, toxicity, advances and applications. Journal of Drug Delivery Science and Technology. 2021;**61**:102308

[47] Kairdolf BA, Smith AM, Stokes TH, Wang MD, Young AN, Nie S. Semiconductor quantum dots for bioimaging and biodiagnostic applications. Annual Review of Analytical Chemistry. 2013;**6**(1):143-162

[48] Song JT, Yang XQ, Zhang XS, Yan DM, Yao MH, Qin MY, et al. Composite silica coated gold nanosphere and quantum dots nanoparticles for X-ray CT and fluorescence bimodal imaging. Dalton transactions (Cambridge, England: 2003). 2015;**44**(25):11314-11320

[49] Ranjbar-Navazi Z, Fathi M, Abdolahinia ED, Omidi Y, Davaran SJMS,

et al. MUC-1 aptamer conjugated InP/ZnS quantum dots/nanohydrogel fluorescent composite for mitochondria-mediated apoptosis in MCF-7 cells. Materials Science and Engineering: C. 2021;**118**:111469

[50] Cai X, Luo Y, Yan H, Du D, Lin Y. pH-responsive ZnO nanocluster for lung cancer chemotherapy. ACS Applied Materials & Interfaces. 2017;**9**(7):5739-5747

[51] Cai X, Luo Y, Zhang W, Du D, Lin Y. pH-Sensitive ZnO quantum dots–doxorubicin nanoparticles for lung cancer targeted drug delivery. ACS Applied Materials & Interfaces. 2016;**8**(34):22442-22450

[52] Matea CT, Mocan T, Tabaran F, Pop T, Mosteanu O, Puia C, et al. Quantum dots in imaging, drug delivery and sensor applications. International Journal of Nanomedicine. 2017;**12**:5421

[53] Li Y, Liu B, Yang F, Yu Y, Zeng A, Ye T, et al. Lobaplatin induces BGC-823 human gastric carcinoma cell apoptosis via ROS-mitochondrial apoptotic pathway and impairs cell migration and invasion. Biomedicine & Pharmacotherapy = Biomedecine & Pharmacotherapie. 2016;**83**:1239-1246

[54] Han W, Wu Z, Li Y, Wang Y. Graphene family nanomaterials (GFNs)—Promising materials for antimicrobial coating and film: A review. Chemical Engineering Journal. 2019;**358**:1022-1037

[55] Dong X, Liang W, Meziani MJ, Sun YP, Yang L. Carbon dots as potent antimicrobial agents. Theranostics. 2020;**10**(2):671-686

[56] Al Awak MM, Wang P, Wang S, Tang Y, Sun YP, Yang L. Correlation of carbon dots' light-activated antimicrobial activities and fluorescence quantum yield. RSC Advances. 2017;7(48):30177-30184

[57] McCollum C, Bertram J, Nagpal P, Chatterjee A. Treatment of Multidrug-Resistant Bacterial Infections Using Quantum Dots. The FASEB Journal. 2022;**36**. DOI: 10.1096/fasebj.2022.36.S1.R4869

[58] Rajendiran K, Zhao Z, Pei D-S, Fu AJ. Antimicrobial activity and mechanism of functionalized quantum dots. Polymers. 2019;**11**(10):1670

[59] Shahshahanipour M, Rezaei B, Ensafi AA, Etemadifar Z. An ancient plant for the synthesis of a novel carbon dot and its applications as an antibacterial agent and probe for sensing of an anti-cancer drug. Materials Science and Engineering: C. 2019;**98**:826-833

[60] Elliott AG, Huang JX, Neve S, Zuegg J, Edwards IA, Cain AK, et al. An amphipathic peptide with antibiotic activity against multidrug-resistant Gram-negative bacteria. Nature Communications. 2020;**11**(1):1-13

[61] Shaw Z, Kuriakose S, Cheeseman S, Dickey MD, Genzer J, Christofferson AJ, et al. Antipathogenic properties and applications of low-dimensional materials. Nature Communications. 2021;**12**(1):1-19

[62] Shikha S, Chaudhuri SR, Bhattacharyya MS. Facile one pot greener synthesis of sophorolipid capped gold nanoparticles and its antimicrobial activity having special efficacy against gram negative vibrio cholerae. Scientific Reports. 2020;**10**(1):1463

[63] Hao X, Huang L, Zhao C, Chen S, Lin W, Lin Y, et al. Antibacterial activity of positively charged carbon quantum dots without detectable resistance for

wound healing with mixed bacteria infection. Materials Science and Engineering: C. 2021;**123**:111971

[64] Yang J, Zhang X, Ma YH, Gao G, Chen X, Jia HR, et al. Carbon dot-based platform for simultaneous bacterial distinguishment and antibacterial applications. ACS Applied Materials & Interfaces. 2016;**8**(47):32170-32181

[65] Lin F, Bao Y-W, Wu F-G. Carbon dots for sensing and killing microorganisms. Journal of Carbon Research. 2019;**5**(2):33

[66] Machado GHA, Marques TR, de Carvalho TCL, Duarte AC, de Oliveira FC, Gonçalves MC, et al. Antibacterial activity and in vivo wound healing potential of phenolic extracts from Jaboticaba skin. Chemical Biology & Drug Design. 2018;**92**(1):1333-1343

[67] Zhao C, Wang X, Yu L, Wu L, Hao X, Liu Q, et al. Quaternized carbon quantum dots with broad-spectrum antibacterial activity for the treatment of wounds infected with mixed bacteria. Acta Biomaterialia. 2022;**138**:528-544

[68] Zmejkoski DZ, Marković ZM, Mitić DD, Zdravković NM, Kozyrovska NO, Bugárová N, et al. Antibacterial composite hydrogels of graphene quantum dots and bacterial cellulose accelerate wound healing. Journal of Biomedical Materials Research Part B, Applied Biomaterials. 2022;**110**(8):1796-1805

[69] Chai S, Zhou L, Pei S, Zhu Z, Chen B. P-Doped Carbon Quantum Dots with Antibacterial Activity. Micromachines. 2021;**12**:1116. DOI: 10.3390/mi12091116

[70] Wu L, Gao Y, Zhao C, Huang D, Chen W, Lin X, et al. Synthesis of curcumin-quaternized carbon quantum dots with enhanced broad-spectrum antibacterial activity for promoting infected wound healing. Biomaterials Advances. 2022;**133**:112608

[71] Wagalgave SM, Birajdar SS, Malegaonkar JN, Bhosale SV. Chapter Eight - Patented AIE materials for biomedical applications. Bhosale RS, Singh V, editors. Progress in Molecular Biology and Translational Science, Academic Press; 2021;**185**:199-223

[72] Fong JFY, Ng YH, Ng SM. Chapter 7 - Carbon dots as a new class of light emitters for biomedical diagnostics and therapeutic applications, Grumezescu AM, editor. Fullerens, Graphenes and Nanotubes. William Andrew Publishing; 2018:227-295. (Swinburne University of Technology Sarawak Campus, Kuching, Sarawak, Malaysia)

[73] Dolatyari M, Aghdam FA, Rostami G, Rostami A, Amiri IS. Introducing new conjugated quantum dots for photothermal therapy in biological applications. Plasmonics. 2020;**15**(6):1565-1575

[74] Zhang M, Zheng T, Sheng B, Wu F, Zhang Q, Wang W, et al. Mn2+ complex-modified polydopamine- and dual emissive carbon dots based nanoparticles for in vitro and in vivo trimodality fluorescent, photothermal, and magnetic resonance imaging. Chemical Engineering Journal. 2019;**373**:1054-1063

[75] Tu X, Wang L, Cao Y, Ma Y, Shen H, Zhang M, et al. Efficient cancer ablation by combined photothermal and enhanced chemo-therapy based on carbon nanoparticles/doxorubicin@SiO2 nanocomposites. Carbon. 2016;**97**:35-44

[76] Xie H, Liu M, You B, Luo G, Chen Y, Liu B, et al. Photothermal Therapy: Biodegradable Bi2O2Se Quantum Dots for Photoacoustic Imaging-Guided Cancer Photothermal Therapy (Small 1/2020). Small, US. 2020;**16**:2070013

[77] Ding D, Guo W, Guo C, Sun J, Zheng N, Wang F, et al. MoO(3-x) quantum dots for photoacoustic imaging guided photothermal/photodynamic cancer treatment. Nanoscale. 2017;**9**(5):2020-2029

[78] Guo T, Tang Q, Guo Y, Qiu H, Dai J, Xing C, et al. Boron quantum dots for photoacoustic imaging-guided photothermal therapy. ACS Applied Materials & Interfaces. 2021;**13**(1):306-311

[79] Chilakamarthi U, Giribabu L. Photodynamic therapy: Past, present and future. The Chemical Record. 2017;**17**(8):775-802

[80] Tabish TA, Scotton CJ, Ferguson J, Lin L, der Veen AV, Lowry S, et al. Biocompatibility and toxicity of graphene quantum dots for potential application in photodynamic therapy. Nanomedicine. 2018;**13**(15):1923-1937

[81] Algorri JF, Ochoa M, Roldán-Varona P, Rodríguez-Cobo L, López-Higuera JM. Photodynamic therapy: A compendium of latest reviews. Cancers. 2021;**13**(17):4447

[82] Niculescu A-G, Grumezescu AM. Photodynamic therapy—An up-to-date review. Applied Sciences. 2021;**11**(8):3626

[83] Martynenko IV, Kuznetsova VA, Orlova AO, Kanaev PA, Maslov VG, Loudon A, et al. Chlorin e6–ZnSe/ZnS quantum dots based system as reagent for photodynamic therapy. Nanotechnology. 2015;**26**(5):055102

[84] Lu D, Tao R, Wang Z. Carbon-based materials for photodynamic therapy: A mini-review. Frontiers of Chemical Science and Engineering. 2019;**13**(2):310-323

[85] Palui G, Aldeek F, Wang W, Mattoussi H. Strategies for interfacing inorganic nanocrystals with biological systems based on polymer-coating. Chemical Society Reviews. 2015;**44**(1):193-227

[86] Volkov Y. Quantum dots in nanomedicine: Recent trends, advances and unresolved issues. Biochemical and Biophysical Research Communications. 2015;**468**(3):419-427

[87] Aizik G, Waiskopf N, Agbaria M, Ben-David-Naim M, Levi-Kalisman Y, Shahar A, et al. Liposomes of quantum dots configured for passive and active delivery to tumor tissue. Nano Letters. 2019;**19**(9):5844-5852

[88] Nasrin A, Hassan M, Gomes VG. Two-photon active nucleus-targeting carbon dots: Enhanced ROS generation and photodynamic therapy for oral cancer. Nanoscale. 2020;**12**(40):20598-20603

[89] Li X, Rui M, Song J, Shen Z, Zeng H. Carbon and graphene quantum dots for optoelectronic and energy devices: A review. Advanced Functional Materials. 2015;**25**(31):4929-4947

[90] Kim J, Song S, Kim Y-H, Park SK. Recent progress of quantum dot-based photonic devices and systems: A comprehensive review of materials. Devices, and Applications. 2021;**2**(3):2000024

[91] Yuan F, Li S, Fan Z, Meng X, Fan L, Yang S. Shining carbon dots: Synthesis and biomedical and optoelectronic applications. Nano Today. 2016;**11**(5):565-586

[92] Wu J, Chen S, Seeds A, Liu H. Quantum dot optoelectronic devices: Lasers, photodetectors and solar cells. Journal of Physics D: Applied Physics. 2015;**48**(36):363001

[93] Fang J, Zhou Z, Xiao M, Lou Z, Wei Z, Shen G. Recent advances in

low-dimensional semiconductor nanomaterials and their applications in high-performance photodetectors. InfoMat. 2020;**2**(2):291-317

[94] Benelmekki M. Zero-dimensional nanostructures. In: Nanomaterials [Internet]. Morgan & Claypool Publishers; 2019. pp. 3-1-3-18. DOI: 10.1088/2053-2571/ab126dch3

[95] Gajjela RSR, Koenraad PM. Atomic-scale characterization of droplet epitaxy quantum dots. Nanomaterials. 2021;**11**(1):85

[96] Mitin V, Antipov A, Sergeev A, Vagidov N, Eason D, Strasser G. Quantum dot infrared photodetectors: Photoresponse enhancement due to potential barriers. Nanoscale Research Letters. 2011;**6**(1):21

[97] Bergeson J, Bommena R, Fahey S, Cowan V, Morath C, Velicu S. Mid and long wavelength infrared HgCdTe photodetectors exposed to proton radiation. In: Proceedings of SPIE. SPIE; 2014;**9226**:92260P-1-10

[98] Blood P. Quantum Efficiency of Quantum Dot Lasers, in IEEE Journal of Selected Topics in Quantum Electronics. 2017;**23**(6):1-8. Art no. 1900608, DOI: 10.1109/JSTQE.2017.2687039

[99] Zhang W, Lim H-C, Taguchi M, Tsao S, Szafraniec J, Movaghar B, et al. High performance In As quantum dot infrared photodetectors (QDIP) on InP by MOCVD. In: Proceedings of SPIE 5732, Quantum Sensing and Nanophotonic Devices II. 25 Mar 2005. pp. 1-8

[100] Rogalski A. Progress in quantum dot infrared photodetectors. In: Tong X, Wu J, Wang ZM, editors. Quantum Dot Photodetectors. Cham: Springer International Publishing; 2021. pp. 1-74

[101] Jiao H, Wang X, Chen Y, Guo S, Wu S, Song C, et al. HgCdTe/black phosphorus van der Waals heterojunction for high-performance polarization-sensitive midwave infrared photodetector. Science Advances. 2022;**8**(19):eabn1811

[102] Chatterjee A, Jagtap A, Pendyala N, Rao K. HgCdTe quantum dot over interdigitated electrode for mid-wave infrared photon detection and its noise characterization. International Journal of Nanoscience. 2019;**19**:1950020

[103] Song H, Lin Y, Zhang Z, Rao H, Wang W, Fang Y, et al. Improving the efficiency of quantum dot sensitized solar cells beyond 15% via secondary deposition. Journal of the American Chemical Society. 2021;**143**(12):4790-4800

[104] Kim Y, Ban K-Y, Honsberg CB. Multi-stacked InAs/GaAs quantum dots grown with different growth modes for quantum dot solar cells. Applied Physics Letters. 2015;**106**(22):222104

[105] Pu J, Takenobu T. Recent advances in light-emitting electrochemical cells with low-dimensional quantum materials. Nihon Gazo Gakkaishi (Journal of the Imaging Society of Japan). 2021;**60**(6):656-672

[106] Wu T, Zhang T, Chen Y, Tang M. Research advances on potential neurotoxicity of quantum dots. Journal of Applied Toxicology. 2016;**36**(3):345-351

[107] Friehs E, AlSalka Y, Jonczyk R, Lavrentieva A, Jochums A, Walter J-G, et al. Toxicity, phototoxicity and biocidal activity of nanoparticles employed in photocatalysis. Journal of Photochemistry and Photobiology C: Photochemistry Reviews. 2016;**29**:1-28

[108] Casals E, Casals G, Puntes V, Rosenholm JM. 1 - Biodistribution, Excretion, and Toxicity of Inorganic Nanoparticles. Cui W, Zhao X, editors. In: Micro and Nano Technologies, Theranostic Bionanomaterials. USA: Elsevier; 2019:3-26

[109] Liu J, Wang Y, Ma J, Peng Y, Wang A. A review on bidirectional analogies between the photocatalysis and antibacterial properties of ZnO. Journal of Alloys and Compounds. 2019;**783**:898-918

[110] Wang Z, Tang M. The cytotoxicity of core-shell or non-shell structure quantum dots and reflection on environmental friendly: A review. Environmental Research. 2021;**194**:110593

[111] Reshma VG, Mohanan PV. Quantum dots: Applications and safety consequences. Journal of Luminescence. 2019;**205**:287-298

[112] Paydary P. Degradation of Quantum Dots in Aqueous Environments. Diss. Northeastern University, US. 2018

[113] Pramanik Sunipa and Hill, Samantha KE. and Zhi, Bo and Hudson-Smith, Natalie V. and Wu, Jeslin J. and White, Jacob N. and McIntire, Eileen A. and Kondeti, V. S. Santosh K. and Lee, Amani L. and Bruggeman, Peter J. and Kortshagen, Uwe R. and Haynes, Christy L.", Comparative toxicity assessment of novel Si quantum dots and their traditional Cd-based counterparts using bacteria models Shewanella oneidensis and Bacillus subtilis. Environmental science: Nano. 2018;**5**(8):1890-1901

[114] Hlavata L, Striesova I, Ignat T, Blaskovisova J, Ruttkay-Nedecky B, Kopel P, et al. An electrochemical DNA-based biosensor to study the effects of CdTe quantum dots on UV-induced damage of DNA. Microchimica Acta. 2015;**182**(9):1715-1722

[115] Rocha TL, Mestre NC, Sabóia-Morais SM, Bebianno MJ. Environmental behaviour and ecotoxicity of quantum dots at various trophic levels: A review. Environment International. 2017;**98**: 1-17. DOI: 10.1016/j.envint.2016.09.021. Epub 2016 Oct 13. PMID: 27745949

[116] Kubicek-Sutherland JZ, Makarov NS, Stromberg ZR, Lenz KD, Castañeda C, Mercer AN, et al. Exploring the biocompatibility of near-IR CuInSe x S2–x/ZnS quantum dots for deep-tissue. Bioimaging. 2020;**3**(12):8567-8574

[117] Bayer M. Bridging two worlds: Colloidal versus epitaxial quantum dots. Advances in Physics of Semiconductors. 2019;**531**(6):1900039

[118] Javanbakht S, Shaabani A. Stimuli-Responsive Bio-Based Quantum Dots in Biomedical Applications. In: Nanoengineering of Biomaterials, Jana S, Jana S, editors. 2022. DOI: 10.1002/9783527832095.ch28

Chapter 7

Recent Advances in Quantum Dots-Based Biosensors

Meysam Safari

Abstract

Biosensors can be developed using quantum dots (QDs). An inorganic nucleus with organic molecules attached to its surface is referred to as a QD, and they are a type of new fluorescent nanomaterial. QDs possess unique excellent optical properties and chemical properties, including broad excitation spectra, adjustable particle sizes, confined emission spectra, emission of multiple fluorescence colors, superior signal brightness, and extended fluorescence lifetime. QDs have abundant functional groups, which make it easy to form hybrid nanomaterials that perform analytically well. With functionalized sensing systems, we can detect metal ions, biomarkers, and antibiotics sensitively and selectively through the hybridization of QDs with nanomaterials. In this chapter, we first introduce the research trends in the application of QDs and then discuss their surface modification for biological applications.

Keywords: biosensors, quantum dots, biological applications, fluorescent nanomaterial

1. Introduction

Semiconductor nanocrystals or quantum dots (QDs) are a type of novel fluorescent nanomaterial consisting of inorganic nuclei with organic molecules in the nanoscale range of 1–10 nm and are typically composed of atoms from groups II–VI (e.g. CdTe, CdSe) and III–V (e.g. InP, InAs) of the periodic table [1]. Their optoelectronic properties change as a function of both shape and size. Colors like orange and red are emitted by larger QDs with diameters of 5–6 nm. QDs with shorter wavelengths (2–3 nm) produce colors such as blue and green [2]. Depending on the exact composition of the QD, the specific colors vary. It is fairly common for these QD cores to be capped with an inorganic layer to boost their quantum yield, which enhances their signal-to-noise ratio [3]. As a result of the size and composition of QDs, they can emit various wavelengths, ranging from ultraviolet (UV) to visible to near-infrared (NIR) [4]. The properties of QDs are intermediate between those of bulk semiconductors and those of discrete atoms or molecules. A hydrophilic material (e.g. mercaptopropionic acid (MPA) or cysteamine) must be added to QD surfaces in order to increase their water solubility [5]. Surface conjugations with synthetic polymers such as polyethylene glycol (PEG) are often useful for preventing aggregation of these nanoparticles [6].

With their unique electronic properties and tunable emission spectrum, quantum dots have attracted considerable interest owing to their resistance to photobleaching, broad excitation wavelength, and high quantum yield [7, 8]. Various applications and

IntechOpen

devices benefit from the unique characteristics of QDs, including solar cells, fluorescent probes, optical switches, and light sources [9, 10].

The quantum dots can be divided into core-shell QDs, doped QDs, and nuclear QDs in terms of structure. Quantum dots can also be easily modified by encapsulating the surface with amphiphilic ligands, salinizing, and exchanging ligands, further expanding their sensor applications [11–14].

Biosensors can be developed using quantum dots (QDs) [15, 16]. Molecular QDs provide researchers with the ability to study cell processes at the molecule level and may be helpful in diagnosing and treating diseases such as cancer [17]. It is possible to use quantum dots as active sensors in high-resolution cellular imaging, by changing their fluorescence properties as they react with the analyte, or by conjugating antibodies to the surface of the dots to act as passive label probes. Due to the presence of highly toxic heavy metal elements such as cadmium in QDs, these materials cannot be used in biomedical applications [18]. Environmental pollution and toxicity have been a concern in the use of nanomaterials for biomedical applications, and the development of a nontoxic and biocompatible nanomaterial is becoming increasingly important. Incorporating QDs into hybrid nanomaterials has proven easy due to their abundant functional groups [19]. Hybridizing QDs with other materials can provide enhanced thermal and chemical stability, high quantum efficiency, longer excited state lifetimes, and minimized toxicity [20].

The biosensor is a micro-analytical system. As part of the development and application of analytical sensors, biosensing incorporates molecular biological recognition entities [21]. Biosensors are sensing systems that leverage biological recognition to confer molecular specificity to analytes. In the chapter, we first introduce the research trends in the application of QDs and then discuss their surface modification for biological applications.

2. Overview of QDs-based biosensors

Biosensors have been developed using QDs as promising tools. In **Table 1**, we summarize the selected methods for detecting different targets using various QDs.

Pourghobadi et al. prepared TGA-capped CdTe QDs for the visual determination of the trace levels of dopamine. The fluorescence intensity of the samples was investigated as a result of interaction between TGA-CdTe QDs and dopamine. Fluorescence intensity decreases dramatically with an increase in dopamine concentration. Taking into account this trend, we developed a straightforward method for the detection of dopamine that is sensitive to fluorescence [22].

According to Wang et al., hydrophilic QDs can be used as molecular probes to detect trace amounts of propafenone. Electrostatic attraction and hydrogen bonds allowed propafenone to combine with CdTe QDs modified with thioglycolic acid in the weak acid. Fluorescence, UV-vis absorption, RRS, spectra, and spectral analysis have been used to study the interaction of CdTe QDs with propafenone. With the RRS method, ppb (ng mL^{-1}) levels of propafenone in serum samples can be detected in less than 30 min [23].

Tetracycline was detected in milk using MoS_2 QDs and CdTe QDs developed by Liang et al. As a first step, MoS_2 QDs emit blue light at 433 nm, while CdTe QDs produce yellow light at 573 nm under 365 nm excitation. We used MoS_2 QDs and CdTe QDs to construct a fluorescent sensor with dual signals at 433 nm and 573 nm. MoS_2/CdTe-based sensors show a lower fluorescence intensity as tetracycline concentration

QD type	Sensor model	Target	Linear range	Real sample type	Ref.
Ni-doped CdTe	Fluorescence	Pyrazinamide	2–100 μM	Plasma samples	4
CdTe	Fluorescence	Dopamine	0.5–10 μM	Biological fluids	22
CdTe	Fluorescence	Propafenone	0.003–7.0 μg mL^{-1}	Human serum	23
MoS_2/CdTe	Fluorescence	Tetracycline	0.1–1 μM	Milk samples	24
CdTe–Con A	Fluorescence	Lipopolysaccharide and Serratia marcescens	10–90 fg/mL	—	25
CdSe	Fluorescence	Urea	1–120 mM		26
CdTe	Fluorescence	Salbutamol	6.27×10^{-8} to 2.09×10^{-7} M	Pig urine samples	27
Pd-doped CdTe	Fluorescence	Diazinon	2.3–100 μM	Environmental water samples	36
CdS@MOF	Electrochemiluminescence	Carcinoembryonic antigen	—	Human serum samples	39
CdTeS @ SiO2	ImageJ software	Folic acid	5–80 μM	Serum samples	29
α-FeOOH@ CdS/Ag	Electrochemiluminescence	17β-estradiol	0.01–10 pg. mL^{-1}		35
ZnCdS QDs@MIP	Fluorescence	Ascorbic acid	1–500 μM	Vitamin C tablets	48
MoS_2/GQD	Electrochemical	Caffeic acid	0.38–100 μM	Red wine samples	52
CdTe	Photoinduced electron transfer	Double-strand DNA	0.0874 μg mL^{-1} and 20 μg mL^{-1}	Synthetic samples	55
Polymer CdTe/CdS	Fluorescence	Glucose	0.2–5 mM	Human body fluids	61

Table 1.
Summary of the QDs-based biosensors.

is increased, and 573 nm is quenched more apparent than 433 nm as tetracycline concentrations are increased [24].

Conjugated CdTe QDs with TGA caps were successfully prepared as a novel biosensor for detecting LPS at concentrations of fg/mL. CdTe QDs showed high crystalline lattice planes with an average size of 4–5 nm in the high-resolution transmission electron photograph. The electrostatic attraction force between positively charged Con A species and negatively charged CdTe QD surfaces was responsible for the adsorption of Con A onto CdTe QD surfaces [25].

The MPA-CdSe QDs were used to detect urea in a case study based on a fluorescence-based “turn-on” probe. MPA-CdSe QDs are sensitive to pH change, and this property makes them useful for detecting urea sensitively and selectively in the presence of urease. By releasing ammonia ions as the enzymatic reaction occurs, the pH of the solution changes, and the fluorescence intensity of the system increases along with the concentration of urea [26].

Researchers conducted experiments to improve the quantum yield of CdTe QDs in 2021 (59.673%). A pH value of 140°C and heating for 40 minutes have been suggested for producing high photoluminescence from CdTe QDs. Salbutamol effectively reduces the fluorescence intensity of CdTe QDs. Fluorescence quenching was linearly related to Salbutamol concentration. Using CdTe QDs as a new fluorescent probe, Salbutamol was successfully measured in pig urine [27].

The main obstacles to their extensive application in biomedicine are their poor biodistribution, low in vivo stability, and high cytotoxicity.

Cd and Se elements in QDs are mostly responsible for their cytotoxicity. The surface conjugations with synthetic polymers such as polyethylene glycol (PEG) are often useful for preventing aggregation of these nanoparticles.

It remains a significant concern about the toxicity of these NPs, despite all the advances in the synthesis of QDs with different coatings. Further research is needed in this area. Researchers are currently working on modifying the surface of QDs to enhance their capacity to capture specific and efficient heavy metals and reduce their release into the environment.

3. Surface modification for biological applications

To maintain the electronic properties and optical of the core material, one must choose the correct passivating agent when considering real-life applications. Surface ligands play a crucial role in the biological applications of quantum dots. The surface of quantum dots can be easily modified with, metal ions doped, metal-organic frameworks (MOFs), molecularly imprinted polymer (MIP), aptamers, multiwalled carbon nanotubes (MWCNTs), graphene quantum dots and carbon quantum dots, silanization, polymer, and transition metal oxide, which further expands the application of quantum dots in the sensor field.

3.1 Silica

To encapsulate QDs, silica nanoparticles have attracted considerable attention due to outstanding biocompatibility, high surface-to-volume ratio, ease of functionalization, and low cost. As labels for signal amplification to detect S. aureus, Ag_2S QDs loaded onto dendritic mesoporous silica nanospheres (DMSNs) were employed by Wang et al. as electrochemical immuno-biosensors.

As signal-amplifying labels, quantum dots and DMSNs have several advantages. A first step in enriching Ag_2S QDs was achieved by taking advantage of high surface area and the extensive pore channels of DMSNs. Each Ag_2S/DMSNs signal amplification label loaded with a tremendous amount of Ag_2S QDs increased the amount of Ag_2S QDs on each S. aureus cell. This resulted in a lower detection limit and a wider detection range for the electrochemical immuno-biosensor. When an electrochemical immuno-biosensor is activated, differential pulse voltammetry can be used to identify S. aureus. From Ab-Ag_2S/DMSNs, Ag (I) ions are liberated by leaching with HNO_3 [28].

Yang et al. developed a fluorescent probe for the visual detection of folic acid based on silica-coated CdTe QDs nanoparticles. The emission intensity of CdTeS QDs@SiO_2 decreased when folic acid was present, and folic acid itself showed fluorescence emission peaks. A gradient color change of nanoparticles was observed with increasing folic acid concentration [29].

For a sensitive NIR immunosensor, Han et al. immobilized CdTe/CdS QDs and antibodies on amino-functionalized SiO_2 [30]. Another case involved the modification of Au@SiO_2 by CdS QDs [31]. ECL signals were greatly enhanced due to the combination of SPR of Au cores and chemical enhancement from coreactants. SiO_2 with quantum dots appears to be a better choice for increasing loading.

CdTe/SiO_2 nanoparticles were compared by Shen et al. for their amplification effects (SiO_2@CdTe and CdTe@SiO_2). This resulted in much stronger ECL emission from CdTe@SiO_2 and lower cytotoxicity than SiO_2@CdTe, suggesting that bioanalysis of CdTe@SiO_2 may prove useful for clinical diagnosis. Compared with solid SiO_2, mesoporous SiO_2 provided higher loading sites and surface area [32].

In their study, Dong et al. experimented with mesoporous SiO_2 as a substitute for solid SiO_2. Their results showed that the ECL intensity and stability were much greater on mesoporous SiO_2 [33].

3.2 Transition metal oxide

QDs-based biosensors can also be assembled using transition metal oxides with variable valence states. ECL reaction could be corrected by using transition metal elements as catalysts. In situ activated CdS QDs/TiO_2 nanocomposites were used by Dai et al. to develop an enhanced PSA aptamer sensor. As a result of the TiO_2 nanotubes' degradation of H_2O_2 reactants, the CdS QDs were significantly more sensitive to the ECL reaction [34].

Miao et al. immobilized the CdS QDs onto the 3D urchin-like α-FeOOH, realizing ultrasensitive quantitation of 17β-estradiol. In contrast, the Fenton-like process produced more $SO^{4\bullet-}$ radicals through the conversion cycle of Fe^{3+} and Fe^{2+}, which supported ECL responses through electron transfer [35].

3.3 Metal ions doped

In recent years, transition metal ions have been investigated as quantum dots because they are not only more efficient than undoped quantum dots, but they also have additional advantages such as chemical stability, longer excited state lifetimes, reduced toxicity, high quantum efficiency, and larger Stokes shift to avoid self-absorption. The doping of transition metal ions with various impurities such as Sm^{3+}, Mn^{2+}, Zn^{2+}, Ag^{+}, Gd^{3+}, Tb^{3+}, Cu^{2+}, Er^{3+}, Eu^{3+}, and Co^{2+} into II–VI QDs has been reported to date. Biological studies can be improved by using QDs doped with cadmium, since they have less toxic cadmium content than their undoped counterparts.

Safari et al. investigated the possibility of using Ni:CdTe d-dots as nanoprobes for the detection of pyrazinamide (PZA) in plasma samples by hydrothermal method with 3-mercaptopropionic acid as the capping reagent **Figure 1**. It has been demonstrated that Ni-doped CdTe QDs have enhanced biocompatibility, photostability, and decreased cytotoxicity when compared to undoped CdTe QDs. For the rapid determination of PZA in plasma samples, a simple, inexpensive, sensitive, and selective method was developed based on fluorescence quenching of Ni:CdTe. Since PZA binds to the surface of Ni-doped CdTe quantum dots, it effectively quenches the emission of the quantum dots [4].

Najafi et al. reported Pd: CdTe QDs by hydrothermal method using palladium ions as a dopant. A linear quenching of fluorescence intensity occurs in the presence of Diazinon. Diazinon was detected in environmental waters using this novel nanoprobe at very high levels of sensitivity and selectivity, as shown in **Figure 2**. Also, the cytotoxicity assay of Pd: CdTe QDs was successfully conducted, which indicates its tremendous potential in biotechnology and medicine [36].

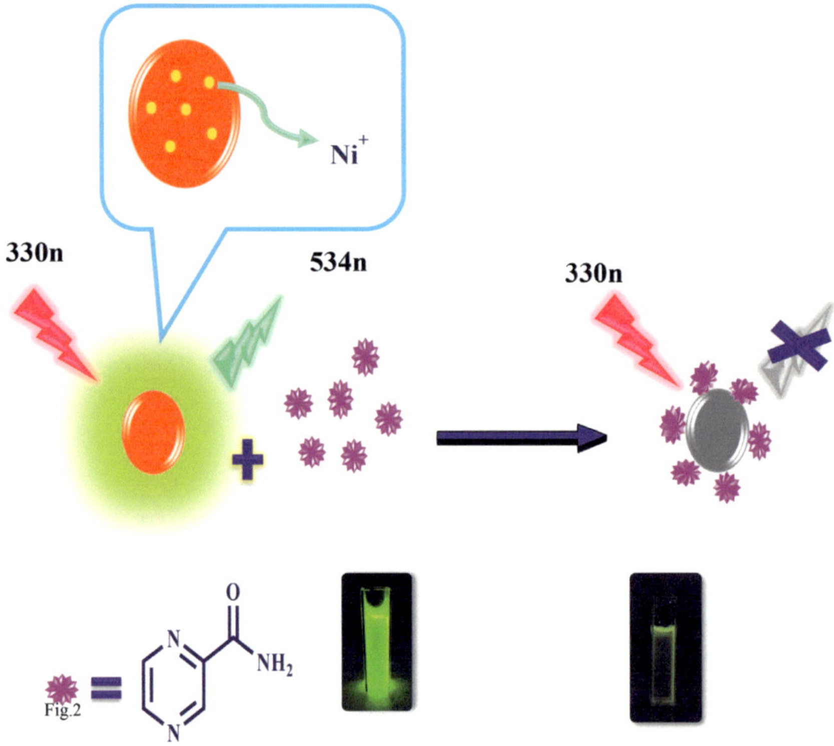

Figure 1.
Schematic illustration of PZA detection mechanism using Ni-doped CdTe [4].

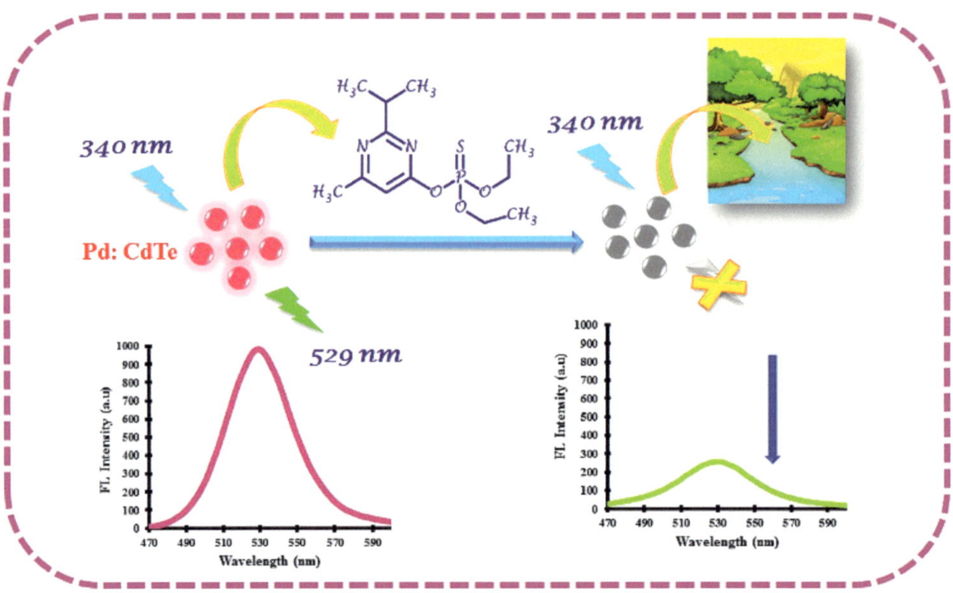

Figure 2.
The schematic illustration for Pd: CdTe QDs sensor for the detection of DZN [36].

3.4 Metal-organic frameworks

With their nice monodisperse structure and large surface area, metal-organic frameworks (MOFs) represent advanced hierarchical nanostructures that have been regarded as suitable QD loading hosts for amplification and stabilization of signals [37].

GQDs/UiO-66 NC were prepared by Abdolmohammad-Zadeh et al. resulting from terephthalic acid (TA) oxidation with H_2O_2 resulting from cholesterol oxidation with ChOx. As a result of the H_2O_2 oxidation of TA, a highly fluorescent product was produced, enabling a direct correlation between the intensity of fluorescence and the amount of cholesterol in solution [38].

The bi-coreactants, TEOA@AuNPs and CdS QDs@MOF, were prepared by Wei et al. to synergistically enhance the ECL signal. A TEOA@AuNPs nanocarrier was used to carry CEA aptamers via Au-S bonds in addition to acting as coreactants. After CEA's aptamer was bound to $Ru(bpy)_3^{2+}$, the ECL signal was also weakened because of increased impedance between $Ru(bpy)_3^{2+}$ and the bi-coreactants. An ECL biosensor for CEA analysis has been successfully constructed using the ECL system [39].

According to Yan et al., a fluorescent aptasensor was constructed by encapsulating CHAs in MOF-5-NH_2 and self-cycling CHAs (scCHAs). Patulin (PAT) in apple juice can be detected sensitively using a dual amplification strategy. As a result of the 17 s CS and 17 s AP hairpins, the 17 s AP includes an aptamer domain, which allows cyclic amplification without the addition of an extra aptamer. Further, hybridization strength between hairpins and aptamer domains was optimized to obtain the strongest signal change following PAT binding to aptamer domains [40].

Using a direct encapsulation method, Cui et al. prepared CdS QDs@MOF-5 composites for ultrasensitive cTnI detection [41].

CdSe nanocomposites were synthesized in situ using MIL-101. The as-prepared composites showed high ECL activity and sensing selectivity because of MIL-101's highly selective adsorption and efficient accumulation abilities [42].

For diverse sensing applications, MIL-53(Al)@CdS, MIL-53(Fe)@CdS, and MIL-53(Al)@CdTe are among the QD-loaded composites described in this study [43, 44].

Some MOFs can also serve as catalysts in the ECL process, in addition to their loading capacity. CdTe QDs were loaded on both the internal framework and external surface of NH_2 MIL-88(Fe) by Zhou et al. for signal amplification [45]. CdTe@IRMOF-3@CdTe composites were used as ECL labels in a sandwich immunosensor developed by Zhuo et al. As a coreactant accelerator, 2-amino terephthalic acid (2-NH_2-BDC) acted as an organic ligand of IRMOF-3 in the ECL process to increase the intensity and sensitivity of the ECL process [46].

According to Du et al., CdS@TiO_2 emits a highly transcendent ECL, whereas curcumin-Au NPs incorporated in ZIF-8 led to an intense reduction in ECL from CdS@TiO_2. ECL-RET acceptors have also been developed from Ag nanoclusters (Ag NCs) [47].

4. Molecularly imprinted polymer

In the presence of template molecules, molecularly imprinted polymers (MIPs) are synthesized through polymerization reactions between functional monomers and crosslinking monomers. These polymers are widely used for many applications, including artificial antibodies, drug delivery, and chemical sensors catalysis. Due to its unique ability to recognize target analytes and chemical stability, MIP has become a research hotspot in chemical sensors. According to Yang et al., ascorbic acid can be quantitatively measured with a ternary QD-based MIP ratiometric fluorescence sensor based on ternary QDs. Under a single excitation wavelength of 380 nm, the sensor exhibited well-resolved emission peaks at 530 and 705 nm, respectively, corresponding to ZnCdS QDs@MIP and CdTeS QDs@SiO_2. CdTe QDs were embedded in the SiO_2 shell to serve as the reference signal, and ZnCdS QDs were encapsulated in the MIP to serve as the response signal [48].

Qi et al. fabricated MIP using L-arginine, which is a part of microcystin, as the segment template for the adsorption and determination of microcystin. After studying a series of structural analogs of MC-LR, we selected L-leucine as the segment template to prepare MIP. Sol-gel polymerization was used to fabricate the MIP@CQDs@SiO_2 fluorescence sensor using 4-aminopropyltriethoxysilane as the monomer and tetraethylorthosilicate as the crosslinker. Polymer layers with specific recognition sites were capable of selectively adsorbing MC-LR molecules, causing CQDs to show fluorescence quenching behavior via electron transfer [49].

Chmangui et al. synthesized a composite material based on the molecularly imprinted technique by using DMC as a dummy template. 5,7-dimethoxycoumarin (DMC) contains coumarin moiety which assists with the specific recognition and imprinting of cavities within the prepared MIP layer. This moiety is identical to that found in AFs. A sensitive and selective detection of total AFs in non-dairy beverages is therefore expected through the anchorage with Mn-doped ZnS QDs. The prepared Mn-doped ZnS QDs were first deposited on a DMC dummy template, which was then coated with polyethyleneglycol (PEG). An efficient fluorescent screening method for non-dairy beverage analysis was found to be reliable with MIP-Mn-doped ZnS QDs composites [50].

5. Graphene quantum dots and carbon quantum dots

As a result of their electronic transport properties, chemical, and exceptional physical, graphene quantum dots (GQDs) and carbon quantum dots (CQDs) are being investigated for the development of next-generation biosensors. An explosion of reports have been published in recent years on biotransducer designs and biosensing applications utilizing these nanomaterials for facilitation, improvement, or otherwise developing novel approaches to analyte detection and monitoring.

Using fluorescent ssDNA probes coupled with CQD sensors, Loo et al. reported complementary DNA strand detection. Fluorescence resonance energy transfer (FRET) is quenched in this scheme thanks to the p-p stacking from DNA base pairs and CQD-conjugated p-systems. Due to electrostatic repulsion, the fluorescence of the ssDNA probe was restored when it hybridized with the target [51].

In order to fabricate a polyphenol index biosensor, Vasilescu et al. developed a molybdenum disulfide (MoS_2)/GQD biotransducer on which they drop-casted laccase. It was reported that the caffeic acid sensor had a linear range of 380 nanomolar to 100 millimolar with a low level of detection of 320 nanomolar. A notable finding of their voltammetric studies was that laccase's response was similar with each electrode formulation, although the MoS_2/GQD electrode contributed to a higher electron transfer efficiency [52].

6. Aptamers

In nucleic acids, aptamers are short, single-stranded sequences that bind specific molecules selectively and strongly. To create them, ligands undergo systematic transformations and exponential evolutions. Several applications rely on aptamers, including cancer cell imaging, biomedicine, therapeutics, gene therapy, biosensing, or targeted drug delivery, despite their high binding affinity, long-term stability, ease of modification, low immunogenicity, and low cost.

An ATmega328P prototype biosensor for the detection of E. coli in water samples using aptamer I and II-conjugated SPIONs and CdTe QDs has been developed by Pandit et al. We then conjugated E. coli-specific aptamers I and II to both types of nanoparticles. In terms of microbial detection, conjugated SPIONs and CdTe-MPA QDs display high quantum yields and significant magnetic and fluorescent properties. It is an effective tool to detect microbes using a conjugated SPION followed by a CdTe-MPA QD that has a high quantum yield, as well as significant magnetic and fluorescent properties. Bioconjugation of *E. coli*-specific aptamer I and II with SPIONs as well as QDs studied using various methods demonstrated selective separation and subsequently, detection of *E. coli*. The ATmega 328P prototype biosensor was further used to capture and detect E.coli using CdTe MPA QDs conjugated with aptamer II. Up to 100 cfu can be detected by the biosensor, demonstrating its sensitivity. We can detect on the spot, it is easy to use and less time-consuming. The ATmega328P prototype biosensor is capable of detecting E. coli in water if it is combined with cadmium-based QDs. As a result of the technology's rational extension and exploration, other pathogenic microorganisms can be detected in a wide variety of food samples. Additionally, the system meets the need for a portable, miniaturized pathogen detection device that can be deployed in the field [53].

Using SPCE/C60/MWCNTs-PEI/PQdot/GLA/APT aptasensors, Jamei et al. analyzed thrombin protein. By modifying SPCE electrodes with C60/MWCNTs-PEI/PQ dot nanocomposites, sensitive electrochemical aptasensors could be manufactured, as shown in **Figure 3**. In addition to their suitable stability, their high number of surface amine groups, their surface-to-volume ratio, their fast electron transfer kinetics, and their high electrical conductivity, these C60/MWCNTs-PEI/PQ dot nanocomposites possess unique characteristics. This nanocomposite's high surface-to-volume ratio and high number of surface amine groups increased the number of thrombin binding sites. A sensitive aptasensor was first developed using polymer quantum dots, which exhibit unique characteristics and are easy to synthesize [55].

A QDs-based biosensor was reported by Liu et al. using Pr^{3+}–rutin complexes as both quenchers and receptors. Using the Pr^{3+}–rutin complex, we were able to detect QDs in a favorable "off" state by efficiently quenching their fluorescence. Our approach relies on the electrostatic association of the cationic Pr^{3+}–rutin complex on the surface of QDs. The static association reduces the fluorescence intensity of QDs due to ultrafast photoinduced electron transfer. In addition, Pr3 + −rutin complexes also serve as receptors for dsDNA. As a result of adding herring sperm DNA, Pr^{3+}–rutin complexes were removed from QD surfaces and the QDs' fluorescence was restored [54].

The FRET method was employed by RezanejadeBardajee et al. in order to detect complementary DNA (as a positive control) and RNA from the COVID-19 virus by high-sensitivity detection of CdTe-ZnS QDs bioconjugates. A ligand exchange process replaces thiolated DNA (capture DNA) with surface-bound TGA molecules to form

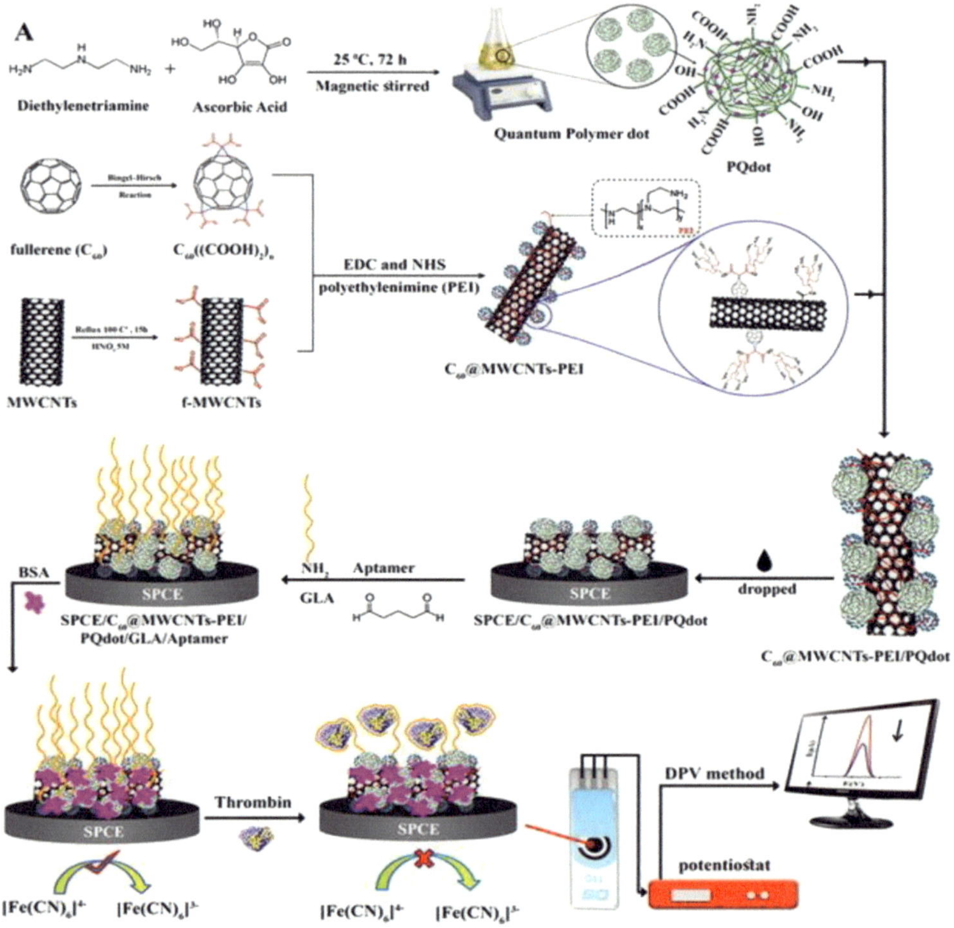

Figure 3.
Schematic of stepwise preparation electrochemical aptasensor and C60/MWCNTsPEI/PQdot [54].

DNA-conjugated quantum dots (QDs-DNA). Besides designing the BHQ_2-DNA oligonucleotide, an oligonucleotide from the viral genome was also designed for the FRET experiment. As a final application, these elements can be used for detecting viral RNAs (via FRET experiment) and for sensing target DNA sequences (via the QDs-DNA bioconjugate nanoprobe) [56].

7. Multiwalled carbon nanotubes

A number of chemically modified QDs have been developed using multiwalled carbon nanotubes (MWCNTs) because of their unique properties, such as high electrical conductivity, high mechanical strength, high thermal conductivity, and large surface area to volume. In analytical sensing, multiwalled carbon nanotubes (MWCNTs) have been shown to reduce detection limits, increase sensitivity, and prevent surface fouling.

As part of their study, Baslak et al. functionalized MWCNT surfaces with silver nanoparticles and encapsulated the silver nanoparticles with poly(glycidyl methacrylate) (pGMA). An initiated chemical vapor deposition (CVD) method was used to coat MWCNT surfaces with pGMA. The chemical vapor deposition (CVD) method produces thin films in a dry environment. By initiating chemical vapor deposition (iCVD), complex geometries can be coated uniformly without solvent-related damages, which are typically observed in conventional wet coating. During the iCVD process, the substrate is cooled so that the species can adsorb and grow on the surface. This prevents the substrate from being damaged by high temperatures, plasma, or light sources, which can alter the substrate's properties physically or chemically. As a result, iCVD was selected as an ideal method for functionalizing and encapsulating MWCNT surfaces [57].

The authors of Vinoth et al. describe a method for integrating ZnO QDs into MWCNTs using ultrasonication to decorate the QDs on the surface of the nanotubes (MWCNT/ZnO QDs), using both electromechanical detection of glucose and photocurrents simultaneously. A chronoamperometry measurement, a differential pulse voltammetry measurement, and a cyclic voltammetry measurement were used to evaluate the performance of the enzyme-free glucose sensor. Ascorbic acid, uric acid, and dopamine all interfere with the glucose sensor based on MWCNT/ZnO QDs, while sucrose exhibits good anti-interference properties. A nanocomposites heterostructure made from MWCNT/ZnO QDs shows excellent photocurrent activity toward visible detection [58].

8. Polymer

Multifunctional polymer layers can enhance QDs' interaction with target analytes by incorporating a variety of functionalities. QDs are usually synthesized with polymers that contain organic cyclic chains such as calixarene and cyclodextrin. The use of polyethylene glycol (PEG) derivatives in the synthesis of QDs has also become increasingly popular in recent years. PEG derivatives are readily available, require simple encapsulation processes, and are readily available.

According to Yi et al., a fluorescence nanoprobe detecting parathion-methyl (MP) through host-guest recognition is the first enzyme-free fluorescent nanoprobe. It is possible to form distinct molecular recognition function on MoS2 QD surfaces by introducing this molecule on the surface of MoS_2 QDs, as shown in **Figure 4**.

With the PET process, p-NP molecules from MP hydrolysis under alkaline conditions can enter into small cavities of molecular beads and further quench their fluorescent properties, thereby providing a fluorescence sensing platform for MP. Based on the optimization of various experimental conditions, the results show that the molecular probe is excellent in terms of selectivity and sensitivity, has a wide linear range, and has been able to detect MP with a low detection limit [59].

Gupta et al. studied interactions between chitosan, the most abundant biopolymer, and luminescent CdSe QDs synthesized by cyclic voltammetry and capped with MPA. MPA-CdSe QDs were found to react dynamically with chitosan (crystal size 2.3 × 0.5 nm, zeta potential 47 × 6 mV). Based on our assessments, the number of chitosan molecules bound and the binding constant Ka by MPA-CdSe QDs are ~1.z 3 and 7.332 × 1015 L Mol-1 at 298 K, respectively. Chitosan's strong affinity for MPA-CdSe QDs is confirmed by the high Ka. Using the above calculations, chitosan-MPA CdSe QDs were found to be stable over a long period of time and to disperse in water homogeneously. Biomolecule interaction can be strongly influenced by the size, shape, and surface chemistry of QDs [60].

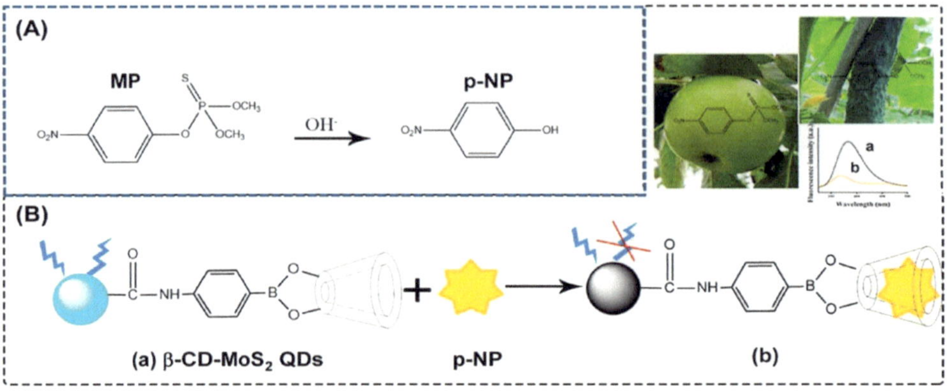

Figure 4.
Schematic illustration of parathion-methyl detection using β-CD-functionalized MoS2 QDs [59].

A ratiometric fluorescence probe developed by Yu et al. detects H_2O_2 and glucose selectively using cationic conjugated polymer (CCP) and cationic transition metal dichalcogenides (CdTe/CdS QDs). H_2O_2 causes fluorescence quenching of QDs due to FRET between CCP and QDs. In order to create new H2O2 and glucose enzymatic assays, these phenomena can be exploited. This method can be applied to other oxidases to quantify substrates such as choline, xanthine, cholesterol, and lacate, since many oxidases generate H_2O_2.

According to the proposed design, it is able to provide two advantages and advantages over conventional systems: (I) we use the signal transformation process to design a dual-emission nanoprobe with excellent properties such as high selectivity, fast response time, and resistance to outside interferences, and (II) because the QDs exhibit excellent NIR properties, we were able to directly measure glucose in whole blood [61].

9. Conclusion

A comprehensive summary of recent advances in surface modification QDs for biological applications is presented in this chapter. Surface ligands are the most important factor in obtaining useful materials from quantum dots for biological applications. The surface of quantum dots can be easily modified with, metal ions doped, metal-organic frameworks, molecularly imprinted polymer, aptamers, multiwalled carbon nanotubes, graphene and carbon quantum dots, silanization, polymer, and transition metal oxide, which further expands the application of quantum dots in the sensor field.

The surface modification QDs have shown that design of QDs to detect, with high sensitivity and selectivity, various analytes was an effective strategy. Traditional QDs have challenges such as being environmentally friendly, biocompatible, having a broad excitation spectrum, and having poor photostability, wavelength tenability dependent on size, and low quantum yield. Modification QDs help overcome these issues by being extremely energy-efficient and highly quantum yielding.

DOI: http://dx.doi.org/10.5772/intechopen.108205

Author details

Meysam Safari
Department of Chemical Engineering, Kermanshah University of Technology, Kermanshah, Iran

*Address all correspondence to: m.safary85@gmail.com

References

[1] Wonci Z, Tsolekile N, Matoetoe MC. Polyvinylpyrrolidone as a polymer template for CuInS quantum dots: Effect on optical properties. Materials Today: Proceedings. 2022;**56**:1989-1994. DOI: 10.1016/j.matpr.2021.11.330

[2] Lee G, Lee SY, Park S, Jang SH, Park H-K, Choi I, et al. Highly effective surface defect passivation of perovskite quantum dots for excellent optoelectronic properties. Journal of Materials Research and Technology. 2022;**18**:4145-4155. DOI: 10.1016/j.jmrt.2022.04.080

[3] Amani-Ghadim AR, Arefi-Oskoui S, Mahmoudi R, Sareshkeh AT, Khataee A, Khodam F, et al. Improving photocatalytic activity of the ZnS QDs via lanthanide doping and photosensitizing with GO and g-C3N4 for degradation of an azo dye and bisphenol-A under visible light irradiation. Chemosphere. 2022;**295**:133917. DOI: 10.1016/j.chemosphere.2022.133917

[4] Safari M, Najafi S, Arkan E, Amani S, Shahlaei M. Facile aqueous synthesis of Ni-doped CdTe quantum dots as fluorescent probes for detecting pyrazinamide in plasma. Microchemical Journal. 2019;**146**:293-299. DOI: 10.1016/j.microc.2019.01.019

[5] Fan G, Wang C, Fang J. Solution-based synthesis of III–V quantum dots and their applications in gas sensing and bio-imaging. Nano Today. 2014;**9**:69-84. DOI: 10.1016/j.nantod.2014.02.007

[6] Tomlinson ID, Gies AP, Gresch PJ, Dillard J, Orndorff RL, Sanders-Bush E, et al. Universal polyethylene glycol linkers for attaching receptor ligands to quantum dots. Bioorganic & Medicinal Chemistry Letters. 2006;**16**:6262-6266. DOI: 10.1016/j.bmcl.2006.09.031

[7] Roy S, Tuinenga C, Fungura F, Dagtepe P, Chikan V, Jasinski J. Progress toward producing n-Type CdSe quantum dots: Tin and indium doped CdSe quantum dots. The Journal of Physical Chemistry C. 2009;**113**:13008-13015. DOI: 10.1021/jp8113946

[8] Nann T, Skinner WM. Quantum dots for electro-optic devices. ACS Nano. 2011;**5**:5291-5295. DOI: 10.1021/nn2022974

[9] Liu Q, Fan Z, Yi X, Chen S, Li B, Luo W. Porous polyimide/carbon quantum dots/ZnS quantum dots material aerogel for efficient visible-light photocatalytic degradation over oxytetracycline. Reactive and Functional Polymers. 2022;**178**:105330. DOI: 10.1016/j.reactfunctpolym.2022.105330

[10] Vinnichenko MY, Makhov IS, Ustimenko RV, Sargsian TA, Sarkisyan HA, Hayrapetyan DB, et al. Doping effect on the light absorption and photoluminescence of Ge/Si quantum dots in the infrared spectral range. Micro and Nanostructures. 2022;**169**:207339. DOI: 10.1016/j.micrna.2022.207339

[11] Shen J, Zhu Q. Stability strategies of perovskite quantum dots and their extended applications in extreme environment: A review. Materials Research Bulletin. 2022;**156**:111987. DOI: 10.1016/j.materresbull.2022.111987

[12] Fu L, Zi Y, Bai X, Xiao D, Haider AA, Qiu J, et al. Preparation and photoluminescence of Cs4PbBr6 perovskite quantum dot embedded in borophosphate glass. Journal of Alloys

and Compounds. 2022;**911**:165004. DOI: 10.1016/j.jallcom.2022.165004

[13] Monika S, Mahalakshmi M, Subha N, Pandian MS, Ramasamy P. Graphene quantum dots and CuS microflowers anchored rGO composite counter electrode for the enhanced performance of quantum dot sensitized solar cells. Diamond and Related Materials. 2022;**125**:109033. DOI: 10.1016/j.diamond.2022.109033

[14] Sahoo Y, Poddar P, Srikanth H, Lucey DW, Prasad PN. Chemically fabricated magnetic quantum dots of InP:Mn. The Journal of Physical Chemistry B. 2005;**109**:15221-15225. DOI: 10.1021/jp050202n

[15] Topcu A, Bağda E, Oymak T, Durmuş M. Development of quantum dot-phthalocyanine integrated G-quadruplex/double-stranded DNA biosensor. Analytical Biochemistry. 2022;**654**:114777. DOI: 10.1016/j.ab.2022.114777

[16] Zhang Q, Zhang X, Ma F, Zhang C-Y. Advances in quantum dot-based biosensors for DNA-modifying enzymes assay. Coordination Chemistry Reviews. 2022;**469**:214674. DOI: 10.1016/j.ccr.2022.214674

[17] Fatima B, Hussain D, Bashir S, Hussain HT, Aslam R, Nawaz R, et al. Catalase immobilized antimonene quantum dots used as an electrochemical biosensor for quantitative determination of H2O2 from CA-125 diagnosed ovarian cancer samples. Materials Science and Engineering: C. 2020;**117**:111296. DOI: 10.1016/j.msec.2020.111296

[18] Zhou W, Hu Z, Wei J, Dai H, Chen Y, Liu S, et al. Quantum dots-hydrogel composites for biomedical applications. Chinese Chemical Letters. 2022;**33**:1245-1253. DOI: 10.1016/j.cclet.2021.09.027

[19] Hamed AA, Saad GR, Abdelhamid IA, Elwahy AHM, Abdel-Aziz MM, Elsabee MZ. Chitosan Schiff bases-based polyelectrolyte complexes with graphene quantum dots and their prospective biomedical applications. International Journal of Biological Macromolecules. 2022;**208**:1029-1045. DOI: 10.1016/j.ijbiomac.2022.03.199

[20] Tufani A, Qureshi A, Niazi JH. Iron oxide nanoparticles based magnetic luminescent quantum dots (MQDs) synthesis and biomedical/biological applications: A review. Materials Science and Engineering: C. 2021;**118**:111545. DOI: 10.1016/j.msec.2020.111545

[21] Khan ME, Mohammad A, Yoon T. State-of-the-art developments in carbon quantum dots (CQDs): Photo-catalysis, bio-imaging, and bio-sensing applications. Chemosphere. 2022;**302**:134815. DOI: 10.1016/j.chemosphere.2022.134815

[22] Pourghobadi Z, Mirahmadpour P, Zare H. Fluorescent biosensor for the selective determination of dopamine by TGA-capped CdTe quantum dots in human plasma samples. Optical Materials. 2018;**84**:757-762. DOI: 10.1016/j.optmat.2018.08.003

[23] Wang X, Liu Z, Gao P, Li Y, Qu X. Surface functionalized quantum dots as biosensor for highly selective and sensitive detection of ppb level of propafenone. Spectrochimica Acta Part A: Molecular and Biomolecular Spectroscopy. 2020;**227**:117709. DOI: 10.1016/j.saa.2019.117709

[24] Liang N, Hu X, Li W, Wang Y, Guo Z, Huang X, et al. A dual-signal fluorescent sensor based on MoS2 and CdTe quantum dots for tetracycline detection in milk. Food Chemistry. 2022;**378**:132076. DOI: 10.1016/j.foodchem.2022.132076

[25] Ebrahim S, Reda M, Hussien A, Zayed D. CdTe quantum dots as a novel biosensor for Serratia marcescens and Lipopolysaccharide. Spectrochimica Acta Part A: Molecular and Biomolecular Spectroscopy. 2015;**150**:212-219. DOI: 10.1016/j.saa.2015.05.042

[26] Sajwan RK, Sharma P, Lakshmi GBVS, Solanki PR. Mercaptopropionic acid capped cadmium selenide quantum dots based urea Biosensor. Materials Letters. 2021;**305**:130794. DOI: 10.1016/j.matlet.2021.130794

[27] Zhang J, Wei Y, Qiu S, Xiong Y. A highly selective and simple fluorescent probe for salbutamol detection based on thioglycolic acid-capped CdTe quantum dots. Spectrochimica Acta Part A: Molecular and Biomolecular Spectroscopy. 2021;**247**:119107. DOI: 10.1016/j.saa.2020.119107

[28] Wang H, Yang L, Tan Y, Deng W, Xie Q. Ag2S quantum dots loaded dendritic mesoporous silica nanospheres as signal amplification labels for ultrasensitive electrochemical immuno-biosensor for Staphylococcus aureus. Journal of Electroanalytical Chemistry. 2022;**919**:116496. DOI: 10.1016/j.jelechem.2022.116496

[29] Yang M, Wang C, Yan Y, Liu E, Hu X, Hao H, et al. Visual detection of folic acid based on silica coated CdTeS quantum dots in serum samples. Materials Research Bulletin. 2021;**144**:111509. DOI: 10.1016/j.materresbull.2021.111509

[30] Wang J, Han H, Jiang X, Huang L, Chen L, Li N. Quantum dot-based near-infrared electrochemiluminescent immunosensor with gold nanoparticle-graphene nanosheet hybrids and silica nanospheres double-assisted signal amplification. Analytical Chemistry. 2012;**84**:4893-4899. DOI: 10.1021/ac300498v

[31] Li X, Xu Y, Chen Y, Wang C, Jiang J, Dong J, et al. Dual enhanced electrochemiluminescence of aminated Au@SiO2/CdS quantum dot superstructures: Electromagnetic field enhancement and chemical enhancement. ACS Applied Materials & Interfaces. 2019;**11**:4488-4499. DOI: 10.1021/acsami.8b14886

[32] Pan D, Chen K, Zhou Q, Zhao J, Xue H, Zhang Y, et al. Engineering of CdTe/SiO2 nanocomposites: Enhanced signal amplification and biocompatibility for electrochemiluminescent immunoassay of alpha-fetoprotein. Biosensors and Bioelectronics. 2019;**131**:178-184. DOI: 10.1016/j.bios.2019.02.022

[33] Dong H, Han T-T, Ren L-L, Ding S-N. Novel sandwich-structured electrochemiluminescence immunosensing platform via CdTe quantum dots-embedded mesoporous silica nanospheres as enhanced signal labels and Fe3O4@SiO2@PS nanocomposites as magnetic separable carriers. Journal of Electroanalytical Chemistry. 2017;**806**:32-40. DOI: 10.1016/j.jelechem.2017.10.038

[34] Dai P, Liu C, Xie C, Ke J, He Y, Wei L, et al. TiO2 nanotubes loaded with CdS nanocrystals as enhanced emitters of electrochemiluminescence: Application to an assay for prostate-specific antigen. Analytical and Bioanalytical Chemistry. 2020;**412**:1375-1384. DOI: 10.1007/s00216-019-02365-1

[35] Liu Y, Li B, Yao Y, Yang B, Tian T, Miao Y, et al. An electrochemilumine scence sensor for 17β-estradiol detection based on resonance energy transfer in α-FeOOH@CdS/Ag NCs. Talanta. 2021;**221**:121479. DOI: 10.1016/j.talanta.2020.121479

[36] Najafi S, Safari M, Amani S, Mansouri K, Shahlaei M. Preparation,

characterization and cell cytotoxicity of Pd-doped CdTe quantum dots and its application as a sensitive fluorescent nanoprobe. Journal of Materials Science: Materials in Electronics. 2019;**30**:14233-14242. DOI: 10.1007/s10854-019-01792-1

[37] Zhou J, Li Y, Wang W, Tan X, Lu Z, Han H. Metal-organic frameworks-based sensitive electrochemiluminescence biosensing. Biosensors and Bioelectronics. 2020;**164**:112332. DOI: 10.1016/j.bios.2020.112332

[38] Abdolmohammad-Zadeh H, Ahmadian F. A fluorescent biosensor based on graphene quantum dots/zirconium-based metal-organic framework nanocomposite as a peroxidase mimic for cholesterol monitoring in human serum. Microchemical Journal. 2021;**164**:106001. DOI: 10.1016/j.microc.2021.106001

[39] Wei X-H, Qiao X, Fan J, Hao Y-Q, Zhang Y-T, Zhou Y-L, et al. A label-free ECL aptasensor for sensitive detection of carcinoembryonic antigen based on CdS QDs@MOF and TEOA@Au as bi-coreactants of Ru(bpy)32+. Microchemical Journal. 2022;**173**:106910. DOI: 10.1016/j.microc.2021.106910

[40] Yan X, Zhao Y, Du G, Guo Q, Chen H, He Q, et al. Magnetic capture of sulfur quantum dots encapsulated in MOF-5-NH2 via a target-driven self-cycling catalyzed hairpin assembly for the sensitive detection of patulin. Chemical Engineering Journal. 2022;**433**:133624. DOI: 10.1016/j.cej.2021.133624

[41] Du D, Shu J, Guo M, Haghighatbin MA, Yang D, Bian Z, et al. Potential-resolved differential electrochemiluminescence immunosensor for cardiac troponin I based on MOF-5-wrapped CdS quantum dot nanoluminophores. Analytical Chemistry. 2020;**92**:14113-14121. DOI: 10.1021/acs.analchem.0c03131

[42] Liu Q, Yang Y, Liu X-P, Wei Y-P, Mao C-J, Chen J-S, et al. A facile in situ synthesis of MIL-101-CdSe nanocomposites for ultrasensitive electrochemiluminescence detection of carcinoembryonic antigen. Sensors and Actuators B: Chemical. 2017;**242**:1073-1078. DOI: 10.1016/j.snb.2016.09.143

[43] Feng D, Tan X, Wu Y, Ai C, Luo Y, Chen Q, et al. Electrochemiluminecence nanogears aptasensor based on MIL-53(Fe)@CdS for multiplexed detection of kanamycin and neomycin. Biosensors and Bioelectronics. 2019;**129**:100-106. DOI: 10.1016/j.bios.2018.12.050

[44] Feng D, Wei F, Wu Y, Tan X, Li F, Lu Y, et al. A novel signal amplified electrochemiluminescence biosensor based on MIL-53(Al)@CdS QDs and SiO2@AuNPs for trichlorfon detection. Analyst. 2021;**146**:1295-1302. DOI: 10.1039/D0AN02158J

[45] Chen P, Liu Z, Liu J, Liu H, Bian W, Tian D, et al. A novel electrochemilumine scence aptasensor based CdTe QDs@NH2-MIL-88(Fe) for signal amplification. Electrochimica Acta. 2020;**354**:136644. DOI: 10.1016/j.electacta.2020.136644

[46] Yang X, Yu Y-Q, Peng L-Z, Lei Y-M, Chai Y-Q, Yuan R, et al. Strong electrochemiluminescence from MOF accelerator enriched quantum dots for enhanced sensing of trace cTnI. Analytical Chemistry. 2018;**90**:3995-4002. DOI: 10.1021/acs.analchem.7b05137

[47] Du Y, Li X, Ren X, Wang H, Wu D, Ma H, et al. Signal-off electrochemiluminescence immunosensors based on the quenching effect between curcumin-conjugated Au nanoparticles encapsulated in ZIF-8 and CdS-decorated TiO2 nanobelts for insulin detection. Analyst. 2020;**145**:1858-1864. DOI: 10.1039/C9AN02288K

[48] Yang M, Wang C, Liu E, Hu X, Hao H, Fan J. A novel ascorbic acid ratiometric fluorescent sensor based on ZnCdS quantum dots embedded molecularly imprinted polymer and silica-coated CdTeS quantum dots. Journal of Molecular Liquids. 2021;**337**:116438. DOI: 10.1016/j.molliq.2021.116438

[49] Qi Z, Lu R, Wang S, Xiang C, Xie C, Zheng M, et al. Selective fluorometric determination of microcystin-LR using a segment template molecularly imprinted by polymer-capped carbon quantum dots. Microchemical Journal. 2021;**161**:105798. DOI: 10.1016/j.microc.2020.105798

[50] Chmangui A, Driss MR, Touil S, Bermejo-Barrera P, Bouabdallah S, Moreda-Piñeiro A. Aflatoxins screening in non-dairy beverages by Mn-doped ZnS quantum dots – Molecularly imprinted polymer fluorescent probe. Talanta. 2019;**199**:65-71. DOI: 10.1016/j.talanta.2019.02.057

[51] Loo AH, Sofer Z, Bouša D, Ulbrich P, Bonanni A, Pumera M. Carboxylic carbon quantum dots as a fluorescent sensing platform for DNA detection. ACS Applied Materials & Interfaces. 2016;**8**:1951-1957. DOI: 10.1021/acsami.5b10160

[52] Vasilescu I, Eremia SAV, Kusko M, Radoi A, Vasile E, Radu G-L. Molybdenum disulphide and graphene quantum dots as electrode modifiers for laccase biosensor. Biosensors and Bioelectronics. 2016;**75**:232-237. DOI: 10.1016/j.bios.2015.08.051

[53] Pandit C, Alajangi HK, Singh J, Khajuria A, Sharma A, Hassan MS, et al. Development of magnetic nanoparticle assisted aptamer-quantum dot based biosensor for the detection of Escherichia coli in water samples. Science of the Total Environment. 2022;**831**:154857. DOI: 10.1016/j.scitotenv.2022.154857

[54] Liu Z, Liu S, Wang X, Li P, He Y. A novel quantum dots-based OFF–ON fluorescent biosensor for highly selective and sensitive detection of double-strand DNA. Sensors and Actuators B: Chemical. 2013;**176**:1147-1153. DOI: 10.1016/j.snb.2012.10.085

[55] Jamei HR, Rezaei B, Ensafi AA. Ultra-sensitive and selective electrochemical biosensor with aptamer recognition surface based on polymer quantum dots and C60/MWCNTs-polyethylenimine nanocomposites for analysis of thrombin protein. Bioelectrochemistry. 2021;**138**:107701. DOI: 10.1016/j.bioelechem.2020.107701

[56] Bardajee GR, Zamani M, Mahmoodian H, Elmizadeh H, Yari H, Jouyandeh L, et al. Capability of novel fluorescence DNA-conjugated CdTe/ZnS quantum dots nanoprobe for COVID-19 sensing. Spectrochimica Acta Part A: Molecular and Biomolecular Spectroscopy. 2022;**269**:120702. DOI: 10.1016/j.saa.2021.120702

[57] Baslak C, Demirel Kars M, Karaman M, Kus M, Cengeloglu Y, Ersoz M. Biocompatible multi-walled carbon nanotube–CdTe quantum dot–polymer hybrids for medical applications. Journal of Luminescence. 2015;**160**:9-15. DOI: 10.1016/j.jlumin.2014.11.030

[58] Vinoth V, Subramaniyam G, Anandan S, Valdés H, Manidurai P. Non-enzymatic glucose sensor and photocurrent performance of zinc oxide quantum dots supported multi-walled carbon nanotubes. Materials Science and Engineering: B. 2021;**265**:115036. DOI: 10.1016/j.mseb.2020.115036

[59] Yi Y, Zeng W, Zhu G. β-Cyclodextrin functionalized molybdenum disulfide

quantum dots as nanoprobe for sensitive fluorescent detection of -methyl. Talanta. 2021;**222**:121703. DOI: 10.1016/j.talanta.2020.121703

[60] Gupta J, Das K, Tanwar A, Rajamani P, Bhattacharya J. An electrochemical study of the binding interaction between chitosan and MPA-CdSe QDs for the development of biocompatible theranostic nanoprobe. Journal of Molecular Liquids. 2022;**358**:119193. DOI: 10.1016/j.molliq.2022.119193

[61] Yu M, Zhao K, Zhu X, Tang S, Nie Z, Huang Y, et al. Development of near-infrared ratiometric fluorescent probe based on cationic conjugated polymer and CdTe/CdS QDs for label-free determination of glucose in human body fluids. Biosensors and Bioelectronics. 2017;**95**:41-47. DOI: 10.1016/j.bios.2017.03.065

Chapter 8

Application of Quantum Dots in Lateral Flow Immunoassays: Non-Communicable and Communicable Diseases

Ncediwe Tsolekile, Noluvuyo Mngcutsha and Nozikumbuzo Vitshima

Abstract

Quantum Dots (QDs) are ideal fluorescent labels for lateral flow assays (LFA) due to their unique optical properties and resistance to chemical and photo-degradation. Lateral flow assay, known as immunochromatography test, is a diagnostic strip test that uses paper or membrane-based devices to detect the presence/absence of an analyte with the pregnancy test, the most known LFA. Diagnostic tools for non-communicable (NCDs) and communicable diseases (CDs) are available. However, other NCDs and CDs do not have LFAs as first-line diagnosis/point-of-care. QDs are promising fluorescent labels as they improve the LFA's colourimetric sensitivity and fluorescence stability. This chapter discusses the tailoring and application of QDs in LFA. Particular focus on the applications of QDs in LFA for detecting or screening NCDs (i.e. cancers etc.) and CDs (i.e. SARS-CoV-2, HIV/AIDs etc.). The book chapter concludes by discussing different challenges and perspectives of QDs in LFA.

Keywords: quantum dots, lateral flow immunoassays, point-of-care, communicable disease, non-communicable disease

1. Introduction

Non-communicable diseases (NCDs) (commonly known as chronic diseases) are diseases that result from environmental, behavioral and genetic reasons. As a result, they tend to linger for extended periods. Some NCDs like cancer, diabetes, and heart disease are preventable through lifestyle and behavior changes [1]. In comparison, communicable diseases (CDs) are diseases that can spread from one person to another (i.e. via bodily fluids and blood, inhaling airborne virus) [2]. Early detection and rapid diagnosis of NCDs and CDs are essential for disease screening and treatment. Several diagnostic tools are available for the detection and prognosis of NCDs. These include: biopsy procedures (i.e. endoscopic, pelvic examination, bronchoscopy, pap test and lumbar puncture.) [1], blood tests (i.e. complete blood count, blood protein testing etc.) [3] and diagnostic tests (i.e. computerized tomography (CT) scan and magnetic resonance imaging (MRI) scans) [4]. Nonetheless, these medical tools are costly, laborious, and require clinical sample preparation and sophisticated instruments with trained operators.

Moreover, they are unsuitable for first-line diagnosis. This has resulted in a need to develop cost-effective, rapid, and reliable methods for clinical diagnoses and prognoses of NCDs and CDs. Point-of-care testing (POCT) tools are medical diagnostic tools used at the time and place of patient care [5]. It employs simple equipment and reduces the time needed to generate clinical results, permitting patients and clinicians to make on-the-spot clinical decisions. POCTs have some advantages over traditional diagnostic tools, i.e. shorter detection time, less costly, limited to no sample processing, basic instrumentation, with minimal operation requirements [6]. Therefore, they permit early detection of diseases, especially in resource-limited areas, thereby improving the time of initial treatment. Nonetheless, the unavailability of clinical POCTs for NCDs such as cancer and the emergency of the COVID-19 pandemic has highlighted the need for developing POCT tools to allow for improved tracking and response time.

Lateral flow assay (LFA) (also known as immunochromatography tests) is one type of POC testing tool that has addressed the challenges of traditional testing streams like the polymerase chain reaction (PCR) and reverse transcription-polymerase Chain Reaction (RT-PCR). The most common LFA example is the pregnancy and human immunodeficiency virus (HIV) and acquired immunodeficiency syndrome (AIDS) test. LFA is a simple paper or membrane-based device that uses a liquid sample to detect the presence or absence of the test analyte (or disease biomarker). LFA has several advantages as an attractive sensing tool because it is portable, low cost, and enables on-site detection (**Table 1**) [7]. In LFA, various materials have been employed to detect the presence/absence of a test analyte in a liquid sample. The use of nanomaterials in the development of LFA has enabled qualitative naked-eye detection, broadened the type of labels used, eased antibody/antigen conjugation, and improved fluorescent and dual colourimetric signal detection [8]. Several nanoparticles used in LFA include magnetic nanoparticles [9] and colored nanoparticles [10], with fluorescent quantum dots serving as promising fluorescent probes. Quantum dots (QDs) are ideal fluorescent labels for immunoassay application. Their unique optical properties, molecular extinction coefficients, and excellent resistance to chemical and photo-degradation have advantaged them significantly over fluoro-dyes [11]. In LFA application, QDs have been used in antigen-antibody reactions to detect a variety of

Advantages
• Low Cost.
• Simple and user-friendly operation.
• Stability in different environments/ settings
• Has long shelf life.
• Less time for analysis.
• Very little to no energy consumption.
• High potential for commercialization.
• Changeable formats, bio-label molecules, and detection zones.

Table 1.
Advantages of lateral flow assay (LFA).

analytes, i.e. tumor/biomarkers [12, 13], bacteria [14], and viruses [15]. Bock et al., 2021 [16] developed a QDs-embedded with silica nanoparticles lateral flow immunoassay to detect prostate-specific antigen (PSA). The study used QDs as a probe for their red-emitting properties. The group coated the test line with anti-PSA antibody and the control line with goat anti-mouse IgG antibody. Bovine serum albumin (BSA), sucrose, Poly(ethylene glycol) (PEG), Tween 20 and QDs-PSA were used to condition the conjugate pad of the LFA. The LFA results were visible within 15 min, with an LOD of 0.138 ng/mL and no false-negative results on the clinical samples.

Moreover, the fluorescence intensity of the strip exhibited no significant decrease for 10 days. Detecting low levels of antibodies at the early stages of diseases has been a significant challenge in clinical diagnosis. Currently used early-stage diagnosis methods are still expensive with long turn-around times. Kim et al., 2022 [17] reported on the highly sensitive LFA kit based on brush-type ligand-coated quantum beads (B-type QBs) with the potential to diagnose several diseases by changing the antibody pair in the LFA kit. The B-type QBs were obtained via the self-assembly of polystyrene-co-poly(acrylic acid) (PS-PAA)and the QDs. The B-type QBs enabled good dispersion and stability and had a high-binding capacity towards antibodies. Controlling the orientation of the antibody created an ultra-high sensitive LFA. This study highlights the effect of antibody orientation and ligand type on LFA performance. Moreover, the human serum spiked hCG detection confirmed the LFA potential application in human body fluid.

This chapter addresses the application of lateral flow assays as a diagnostic tool for communicable and non-communicable diseases. The chapter discusses the basic principle and mechanisms of lateral flow assays. This is followed by discussions on the tailoring and application of quantum dots (QDs) in LFA to detect NCDs and CDs. The chapter then focuses on applying QDs in the LFA detection of various non-communicable and communicable diseases. The chapter further addresses some of the challenges within LFA application and concludes with future perspectives on QDs application in LFA.

2. Principle of lateral flow immunoassays

Lateral flow assays (LFA) are fabricated on a series of capillary beds made of a specific material. The material used (i.e. glass fiber or nitrocellulose membrane) contributes to the LFA performance. It ensures that the samples migrate towards the detector bed, which hosts the test analyte. The central beds of the LFA are (i) sample application bed, (ii) conjugate pad, (iii) substrate membrane, and (iv) absorbent bed (**Figure 1**).

2.1 Mechanism of lateral flow assay

The sample is presented at the sample bed (made from cellulose fiber) and migrates to the conjugate pad, which consists of material labeled with analyte-specific antibodies. The conjugate antibodies bind and move up with the sample towards the test line. The test and control lines (also known as the detection compromise of nitrocellulose membrane immobilized with biological moieties (primarily antibodies or antigens). When flowing into the detection zone, the conjugated sample reacts with the analyte bound to a specific antibody. Upon detection, visual changes appear as a line (for a positive test result), while the control line appears whether the target analyte is present or not [6, 7]. The conjugation of the label materials (i.e. quantum dots) to biomolecules such as antibodies allows for the improvement in the capture

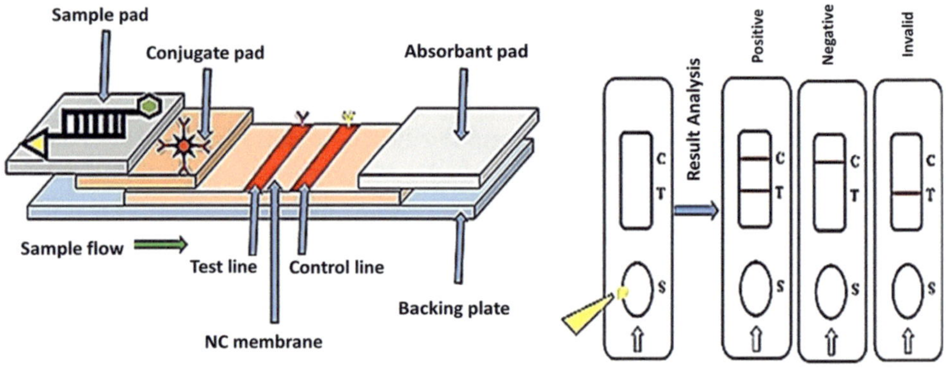

Figure 1.
Schematic illustration of the lateral flow assay (LFA) . Reproduced from Huo C, Li D, Hu Z, Li G, Hu Y and Sun H (2021) a novel lateral flow assay for rapid and sensitive nucleic acid detection of Avibacterium paragallinarum. Front. Vet. Sci. 8:738558 under license https://creativecommons.org/licenses/by/4.0/ [18].

and release of the target analyte in the LFA beds (majorly in the conjugate, test, and control beds). This improves the detection of disease biomarkers.

Although LFAs enjoy the above advantages, traditional lateral flow immunoassays do suffer from limitations which include but are not limited to:

i. Lack the adequate sensitivity that enables color changes that the human eye can view,

ii. Loss of conformation and recognition functionality in non-aqueous media by antibody biosensors,

iii. Challenges in the maintenance of colloidal and fluorescence stability in aqueous biological environments, particularly in enzyme-mediated conditions,

iv. Challenges of ligand exchange and

v. Qualitative results that suffer from obstruction of pores due to matrix components.

To address these challenges, optimization of LFA using various experimental approaches has been reported. Optimizing multiple parameters on the bed is necessary for optimal LFA performance enabling adequate flow, release, and stability. **Table 2** lists the components (i.e. buffers, stabilizers, detergents/surfactants/wetting agents and blocking agents) commonly used to optimize the efficiency, reproducibility, sensitivity, and minimize non-specific binding of the LFA. To further address some of the above LFA shortcomings, research has shifted towards applying QDs as signal-generating material (SGM) [17] (Kim et al., 2022). For example, Liang et al., 2020 [19] reported on a QDs-based LFA used in conjunction with a portable fluorescence immunoassay chip for detecting exact IgE for mite allergens obtained from patients with allergic rhinitis. Clinical samples were analyzed with good reproducibility, fluorescent test, and control line.

The brightness of QDs has therefore improved LFA sensitivity and *in-situ* monitoring [20]. Moreover, QDs have become essential in LFA due to their ease of bio-functionalization, which enables specific and stable recognition of antibodies,

Component	Commonly used reagent
Buffers • for ionic strength and pH • reproducibility • sensitivity and specificity	• Tris(hydroxymethyl)aminomethane (Tris) • Glycine • 4-(2-hydroxyethyl)-1-piperazineethanesulfonic acid) (HEPES) • Bicarbonate • Borate • Phosphate
Stabilizers • stabilizes capture bioreceptor	• Trehalose • Sucrose • Lactose
Detergents/ Surfactants/ Wetting Agents • prevent non-specific binding • Assist in the flow of detection labels along the LFA	• Polyoxyethylene sorbitan monolaurate (Tween) • Sodium dodecyl sulfate (SDS) • 3-[(3-cholamidopropyl)dimethylammonio]propane--1-sulfonate (CHAPS) • Polyvinyl alcohol (PVA) • Polyethylene glycol (PEG) • Sodium deoxycholate
Blocking agents • prevent non-specific binding • improve specificity • faster flow	• PEG • Bis(trimethylsilyl)acetamide (BSA) • Casein • Milk • PVA • Chaotropic agents

Table 2.
Optimization parameters.

addressing the lack of commonly reported conformation. QDs as fluorescent probes have thus become useful for the immune recognition of target analytes [21] (Wilkins et al., 2018), and allow for quantitative detection of biomarkers. In a study by Rong et al., 2021 [22], red-colored quantum dot nanobeads (i.e. silica or polymer nanobeads embedded with hundreds of quantum dots) were used for the multiplex and simultaneous detection of four infectious disease biomarkers. The group were able to detect human immunodeficiency virus (HIV), *Treponema pallidum* (TP), hepatitis C virus (HCV), and hepatitis B virus (HBV) at LOD levels of 0.11 NCU/mL, 0.62 IU/L, 0.14 NCU/mL and 0.22 IU/mL respectively within 20 min. The direct aqueous synthesis of QDs has further eliminated the challenges of colloidal and fluorescence stability of LFA when biomolecule (i.e. blood, serum, plasma etc.) samples are analyzed. While the small sizes and good size distribution of QDs prevents the obstruction/blocking of membrane pores in the sample pad and detection zone.

3. Engineering quantum dots for lateral flow immunoassays

Quantum dots are semiconductor materials with shape and size-dependent optical properties. They have found increased application in the diagnosis and prognosis of infectious diseases [23]. QDs conjugated to biomolecules have improved the binding efficiency within the conjugate pad. QDs bonded to bio-recognition antibodies offer a viable system to enhance specificity and sensitivity in the detection zone. Optimizing multiple parameters for adequate flow, release, and stability is necessary for optimal LFA performance.

Most importantly, the label material must be able to detect antibodies to capture and release the test analyte sufficiently. Antibodies are used as bio-recognition components in LFA's detection zone (test and control lines), and they co-join the target analyte via immunochemical exchange [6, 24]. It is, therefore, essential that the QDs used are also able to bind with the conjugate pad material. The conjugate pad plays an essential role in running the LFA. It serves as the location for chemicals and ensures the constant transfer of detector reagents and analytes to the detector zone. It is, therefore, necessary to ensure QDs exhibit non-specific binding to conjugate pad material to ensure sufficient target analytes reach the detection zones. The conjugate pad also houses the detector particles and ensures they are kept functionally stable until the test is used [8, 9, 25]. In instances where the QDs as fluorescent probes are not well coated or unstable, thus losing their properties over time may result in insufficient target analytes reaching the test line, reducing LFA sensitivity and signal intensity. Furthermore, delays in QDs bound to an analyte of interest due to extended interaction with conjugate pad material may cause inconsistent flow. This may result in the test and control lines appearing as fluorescent streaks due to inconsistent flow [9, 25].

To improve the binding efficiency of QDs, conjugation of QDs via carboxyl-reactive crosslinker reactive groups is commonly used for the labelling/crosslinking of QDs. Water-soluble 1-ethyl-3-(−3-dimethyl aminopropyl) carbodiimide hydrochloride (EDC) is used for aqueous crosslinking in conjunction with *N*-hydroxy succinimide (NHS). NHS is commonly added to EDC coupling to improve its efficiency and stability. For non-aqueous organic synthetic methods, N′, N′-Cyclohexyl carbodiimide (DCC) crosslinker is widely used. Carbodiimide crosslinkers function by activating the -COOH functional group of the QDs, allowing for direct conjugation to primary amines ($-NH_2$) via amide bonds [26]. The functionalization of QDs in LFA enables increased sensitivity and specificity for a 2-part recognition scheme whereby the label and capture agent become specific to the target analyte [21].

QDs are often functionalized with different capping ligands, which provide colloid solubility (i.e. convert hydrophobic material to hydrophilic), inherent stability and assist in chemical binding for biomolecules (i.e. DNA, proteins, peptides, and enzymes) [27, 28]. **Figure 2** shows various generic QDs solubilization and biofunctionalization often performed to condition QDs. QDs synthesized using an organic medium are insoluble and require phase transfer via ligand exchange (**Figure 2a**, **b**) to improve their water solubility and enable their bio-application. Ligand exchange, therefore, converts the hydrophobic ligands with bifunctional (viz., drugs, antibodies, or proteins) moieties enabling direct bio-application of the QDs [28, 29]. **Figure 2c** shows the use of polymers to insulate the hydrophilic surface of the QDs. In contrast, **Figure 2d-f** and **g** depicts the aqueous dispersion and derivatization of the QDs surface and its interaction with amphilic 'diblock' and 'triblock' copolymers and phospholipids. **Figure 2h** shows the attachment of QDs with amino acids (protein or peptides) as a bio-functionalization technique, therefore solubilizing and converting the surface of the QDs.

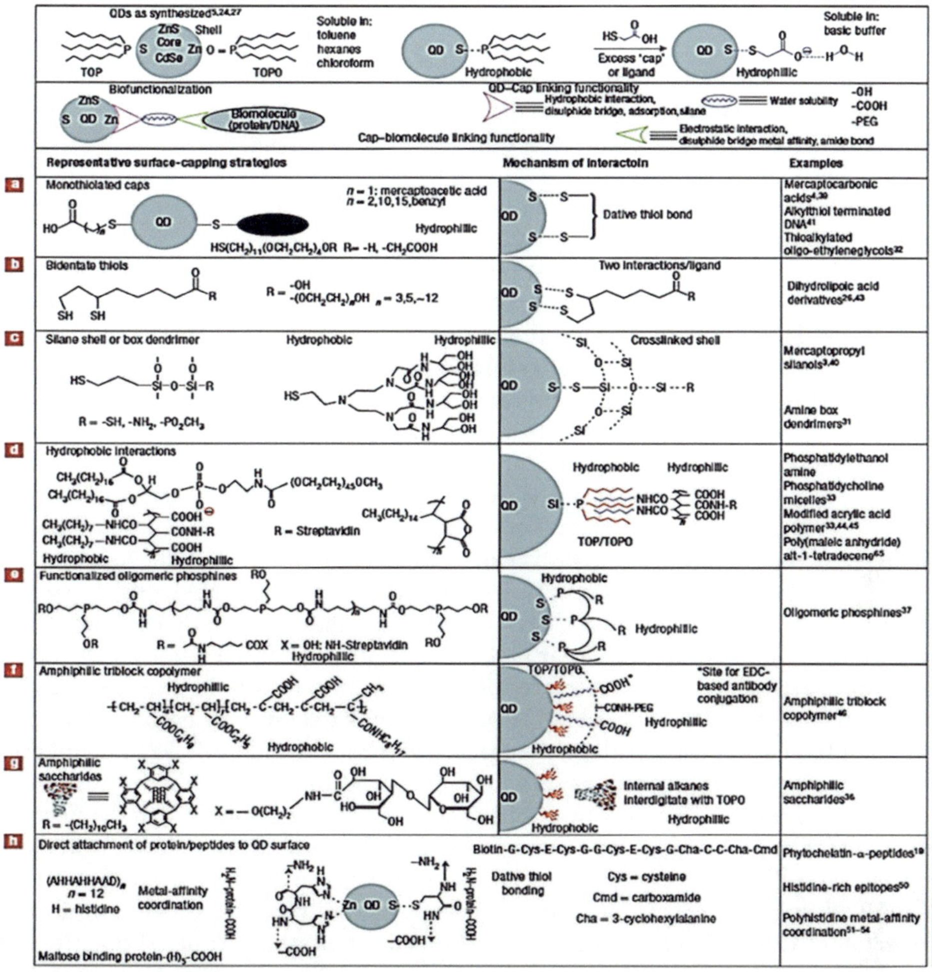

Figure 2.
Schematic of generic QDs solubilization and biofunctionalization (a–h). Reprinted by permission from springer nature customer service Centre GmbH: Nature, nature materials IGor L. Medintz, H. Tetsuo Uyeda, Ellen R. Goldman and Hedi MattoussI), [COPYRIGHT] (2005) [28].

4. Application of quantum dots in LFA to detect communicable (infectious) diseases

Quantum dots have superior fluorescent properties that render them ideal labels. Their application in LFA has improved the detection sensitivity of LFA as a result of their chemical-physical properties (i.e. high signal-to-noise ratio, narrow emission, high fluorescent quantum yields, and broad excitation) [30]. The bright luminescence and stability of QDs qualify their use in LFA for the sensitive detection of communicable (infectious diseases) and non-communicable (chronic) diseases.

4.1 Severe acute respiration syndrome coronavirus 2 (SARS-CoV-2)

Severe acute respiration syndrome coronavirus 2 (SARS-CoV-2) originated in 2019 as the source of the coronavirus disease 2019 (COVID-19). Real-time Reverse Transcription-Polymerase Chain Reaction (RT-PCR) has been the primary test used for the molecular detection of SARS-CoV-2. Its application requires a certified laboratory infrastructure and trained personnel [31]. Nonetheless, for large-scale testing, quick results and contact tracing, RT-PCR is not a suitable diagnosis technique. COVID-19 diagnostic LFIA tests have provided fast, low-cost, and are a widely available tool for COVID-19 testing. Generally, these essential requirements are performed by serological tests based on lateral flow immunoassay (LFIA) [32]. The mechanism of COVID-19 detection/sensing using QDs is based on various paths that include; (i) inhibition of the binding of S protein receptors of the virus to the host cells, (ii) the generation of antiviral radicals upon exposure to light, (iii) reducing of viral RNA genome amplification and (iv) application of QDs as bright labels for COVID-19 [33, 34]. The unique properties (i.e. water solubility, small size ~10 nm, high sustainability, alluring photoluminescence) of carbon QDs (CQDs) have seen their application in the detection, inhibition, and treatment of viral infections [35]. Research on the interaction of QDs with genetic material (i.e. RNA and DNA), and cellular receptors to inhibit biological processes is still at its infancy. Nonetheless, their small sizes, ability to interact with cellular material, biocompatible nature and tuneable optical properties has seen growing reports on their importance in disease/virus inhibition processes. Loczechin et al., 2019 [33] reported on synthesizing carbon dots and their concentration-dependent interference with HCoV-229E-Luc infection. The mechanism of inhibition was said to be based on the CQD's ability to act in the early stages of virus infection by inhibiting the protein S-receptors with the host cell membrane. The group reported that the CQDs inhibited the HCoV-229E from entering the host cells infection as the functional groups of the CQDs interacted with the HCoV-229E entry receptors [36]. Another study by [19] reported on viral inhibition properties of CQDs via type 1 interferon responses. The study reported that the CQDs functioned by expressing IFN-stimulating genes (ISGs) and inducing interferon-α (IFN-α) production, which are responsible for inhibiting virus replication. Recently, Li et al., 2022 [36] reported on a smartphone-based QD-LFIA to detect IgG and Nab specific to SARS-CoV-2 in biological samples. The QDs were conjugated to anti-His mAb via covalent attachment (i.e. EDC/NHS). To enhance the performance of the LFIA, parameters such as the quantity of anti-His mAb-labeled QDs, the type of sample pads, the sample and the treatment buffer to reduce non-specific interactions within the QDs-LFIA were optimized. The developed QD-LFIA quantitatively detected IgG and Nab. The accurate and sensitive detection of the immune status of the clinical samples compared well with other methods. This highlights the progression in QD-LFIA application for the clinical detection of COVID-19.

4.2 Human immunodeficiency virus (HIV)

Human Immunodeficiency Virus (HIV) is an immunodeficiency disease that progresses into acquired immune deficiency syndrome (AIDS) at low CD4 levels (<200 cells/uL) [37, 38]. Currently, the most used approach to detect HIV is a dip-stick test which detects the virus antibodies [39]. The growing interest in the application of QDs for the detection of HIV is a result of QDs (i) small size that permits ease of flow over the strip, (ii) fluorescence lifetime and photostability, which enables QDs to be signifiers for detection, (iii) a surface area-to-volume ratio that enables increasing binding sites and (iv) ability to combine with bio-molecules with ease as a result of their compatibility and stability [40, 41]. Deng et al., 2018 [41] used a fluorescent LFA strip to detect HIV-DNA quantitatively. The group used a one-step synthetic method to assemble the CdTe QDs with hairpin DNA as the reporter probe. At the same time, strand displacement amplification (SDA) worked as an amplification tool. The LFA consisted of a test and control zone on a nitrocellulose membrane immobilized with streptavidin in biotinylated t-DNA and cDNA, respectively. The developed LFA strip reproducibly detected HIV-DNA using CdTe-dsDNA (double-strand DNA) at a LOD of 0.76 pm.

4.3 Influenza

Influenza is seasonal endemic viruses that circulate yearly, causing mild to severe illnesses. Type A viruses such as (H1N1) and A (H3N2) are seasonal influenza caused by human infection. While Avian influenza viruses, A (H5N1) A(H5N6), A(H7N9), and A(H9N2), usually passed to domiciliary poultry leading to high disease surges resulting in life-threatening infections among humans [42]. Therefore, the accurate and fast diagnosis of influenza viruses is essential for early detection, hospital care and prevention of influenza infection outbreaks. The brightly fluorescent nature of QDs and bio-labelling properties have enabled the specific, accurate, quick, and simultaneous detection of *influenza* viruses. The sensitive, reproducible, and simultaneous detection of influenza A virus subtypes H5 and H9 using water-soluble COOH- functionalized QDs has been reported [43]. The QDs were covalently bounded to the influenza A virus antibodies and used as fluorescent tags in the LFA. The study harnessed the fluorescent signal produced by the QDs on the LFA strips to quantify the influenza A viruses (at LOD =0.016 HAU and 0.25 HAU for influenza virus H5 and H9 subtypes, respectively). In another study, Nguyen and co-workers [42] produced a CdSe/CdS/ZnS QDs LFA for sensitive and quick analysis of two influenza subtypes (H1N1 and H3N2). The test and control lines were coated with anti-influenza monoclonal and goat anti-mouse IgG antibodies. Although the number of positive samples was limited, the study could detect A/H1N1 64-folds higher than rapid diagnostic test (RDT; Standard Diagnostics BIOLINE Influenza A/B). These works highlight the potential of fluorescent QDs-based LFAs for clinical use in detecting influenza strains.

4.4 Hepatitis B

Hepatitis B virus (HBV) is a viral infection that can be transmitted through contact with infected body fluids (i.e. infected blood) [44, 45]. Hepatitis B surface antigens (HBsAg) and envelope antigens (HBeAg) are recognized clinical indicators used in the diagnosis of HBV [45]. Therefore, the accurate and rapid detection of HBsAg and HBeAg antigens is essential for the clinical diagnosis of HBV. Gold nanoparticles (GNP) coated with streptavidin (SA – GNP conjugate pad), biotin and antibody (Bio – GNP – Ab conjugate bed) have been employed to advance the sensitivity of LFA in the detection of HBV [46]. However, the expensive synthesis of GNPs, the dependence of the LFA on the localized surface plasmon resonance effect of GNPs [47] and their low sensitivity towards HBsAg have resulted in the exploration of QDs for HBV detection [48]. The ease in size tuning, biocompatibility and stable fluorescence has further advantaged QDs over other nanoparticles in LFA development. Several studies have reported the effect of GNPs size on the sensitivity and reproducibility of LFA [45, 46, 49, 50]. For optimum performance of LFA and for enhanced LFA sensitivity, the narrow size distribution of GNPs has been proposed. Kim et al., 2016 [45] reported a reduction in signal intensity with increased GNP size. The group attributed the results to reduced conjugation efficiency of GNP at sizes >42.7 ± 0.8 nm. In another study, Chotithammakul et al., 2021 [49] reported the synthesis of single and mixed-size GNPs for LFA detection of Albumin. Their results showed mixed-sized GNP conjugates exhibited enhanced signals with higher stability and narrower resonance peaks than GNPs with 20 nm sizes. The results suggest that the combination of different sizes of GNP can affect their plasmon resonance. Moreover, in the study, the conjugated mixed-size GNPs - antibodies (anti-BSA) accumulated at the test spot, which improved localized surface plasmon transduction resulting in enhanced optical labelling in the LFA.

4.5 Tuberculosis

Tuberculosis (TB) is recorded as the deadliest infectious communicable disease caused by *Mycobacterium tuberculosis* and has become a significant public health problem worldwide. Experienced physicians use chest radiographs in clinical practice to detect TB [51, 52]. The clinical diagnosis at reduced time and low cost via LFA is still very underdeveloped for TB diagnosis. Detection of *mycobacterial free tuberculosis* proteins (fprA) using QDs conjugated to two monoclonal antibodies has been reported. The fluorescent QDs were conjugated to the antibodies via a biotin-streptavidin bridge. In this specific LFA, the antibodies were used as capture probes and immobilized onto the test lines of the nitrocellulose strip. The double-antibody sandwich LFA format allowed recombinant fprA protein detection at ~12.5 pg./μL. Compared to traditional immune-based testes (i.e. dot immunobinding assay and Elisa), the antibody labeled QDs LFIA improved the sensitivity, speed (results within 10 min) and simplicity [53]. In another study, Kabwe et al., 2022 [54] explored a paper-based lateral for detecting anti-mycolic acid antibodies. The group used mycolic acid (MA) coated CdSe QDs as a luminescent probe to detect anti-MA antibodies, biomarkers for tuberculosis. The study's results supported test strip-based immunoassays for TB at low-cost and optimum speed.

5. Application of quantum dots in LFA for detecting non-communicable (chronic) diseases

5.1 Cardiovascular diseases

Cardiovascular diseases (CVDs) are disorders related to the heart, such as ischemic heart disease, heart failure, stroke, cerebrovascular diseases, thrombosis, arrhythmia, and cardiomyopathies [55]. N-terminal pro-B-type natriuretic peptide (NT-proBNP) is a diagnostic cardiac biomarker for acute heart failure. However, inadequate products are available in the market for the diagnostic quantification of NT-proBNP. Wilkins and co [21] recently reported on sandwich LFA for NT-pro-BNP. The group used monoclonal antibodies conjugated to QDs via reductive amination. The study showed that conjugating QDs to the antibodies using a site-specific reductive amination strategy exhibited a higher signal at all tested concentrations than LFAs run using EDC-conjugated antibodies. Along with other studies [56, 57], this work verified the potential use of QDs conjugated to CVDs antibodies as a viable means for risk assessment of human cardiovascular diseases.

5.2 Cancer

Cancer is one of the leading causes of death all over the world. An early cancer diagnosis is meaningful and helpful for extending the overall survival of patients. Despite developing highly sophisticated diagnostic tools, these are expensive, time-consuming, and require complex operations and skilled personnel. Moreover, they are not suitable for the first-line diagnosis. One of the early diagnostic approaches for cancer is to test tumor markers. Carbohydrate antigen 72–4 (CA72–4) is a biomarker linked with the diagnosis and prognosis of early stages of gastric cancer. A CdSe QD labeled sandwich LFA coupled with a charge-coupled device (CCD)-based reader has been reported to detect CA72–4 in clinical samples. Monoclonal CC49 and B72.3 antibodies are often used to detect CA72–4; therefore, to improve the sensitivity of LFA, CC49 was conjugated to the CdSe QDs. The QD-labeled CC49 exhibited narrower emissions with strong fluorescence signals than the QDs alone and proved to be a suitable sensor probe for CA72–4 detection. The QDs-based LFA could sensitively detect CA72–4 at 2 IU/mL levels within 10 min. Furthermore, compared to the Roche electrochemiluminescence assay, the developed strip method has shown excellent reproducibility and specificity [58]. Magnetic-quantum dot nanobeads (MQBs) are composite nanomaterials with distinctive fluorescent and magnetic properties. Antibody-conjugated MQBs are employable in recognizing the immune and can magnetically separate analytes under complex samples. Rong et al., 2019 [22] conjugated MQBS to free prostate-specific antigen (f-PSA) and complexed prostate-specific antigen (c-PSA) as fluorescent probes for prostate cancer. The conjugates were used for immune recognition and simultaneous detection of f-PSA and c-PSA analytes in a complex biological matrix. The developed sandwich LFA detected both antigens (-PSA and c-PSA) on a single test line with high sensitivity and specificity.

6. Current challenges

Current LFAs use label materials (in the conjugate pad) that are synthesized in the hydrophobic state, unstable, lack colourimetric fluorescence stability and are incompatible with the protein–particle binding. For example, using enzymes results in detection sensitivity that depends on an enzyme-substrate combination [59]. Colloidal carbon nanoparticles as labels suffer from the non-specificity and the presence of irregular shapes [59]. Although quantum dots seemingly addressed some of these issues. The use of binary QDs (composed of groups I-VI and VII) is of concern as they are toxic by their very nature due to the presence of heavy metals such as Cd and Pb. Moreover, the risk of toxic heavy metals leaking within the LFA is a significant concern, raising toxicity, health, and environmental concerns. Additionally, the synthesis of currently used QDs has been via organic reagent, thus requiring ligand exchange which compromises the optoelectrical properties of the QDs required for ideal LFA development. Another shortcoming of current LFAs is that they: (i) lack the adequate sensitivity that enables color changes that the human eye can view, (ii) loss of conformation and recognition functionality in non-aqueous media by antibody biosensors, and (iii) challenges in the maintenance of colloidal and fluorescence stability in aqueous biological environments, particularly in enzyme-mediated conditions, (iv) challenges of ligand exchange and (v) qualitative results of that suffer from obstruction of pores due to matrix components. To address these challenges, there is a need to develop multi-colored ternary and quaternary QDs-based LFA for the clinical detection of single and multi-disease/virus biomarkers.

7. Future perspective and conclusion

Quantum dots' unique optoelectrical properties have significantly advantaged them as ideal labels in LFA. Their high quantum yields, molecular extinction coefficients and excellent r chemical and photo-degradation further enable their broader application in LFA sensing and diagnosis. An ideal label for application in LFA includes colloidal stability under various conditions and temperatures, suitability for detection over large dynamic ranges, reproducibility, efficiency without losing biological and chemical integrity or activity, specific binding, and sensitivity. These are often achieved via bio-functionalization of QDs with biomolecules (i.e. DNA, protein, and peptide) which further enables the binding and capture of target antibodies within the conjugate pad. In the clinical diagnosis and treatment of communicable and non-communicable diseases, QDs' ability to enter cells, thus inhibiting virus replication, has been shown as a treatment strategy. However, the lack of fast, specific, reproducible, and stable diagnosis tools still hinders early disease detection. The development of QDs-based LFA (for NCDs and CDs) for clinical application has been reported in the literature. Their clinical use is, however, still not approved. Therefore, the exploration of clinical trials for QDs -based LFA as an alternative POC for NCDs and CDs is greatly needed.

Acknowledgements

The authors would like to thank the National Research Foundation (NRF), South Africa under the Thuthuka Grant (Grant No: 121986), Chemistry department, Faculty of Applied Science at the Cape Peninsula University of Technology, South Africa.

Conflict of interest

"The authors declare no conflict of interest."

Author details

Ncediwe Tsolekile*, Noluvuyo Mngcutsha and Nozikumbuzo Vitshima
Department of Chemistry, Cape Peninsula University of Technology, Cape Town, South Africa

*Address all correspondence to: tsolekilen@cput.ac.za

References

[1] Budreviciute A, Damiati S, Sabir DK, Onder K, Schuller-Goetzburg P, Plakys G, et al. Management, and prevention strategies for non-communicable diseases (NCDs) and their risk factors. Frontiers in Public Health. 2020;**8**:788. DOI: 10.3389/fpubh.2020.574111

[2] World Health Organization. Regional Office for South-East Asia. Department of communicable diseases: profile and vision. WHO Regional Office for South-East Asia. Document number: SEA-CD-145. 2007. Available from: https://apps.who.int/iris/handle/10665/204952

[3] Tsolekile N, Nelana S, Oluwafemi OS. Porphyrin as a diagnostic and therapeutic agent. Molecules. 2019;**24**:2669. DOI: 10.3390/molecules24142669

[4] Hong B, Zu Y. Detecting circulating tumor cells: Current challenges and new trends. Theranostics. 2013;**3**:377. DOI: 10.7150/thno.5195

[5] Chang AE, Matory YL, Dwyer AJ, Hill SC, Girton ME, Steinberg SM, et al. Magnetic resonance imaging versus computed tomography in the evaluation of soft tissue tumors of the extremities. Annals of Surgery. 1987;**205**:340. DOI:10.1097/00000658-198704000-00002

[6] Hayes B, Murphy C, Crawley A, O'Kennedy R. Developments in point-of-care diagnostic technology for cancer detection. Diagnostics. 2018;**8**:39. DOI: 10.3390/diagnostics8020039

[7] Bahadır EB, Sezgintürk MK. Lateral flow assays: Principles, designs, and labels. TrAC Trends in Analytical Chemistry. 2016;**82**:286-306. DOI: 10.1016/j.trac.2016.06.006

[8] Koczula KM, Gallotta A. Lateral flow assays. Essays in Biochemistry. 2016;**60**:111-120. DOI: 10.1042/EBC20150012

[9] Parolo C, Sena-Torralba A, Bergua JF, Calucho E, Fuentes-Chust C, Hu L, et al. Tutorial: Design and fabrication of nanoparticle-based lateral-flow immunoassays. Nature Protocols. 2020;**15**:3788-3816. DOI: 10.1038/s41596-020-0357-x

[10] Rong Z, Bai Z, Li J, Tang H, Shen T, Wang Q, et al. Dual-color magnetic-quantum dot nanobeads as versatile fluorescent probes in test strip for simultaneous point-of-care detection of free and complexed prostate-specific antigen. Biosensors and Bioelectronics. 2019;**145**:111719. DOI: 10.1016/j.bios.2019.111719

[11] Qi H, Sun Q, Ma Y, Wu P, Wang J. Advantages of lateral flow assays based on fluorescent submicrospheres and quantum dots for Clostridium difficile toxin B detection. Toxins. 2020;**12**:722. DOI: 10.3390/toxins12110722

[12] Yang X, Liu X, Gu B, Liu H, Xiao R, Wang C, et al. Quantitative and simultaneous detection of two inflammation biomarkers via a fluorescent lateral flow immunoassay using dual-color SiO_2@QD nanotags. Microchimica Acta. 2020;**187**:570. DOI: 10.1007/s00604-020-04555-6

[13] An J, Huynh KH, Ha Y, Jung HS, Kim HM, Kim DM, et al. Surface modification of a stable CdSeZnS/ZnS alloy quantum dot for immunoassay. Journal of Nanomaterials. 2020;**2020**:1-9. DOI: 10.1155/2020/4937049

[14] Kim HM, Oh C, An J, Baek S, Bock S, Kim J, et al. Multi-quantum dots-embedded silica-encapsulated

nanoparticle-based lateral flow assay for highly sensitive exosome detection. Nanomaterials. 2021;**11**:768. DOI: 10.3390/nano11030768

[15] Xu LD, Zhang Q, Ding SN, Xu JJ, Chen HY. Ultrasensitive detection of severe fever with thrombocytopenia syndrome virus based on immunofluorescent carbon dots/SiO2 nanosphere-based lateral flow assay. ACS Omega. 2019;**4**:21431-21438. DOI: 10.1021/acsomega.9b03130

[16] Bock S, Kim HM, Kim J, An J, Choi YS, Pham XH, et al. Lateral flow immunoassay with quantum-dot-embedded silica nanoparticles for prostate-specific antigen detection. Nanomaterials. 2022;**12**:33. DOI:10.3390/nano12010033

[17] Kim S, Ahn Y, Bae Y, Woo S, Park J, Han IK, et al. Highly sensitive pregnancy test kit via oriented antibody conjugation on brush-type ligand-coated quantum beads. Biosensors and Bioelectronics. 2022;**213**:114441. DOI: 10.1016/j.bios.2022.114441

[18] Huo C, Li D, Hu Z, Li G, Hu Y, Sun H. A novel lateral flow assay for rapid and sensitive nucleic acid detection of Avibacterium paragallinarum. Frontiers in Veterinary Science. 2021;**8**:738558

[19] Du T, Liang J, Dong N, Liu L, Fang L, Xiao S, et al. Carbon dots As inhibitors of virus by activation of type I interferon response. Carbon. 2016;**110**:278-285. DOI: 10.1016/j.carbon.2016.09.032

[20] Wang S, Lu H, Wang L, Zou J, Zhang R. Rapid determination of serum vascular endothelial growth factor (VEGF) by a fluorescence. Immunochromatographic Assay, Analytical Letters. 2020;**54**:1233-1241. DOI: 10.1080/00032719.2020.1798455

[21] Wilkins MD, Turner BL, Rivera KR, Menegatti S, Daniele M. Quantum dot enabled lateral flow immunoassay for detection of cardiac biomarker NT-proBNP. Sensing and Bio-Sensing Research. 2018;**21**:46-53. DOI: 10.1016/j.sbsr.2018.10.002

[22] Rong Z, Xiao R, Peng Y, Zhang A, Wei H, Ma Q, et al. Integrated fluorescent lateral flow assay platform for point-of-care diagnosis of infectious diseases by using a multichannel test cartridge. Sensors and: B. Chemical. 2021;**329**:129193. DOI: 10.1016/j.snb.2020.129193

[23] Sohrabi H, Majidi MR, Fakhraei M, Jahanban-Esfahlan A, Hejazi M, Oroojalian F, et al. Lateral flow assays (LFA) for detection of pathogenic bacteria: A small point-of-care platform for diagnosis of human infectious diseases. Talanta. 2022;**243**:123330. DOI: 10.1016/j.talanta.2022.123330

[24] Lox JF, Dang Z, Dzhagan VM, Spittel D, Martín-García B, Moreels I, et al. Near-infrared Cu–In–Se-based colloidal nanocrystals via cation exchange. Chemistry of Materials. 2018;**30**:2607-2617. DOI: 10.1021/acs.chemmater.7b05187

[25] Borse V, Srivastava R. Process parameter optimization for lateral flow immunosensing. Materials Science for Energy Technologies. 2019;**2**:434-441. DOI: 10.1016/j.mset.2019.04.003

[26] Song F, Chan WC. Principles of conjugating quantum dots to proteins via carbodiimide chemistry. Nanotechnology. 2011;**22**:494006. DOI: 10.1088/0957-4484/22/49/494006

[27] Tsolekile N, Parani S, Maluleke R, Joubert O, Matoetoe M, Songca SP, et al. Green synthesis of amino acid functionalized CuInS/ZnS- mTHPP

conjugate for biolabeling application. Dyes and Pigments. 2021;**185**:108960. DOI: 10.1016/j.dyepig.2020.108960

[28] Medintz IL, Uyeda HT, Goldman ER, Mattoussi H. Quantum dot bioconjugates for imaging, labelling and sensing. Nature Materials. 2005;**4**:435-446. DOI: 10.1038/nmat1390

[29] Tsolekile N, Parani S, Matoetoe MC, Songca SP, Oluwafemi OS. Evolution of ternary I–III–VI QDs: Synthesis, characterization, and application. Nano-Structures & Nano-Objects. 2017;**12**:46-56. DOI: 10.1016/j.nanoso.2017.08.012

[30] Di Nardo F, Anfossi L, Giovannoli C, Passini C, Goftman VV, Goryacheva IY, et al. A fluorescent immunochromatographic strip test using quantum dots for fumonisins detection. Talanta. 2016;**150**:463-468. DOI: 10.1016/j.talanta.2015.12.072

[31] Tahamtan A, Ardebili A. Real-time RT-PCR in COVID-19 detection: Issues affecting the results. Expert Review of Molecular Diagnostics. 2020;**20**:453-454. DOI: 10.1080/14737159.2020.1757437

[32] Conte DD, Carvalho JM, de Souza Luna LK, Faíco-Filho KS, Perosa AH, Bellei N. Comparative analysis of three point-of-care lateral flow immunoassays for detection of anti-SARS-CoV-2 antibodies: Data from 100 healthcare workers in Brazil. Brazilian Journal of Microbiology. 2021;**52**:1161-1165. DOI: 10.1007/s42770-021-00498-z

[33] Łoczechin A, Séron K, Barras A, Giovanelli E, Belouzard S, Chen YT, et al. Functional carbon quantum dots as medical countermeasures to human coronavirus. ACS Applied Materials & Interfaces. 2019;**11**:42964-42974. DOI: 10.1021/acsami.9b15032

[34] Rasmi Y, Li X, Khan J, Ozer T, Choi JR. Emerging point-of-care biosensors for rapid diagnosis of COVID-19: Current progress, challenges, and future prospects. Analytical and Bioanalytical Chemistry. 2021;**413**:4137-4159. DOI: 10.1007/s00216-021-03377-6

[35] Xue Y, Liu C, Andrews G, Wang J, Ge Y. Recent advances in carbon quantum dots for virus detection, as well as inhibition and treatment of viral infection. Nano Convergence. 2022;**9**:1-31. DOI: 10.1186/s40580-022-00307-9

[36] Li J, Liu B, Tang X, Wu Z, Lu J, Liang C, et al. Development of a smartphone-based quantum dot lateral flow immunoassay strip for ultrasensitive detection of anti-SARS-CoV-2 IgG and neutralizing antibodies. International Journal of Infectious Diseases. 2022;**121**:58-65. DOI: 10.1016/j.ijid.2022.04.042

[37] Waymack JR, Sundareshan V. Acquired Immune Deficiency Syndrome. [Updated 2022 Sep 5]. In: StatPearls [Internet]. Treasure Island (FL): StatPearls Publishing; Jan 2022. Available from: https://www.ncbi.nlm.nih.gov/books/NBK537293/

[38] Wang Y, Bai X, Wen W, Zhang X, Wang S. Ultrasensitive electrochemical biosensor for HIV gene detection based on graphene stabilized gold nanoclusters with exonuclease amplification. ACS Applied Materials & Interfaces. 2015;7:18872-18879. DOI: 10.1021/acsami.5b05857

[39] Zhou Y, Chen Y, Liu W, Fang H, Li X, Hou L, et al. Development of a rapid and sensitive quantum dot nanobead-based double-antigen sandwich lateral flow immunoassay and its clinical performance for the detection of SARS-CoV-2 total antibodies. Sensors and Actuators B: Chemical. 2021;**343**:130139. DOI: 10.1016/j.snb.2021.130139

[40] Deng X, Wang C, Gao Y, Li J, Wen W, Zhang X, et al. Applying strand

displacement amplification to quantum dots-based fluorescent lateral flow assay strips for HIV-DNA detection. Biosensors and Bioelectronics. 2018;**105**:211-217. DOI: 10.1016/j.bios.2018.01.039

[41] Nguyen AV, Dao TD, Trinh TT, Choi DY, Yu ST, Park H, et al. Sensitive detection of influenza a virus based on a CdSe/CdS/ZnS quantum dot-linked rapid fluorescent immunochromatographic test. Biosensors and Bioelectronics. 2020;**155**:112090

[42] Wu F, Mao M, Liu Q, Shi L, Cen Y, Qin Z, et al. Ultra sensitive detection of influenza a virus based on Cdse/Zns quantum dots immunoassay. SOJ Biochem. 2016;**2**:2-6. DOI: 10.15226/2376-4589/2/3/00108

[43] Martiskainen I, Talha SM, Vuorenpää K, Salminen T, Juntunen E, Chattopadhyay S, et al. Upconverting nanoparticle reporter–based highly sensitive rapid lateral flow immunoassay for hepatitis B virus surface antigen. Analytical and Bioanalytical Chemistry. 2021;**413**(4):967-978. DOI: 10.1007/s00216-020-03055-z

[44] Kim DS, Kim YT, Hong SB, Kim J, Heo NS, Lee MK, et al. Development of lateral flow assay based on size-controlled gold nanoparticles for detection of hepatitis B surface antigen. Sensors. 2016;**16**:2154. DOI: 10.3390/s16122154

[45] Shen Y, Shen G. Signal-enhanced lateral flow immunoassay with dual gold nanoparticle conjugates for the detection of hepatitis B surface antigen. ACS Omega. 2019;**4**(3):5083-5087

[46] Ling Y, Xu SB, Lin YX, Tian D, Zhu ZQ, Dai FH, et al. Persistence and clearance of viral RNA in 2019 novel coronavirus disease rehabilitation patients. Chinese Medical Journal. 2020;**133**:1039-1043. DOI: 10.1097/CM9.0000000000000774

[47] Manivannan S, Ponnuchamy K. Quantum dots as a promising agent to combat COVID-19. Applied Organometallic Chemistry. 2020;**34**:2-6. DOI: 10.1002/aoc.5887

[48] Chotithammakul S, Cortie MB, Pissuwan D. Comparison of single- and mixed-sized gold nanoparticles on lateral flow assay for albumin detection. Biosensors. 2021;**11**:209. DOI: 10.3390/bios11070209

[49] Liu Y, Zhang N, Li P, Yu L, Chen S, Zhang Y, et al. Low-cost localized surface Plasmon resonance biosensing platform with a response enhancement for protein detection. Nanomaterials. 2019;**9**:1019. DOI: 10.3390/nano9071019

[50] Ariffin N, Yusof NA, Abdullah J, Abd Rahman SF, Ahmad Raston NH, Kusnin N, et al. Lateral flow immunoassay for naked eye detection of mycobacterium tuberculosis. Journal of Sensors. 2020;**2020**:1-10. DOI: 10.1155/2020/1365983

[51] Rahman T, Khandakar A, Kadir MA, Islam KR, Islam KF, Mazhar R, et al. Reliable tuberculosis detection using chest X-ray with deep learning, segmentation and visualization. IEEE Access. 2020;**8**:191586-191601

[52] Cimaglia F, Aliverti A, Chiesa M, Poltronieri P, De Lorenzis E, Santino A, Sechi LA. Quantum dots nanoparticle-based lateral flow assay for rapid detection of Mycobacterium species using anti-FprA antibodies. Nanotechnology Development. 2012;**2**:26-30. DOI: 10.4081/nd.2012.e5

[53] Kabwe KP, Nsibande SA, Lemmer Y, Pilcher LA, Forbes PB. Synthesis and

characterization of quantum dots coupled to mycolic acids as a water-soluble fluorescent probe for potential lateral flow detection of antibodies and diagnosis of tuberculosis. Luminescence. 2022;**37**(2):278-289

[54] Lopez EO, Ballard BD, Jan A. Cardiovascular Disease. [Updated 2022 Aug 8]. In: StatPearls [Internet]. Treasure Island (FL): StatPearls Publishing; Jan 2022. PMID: 30571040. Available from: https://pubmed.ncbi.nlm.nih.gov/30571040/

[55] Wu R, Zhou S, Chen T, Li J, Shen H, Chai Y, et al. Quantitative and rapid detection of C-reactive protein using quantum dot-based lateral flow test strip. Analytica Chimica Acta. 2018;**1008**:1-7. DOI: 10.1016/j.aca.2017.12.031

[56] Cheng X, Pu X, Jun P, Zhu X, Zhu D, Chen M. Rapid and quantitative detection of C-reactive protein using quantum dots and immunochromatographic test strips. International Journal of Nanomedicine. 2014;**9**:5619. DOI: 10.2147/IJN.S74751

[57] Yan X, Wang K, Lu W, Qin W, Cui D, He J. CdSe/ZnS quantum dot-labeled lateral flow strips for rapid and quantitative detection of gastric cancer carbohydrate antigen 72-4. Nanoscale Research Letters. 2016;**11**:1-8. DOI: 10.1186/s11671-016-1355-3

[58] Christodouleas DC, Kaur B, Chorti P. From point-of-care testing to eHealth diagnostic devices (eDiagnostics). ACS Central Science. 2018;**4**:1600-1616. DOI: 10.1021/acscentsci.8b00625

[59] Sajid M, Kawde AN, Daud M. Designs, formats and applications of lateral flow assay: A literature review. Journal of Saudi Chemical Society. 2015;**19**:689-705. DOI: 10.1016/j.jscs.2014.09.001

Chapter 9

Quantum Dots in Cancer Cell Imaging

Salar Khaledian, Mohadese Abdoli, Reza Fatahian and Saleh Salehi Zahabi

Abstract

Research on quantum dots (QDs) as zero-dimensional nanostructures whose size is not more than a few nanometers has accelerated in the last two decades, especially in the field of medicine. These nanostructures have attracted much attention due to their unique features such as broad excitation range, narrow emission, strong fluorescence, and high resistance to photobleaching. In this chapter, besides common QDs such as cadmium (Cd)-containing semiconductor QDs, other QDs including carbon-based QDs, chalcogenide QDs, and black phosphorus QDs will be discussed. In addition to describing the optical characteristics of these nanostructures, the usual synthesis methods, their modification and cytotoxicity will be reviewed. Finally, the application of each category of QDs in cancer cell imaging will prospect in more detail.

Keywords: cell imaging, semiconductor QDs, carbon-based QDs, MoS2 QDs, black phosphorus QDs

1. Introduction

Cancer is one of the main causes of death all over the world, and after cardiovascular diseases, it is the most common cause of death. This disease is usually caused by defects in the functioning of the regulatory mechanisms of the process of cell growth and division [1]. In the United States alone, 600,000 people die from cancer each year and 1.7 million new cases are diagnosed [2]. Reducing mortality, increasing survival, improving patients, quick diagnosis, and then timely and specific treatments are the keys to success in cancer treatment [3]. Early diagnosis of cancer is important because the length of the treatment period is shortened and treatment costs are reduced. In addition, some cancers are aggressive and asymptomatic in their early stages, and their rapid detection can be very critical [4]. Microscopic imaging techniques (optical and fluorescence), despite having advantages such as inexpensiveness, non-invasive, and ease of use, also have limitations, among which the emission of fluorescence from cancer cell proteins and also the short time of fluorescence emission. The use of small fluorescent dyes (such as 5-aminolevulinic acid, methylene blue, and indocyanine green) can be an entrepreneur in this regard, but cytotoxicity, photobleaching, and rapid clearance through the lymph system limit their use [5]. Clinical methods of cancer imaging include X-ray computed tomography (CT), magnetic resonance imaging (MRI), positron emission tomography

(PET), etc., which, despite providing appropriate morphological data of cancer cells and tissue, but in cases where the cancer tissue is very similar to healthy tissue, the resolution and contrast are not appropriate [6, 7]. Therefore, the development of probes that overcome the challenges associated with signal intensity, stability, and tissue penetration will be essential for more extensive clinical implementation. QDs, often described as "artificial atoms," are semiconductor nanocrystals that are in the nanoscale region in all dimensions (so-called zero-dimensional nanostructures) and have a size less than 10 nm [8]. Tunable emission into the NIR region, broad excitation range, narrow emission, and a large Stokes shift, high photoluminescence quantum yield, long photoluminescence lifetime and compatible with biomolecular functionalization and the EPR effect among the attractive optical characteristics of QDs as fluorescent imaging probes [2]. Participation in reactions as catalysts or energy acceptor, or direct oxidation and simultaneous energy transfer are among the mechanisms of QDs to enhance signal intensity [9]. Furthermore, light emission for broad-spectrum excitation of the QDs, emits light at a longer wavelength and enhances tissue penetration [10]. In addition, due to the large surface area of QDs, they can covalently link to biorecognition molecules, such as peptides, antibodies, nucleic acids, or small-molecule ligands for further application as fluorescent probes [11, 12]. Molecular beam epitaxy (MBE), ion implantation, electron-beam lithography, X-ray lithography, wet-chemical, and vapor-phase methods are common methods of quantum dot synthesis, which fall into two general categories: top-down and bottom-up [5]. In the last years, QDs have been extensively studied in biosensing, in vitro diagnosis, cancer treatment, bioimaging, drug delivery, etc. [13–16]. Currently, QDs are widely used for optical imaging of cancer cells, which has great significance for clinical diagnosis. Herein, we briefly introduce typical QDs (semiconductor QDs, carbon-based QDs, chalcogenide QDs and black phosphorus QDs), review the different synthesis methods and modifications, and analyze their cytotoxicity. Finally, the applications of QDs in cancer cell imaging are prospected.

2. Semiconductor QDs (SQDs): synthesis methods, modification, cytotoxicity, and application in cancer cell imaging

This class of QDs is usually core-shell (**Figure 1**), in the primary classification, the biocompatible fluorescent core consists of elements from groups II–IV, for example, CdSe, or groups III–V, for example, InP and is covered by a semiconductor shell which improves efficiency and photostability (ZnS, ZnSe). Finally, a variable outer layer such as silica can offer a large surface area for surface functionalization that enables their dispersion in water, and functionalized with other biomolecules (like peptides, antibodies, nucleic acids, and small-molecule ligands) to target specific proteins expressed on the surface of the cancer cells [12, 18, 19].

The methods of synthesizing SQDs are different depending on the used material, desired size, quantum yield and their applications [20]. Synthesis through high temperature, synthesis through γ-Irradiation, microwave-assisted method, sol–gel technique, and core-shell technique are among the common methods for synthesis of SQDs. In the high-temperature method, the NaHTe solution (mixture of potassium tellurite and sodium borohydride which were previously heated under N2 protection) was injected into the N2-saturated $Cd2^{+}$–MSA–alginate precursor. Then, the reaction mixture was heated to reflux (100°C) under atmospheric conditions with a condenser attached for different time intervals. Finally, the as-prepared CdTe alginate QDs were precipitated with ethanol, collected via centrifugation and dispersed in distilled water [21]. In the

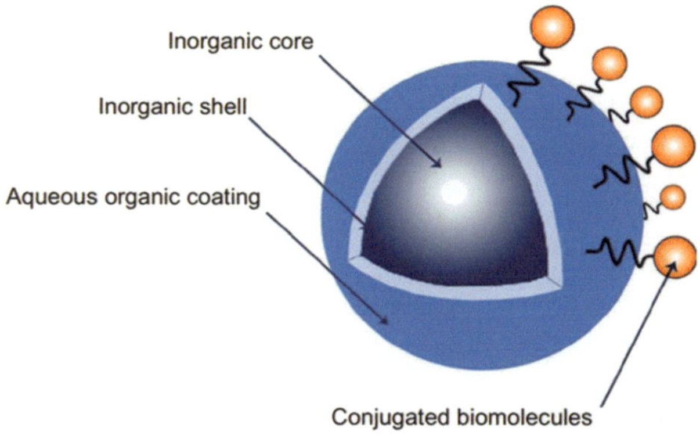

Figure 1.
Schematic diagram of SQDs structure; QDs consist of an inorganic (fluorescent) core, an inorganic shell, and aqueous organic coating [17].

gamma irradiation method, the aqueous solution containing CdCl2 and SeO2 is aerated under N2. Then the solution is irradiated with gamma ray. After irradiation, the synthesized quantum dots are separated by centrifugation at 10,000 for 5 min [22]. In the microwave-assisted method, the 3-mercaptopropionic acid is added to an N2-saturated CdCl2 solution (solution 1). Then the desired pH is adjusted using NaOH (about 11). After that, the freshly prepared NaHTe solution is added to solution 1 and exposed to microwave irradiation at 120°C for 2–5 min to obtain monodispersed CdTe core QDs. Next, the GSH is added to an N2-saturated CdCl2 solution (solution 2) and adjusted pH to 11 by NaOH. Then freshly prepared NaHSe solution is added to solution 2 and exposed to microwave irradiation at 60°C for 3–20 min to obtain CdTe/CdSe core/shell QDs [23]. In the sol–gel technique, the materials exist in solution form and grow at low temperature to form a wet gel. This technique is utilized for the formation of Au, Cu, and Zn QDs [20]. In the core-shell technique, synthesis is done by the formation of metal droplet precursor and intermediate growth on the surface of quantum dots [20].

In order to increase the luminous efficiency and stability of QDs, some modifications are made on their surface. Organic modification of SQDs which include organic ligand, organic polymer, proteins, and other organics modifiers. Inorganic modifications that include elements such as Co^{2+}, Ni^{2+}, Mn^{2+}, and Cu^{2+} doped into SQDs [24]. SQDs, especially Cadmium (Cd)-containing QDs, despite having effects such as high sensitivity and strong stability, but due to cytotoxicity, especially in normal cells, they are not very effective in cell imaging [25]. The released Cd ions are highly toxic and lead to cell damage. In addition, the small size of nanoparticles is also very important, where studies have shown that CdTe-QDs with a size of 2.2 nm are more toxic than large particles with a size of 5.2 nm [26]. Furthermore, different studies indicated that CdTe QDs cause damage to the DNA of HUVEC cells by inducing the production of reactive oxygen species (ROS) [27]. Thus, this type of QDs due to having heavy metals cannot be used for clinical treatment. Therefore, Copper indium sulfide ($CuInS_2$, CIS) QDs proposed as a non-toxic and potential alternative. CIS QDS with a ZnS shell (CIS/ZnS) have a high photoluminescence quantum yield that can be easily transferred in aqueous medium and facilitate their application in biological imaging [25]. Liu et al. synthesized CuInS2/ZnS QDs as a near-infrared (NIR) fluorescence nanoprobe. They used Arg-Gly-Asp (RGD)-labeled bovine serum albumin-poly (ε-caprolactone)-coated CuInS2/ZnS QDs to further evaluate the cytotoxicity and in

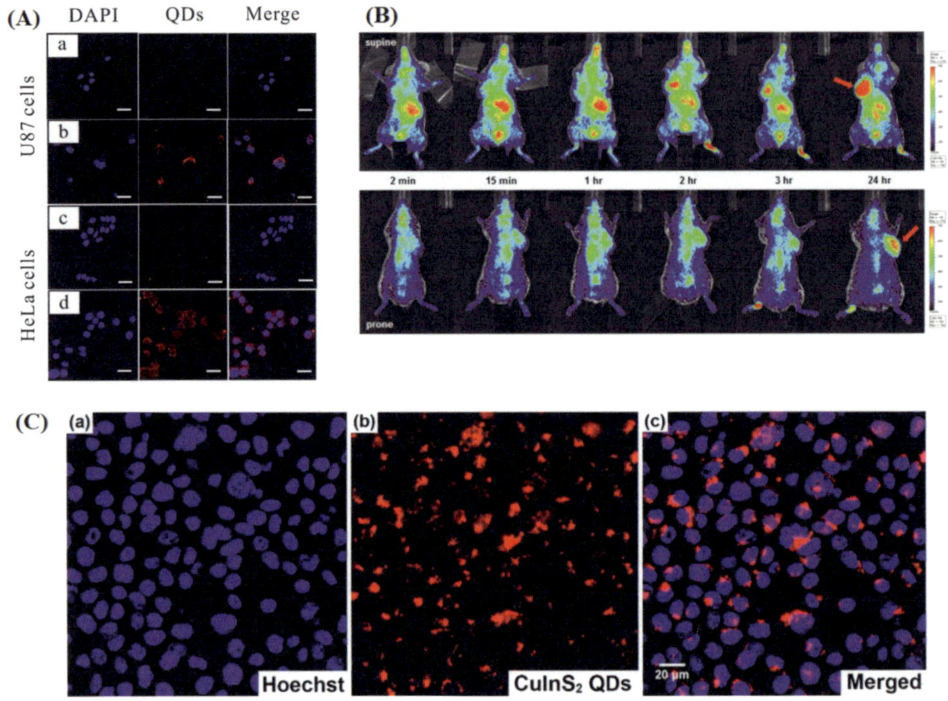

Figure 2.
Application of $CuInS_2$ QDs in cancer cell imaging. (A) Confocal fluorescence image of U87 and HeLa cells treated with nontargeting nanoprobe (a, c), and treated with cRGD-functionalized nanoprobe (b, d). reprinted with permission from [26]. (B) NIR fluorescence images of RR1022 tumor-bearing mouse after intravenous injection with cRGDyk-GCM-QDs. reprinted with permission from [27]. (C) Confocal fluorescence image of MCF10CA1a breast tumor cells treated with GSH-capped CuInS2/ZnS QDs. reprinted with permission from [28].

vitro tumor targeting in U87 and HeLa cells. As shown in **Figure 2A**, the presence of cRGD significantly increases the internalization of QDs in two kinds of cancer cells. In addition, BSA as an outer shell QDs, significantly reduced non-specific cellular binding and improved biocompatibility [28]. In another study, Kim et al. synthesized Cd-free high-quality CuInS2/ZnS core/shell QDs and used for in vivo tumor targeting in RR1022 cancer cell xenograft mice (**Figure 2B**). Their results showed that after intravenous injection of cRGDyk-GCM-QDs, the NIR fluorescence emission increased with time in the supine and prone positions of the mice which indicates high tumor uptake [29]. In the study conducted by Zhao et al., the ability to use GSH-capped CuInS2/ZnS QDs in cancer cell imaging was investigated (**Figure 2C**). Their results indicated that the CuInS2/ZnS QDs can enter the cell and most of them were in the cytoplasm [30]. Previous studies have shown that CIS QDs without ZnS shells rapidly degrade and cause significant toxicity in blood chemistry, organ weight, and histology. Therefore, more caution should be taken in clinical practice [31].

3. Carbon-based QDs: synthesis methods, modification, cytotoxicity, and application in cancer cell imaging

Carbon-based quantum dots include a wide group, which is discussed more in this chapter on carbon quantum dots (CQDs) and graphene quantum dots (GQDs)

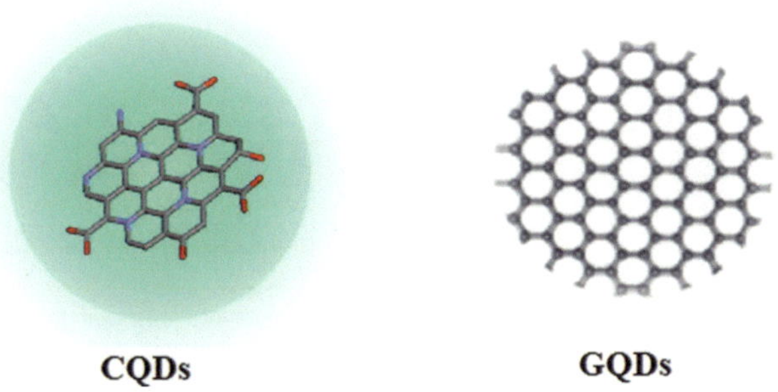

Figure 3.
Schematic diagram of the structure of carbon-based QDs.

(**Figure 3**). CQDs, also called carbon dots, are a new category of carbon nanomaterials with a size below 10 nm that were first discovered in 2004 [32]. The shape of CQDs, is spherical and has a crystal lattice with surface chemical groups, which possess quantum confinement effect (QCE) and intrinsic state luminescence [33]. A major feature of quantum dots is the QCE, which occurs when quantum dots are smaller than their exciton Bohr radius. According to the results of previous studies, it seems that the small size (below 10 nm) and the increase in the thickness of the shell in quantum dots create a strong confinement effect, which ultimately increases the amount of luminescence [34–36]. In addition to having excellent optical properties, CQDs are less cytotoxic, environmental, and biohazardous than traditional semiconductor quantum dots. Moreover, in addition, CQDs have good water solubility, chemical stability, and photobleaching resistance, ease of surface functionalization, and large-scale preparation [37]. This group of quantum dots is widely used in the field of biosensing and bioimaging [38, 39]. GQDs is one of the most attractive and newest members of the graphene family, which have an exposed graphene network

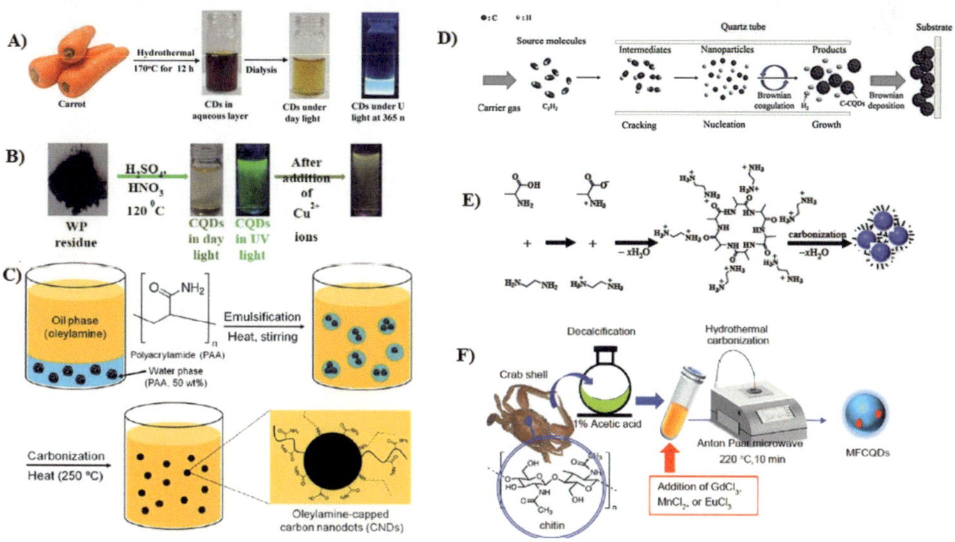

Figure 4.
The typical approaches for the synthesis of CQDs. (A) hydrothermal method, (B) chemical oxidation, (C) emulsion-templated carbonization, (D) chemical vapor deposition, (E) hydrothermal method, (F) microwave-assisted pyrolysis synthesis method. Reprinted with permission from [36].

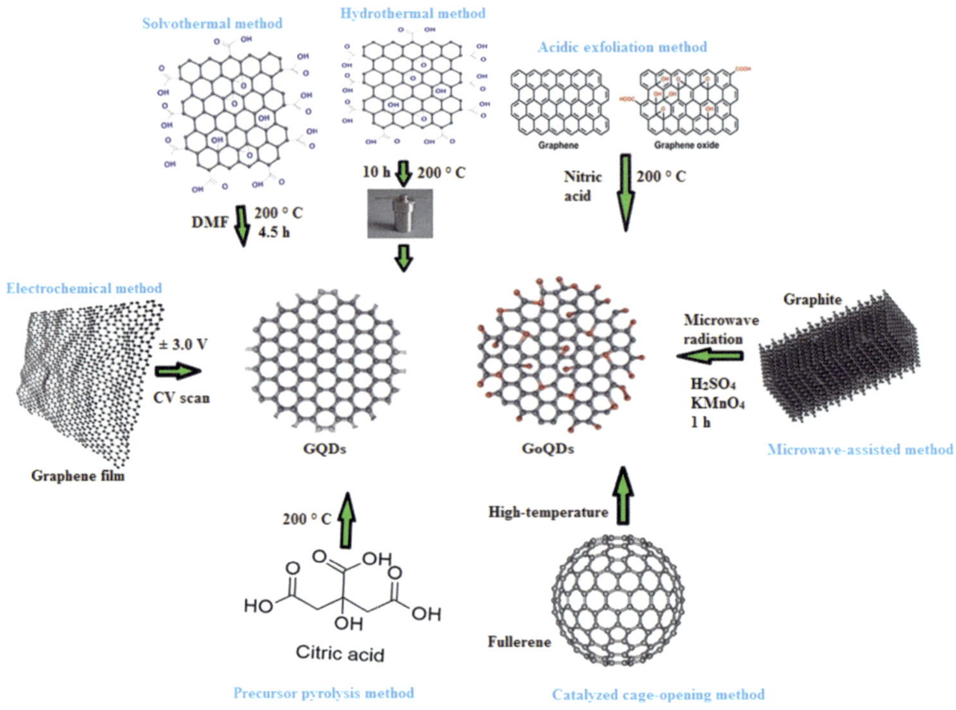

Figure 5.
The typical approaches for the synthesis of GQDs [5].

and are composed of single or multiple sheets of graphene fragments [40]. This type of QDs has many applications in the field of biomedical, including cell imaging [41].

The synthesis methods of both quantum dots are almost similar and include: acidic exfoliation method, laser ablation, hydrothermal method, solvothermal method, electrochemical method, precursor pyrolysis, microwave-assisted synthesis chemical vapor deposition (CVD), etc. [5, 42]. The most common and important synthesis methods are shown in **Figures 4** and **5**. In CQDs, due to having oxygen-containing groups, the possibility of covalent bonding with other functional groups is very high. Covalent bonding by chemical agents such as amine groups is a current approach for surface modification of CQDs [43]. In the case of GQDs, surface functionalization with folic acid, arginine-glycine-aspartic acid (RGD) and polyethyleneimine (PEI) were performed in recent years [5].

Previous studies in the in vitro and in vivo showed that both QDs at low concentrations have little cytotoxicity even if are synthesized from toxic ingredients [44]. The size and concentration of quantum dots are two important factors that affect cytotoxicity [45]. For example, it has been reported that GQDs with a concentration of less than 50 μg/ml have less than 10% cytotoxicity, while at a concentration of 200 μg/ml, they have more than 20% cytotoxicity [46].

Carbon-based quantum dots are used in various fields such as environment, energy, sensing, and imaging. Wang et al. synthesized polymer-coated nitrogen-doped carbon nanodots by direct solvothermal reaction. The CQDs were stable and water soluble with a particle size in the range of 5–15 nm. The prepared quantum dots did not show obvious cytotoxicity. In the in vivo study that was performed on the glioma-bearing nude mice,

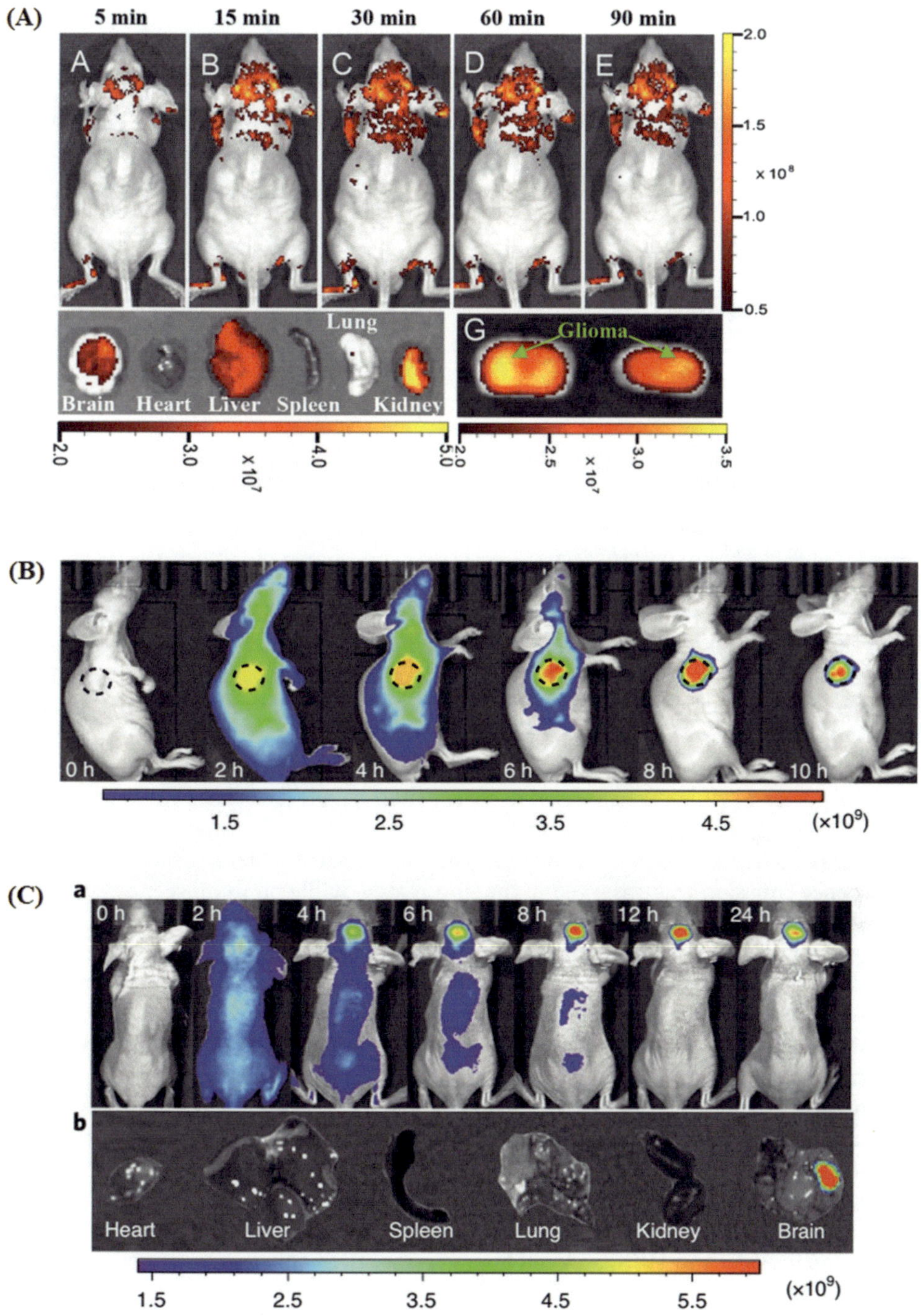

Figure 6.
Application of CQDs in cancer cell imaging. (A) In- and ex-vivo imaging of glioma-bearing mice intravenously administered with the pN-CNDs. reprinted with permission from [41]. (B) NIR fluorescence images of a representative mouse bearing a HeLa tumor that received intravenous injection of LAAM TC-CQDs at the indicated time points. (C) In- and ex-vivo NIR fluorescence images of a representative U87-tumor-bearing mouse (a) and indicated organs and tumor (b) after intravenous injection of LAAM TC-CQDs at the indicated time points. Reprinted with permission from [42].

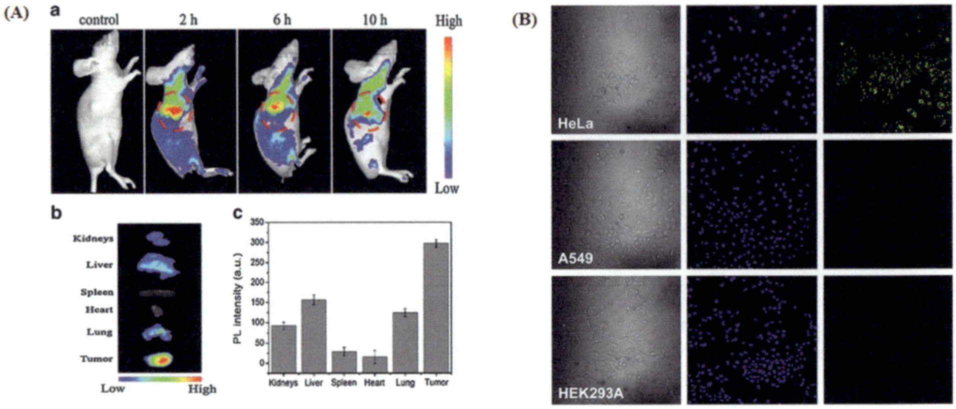

Figure 7.
Application of GQDs in cancer cell imaging. (A) In vivo imaging of HeLa tumor-bearing nude mice after injection of (GQD/DBM)3EuPhen/GQD (5 mg/kg) (a). Ex vivo images of isolated organs of mice at 10 h after injection (b) and PL intensities of (GQD/DBM)3EuPhen/GQD from isolated organs (c). Reprinted with permission from [43]. (B) Confocal laser scanning microscopy of Hela, A549, and HEK293A cells incubated with GQD–FA. Reprinted with permission from [44].

the high fluorescence emission was observed at 30-min post injection of the pN-CNDs (**Figure 6A**). They stated that surface coating of the QDs with a hydrophilic polymer, in addition to increasing the accumulation of the particles in the glioma, extended the circulation time in the bloodstream, which increases the chance of binding to the target tumor site [47]. In another study Li and co-worker showed that CQDs functionalized with multiple paired α-carboxyl and amino groups that bind to the large neutral amino acid transporter 1 (which is expressed in most tumors), selectively accumulate in human tumor xenografts in mice and in an orthotopic mouse model of human glioma. They reported that 8 h after intravenous injection into mice bearing HeLa tumors, maximum fluorescence emission was observed in the tumor area, while no fluorescence was observed in other areas (**Figure 6B**). In order to investigate the capability of LAAM TC-CQDs for brain cancer imaging and treatment, the prepared CQDs injected intravenously into mice bearing U87 gliomas. Their results showed the maximum accumulation of carbon quantum dots in the brain occurs 8–12 h after injection. In addition, by euthanizing the mice after 12 h, ex vivo results showed that LAAM TC-CQDs were significantly accumulated in the brain tumor compared to other organs (**Figure 6C**) [48]. In one study, the GQD–europium complex composites was used as a probe for in vivo fluorescence imaging of HeLa tumor-bearing nude mice. They stated that the maximum amount of fluorescence emission occurs at the tumor site 2 h after injection. While fluorescence is not observed in other organs (**Figure 7A**) [49]. Wang et al. reported that Folic acid (FA)-conjugated GQDs were capable to selective imaging of Hela cancer cell in comparison to A549 and HEK293A cell line (**Figure 7B**). Their result showed that GQD–FA enter to Hela cells with FR-induced endocytosis, which is consistent with the fact that HeLa cells overexpress FR while A549 and HEK293A express FR at a low level [50].

4. Chalcogenide QDs: synthesis methods, modification, cytotoxicity, and application in cancer cell imaging

Beside to graphene and graphene oxide [51], transition metal dichalcogens (TMDs), which belong to a large family of layered compounds, have attracted a lot

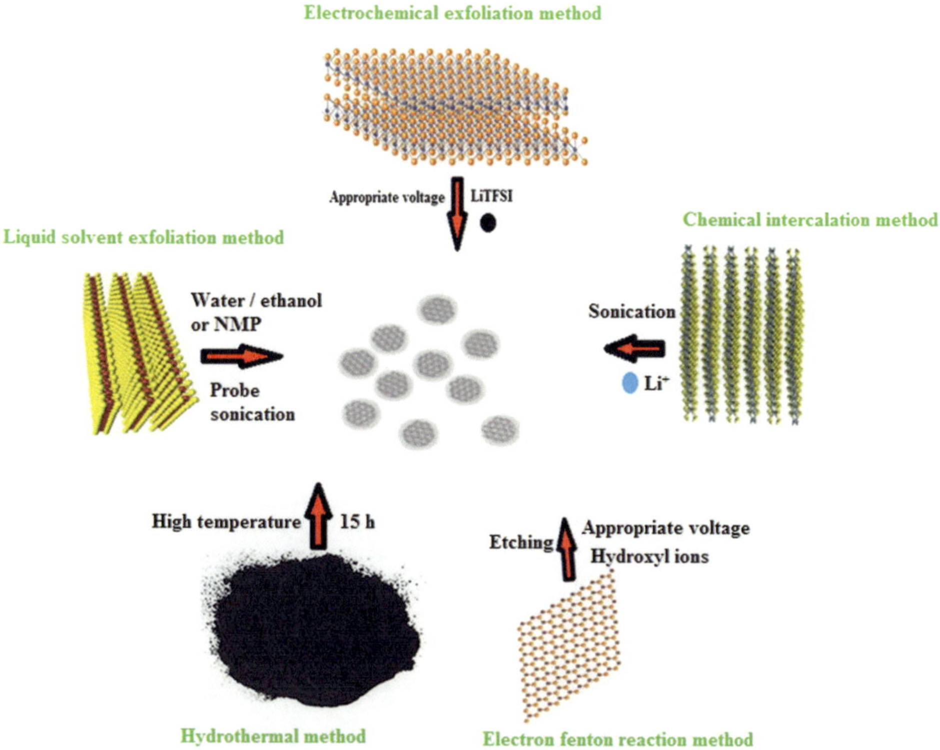

Figure 8.
The typical approaches for the synthesis of MoS2 QDs.

of attention in the field of drug delivery and imaging [52, 53]. Among TMDs, molybdenum disulfide (MoS2) is used more than in the fields of drug delivery, sensing and imaging [54, 55]. MoS2 QDs due to features such as water-solubility, biocompatibility, high stability, less toxicity, and high surface-area has attracted enormous interest [56, 57]. There are many methods for the preparation of MoS2 QDs, including chemical intercalation, electrochemical exfoliation, liquid solvent exfoliation, electron Fenton reaction, and hydrothermal reaction which are shown in **Figure 8**.

Several studies have investigated the use of MoS2 QDs in cancer cell imaging. Roy et al. used free folic acid-sensitive MoS2 QDs based "turn-off" nanoprobes for bioimaging of cancer cells. Their results showed that the FA-pretreated B16F10 cancer cells show higher population of dimmed fluorescence compared to untreated cancer cells and HEK-293 normal cells (**Figure 9A**) [58]. In another study, Liu and colleagues produced MoS2 QDs through a facile one-step and low cost and green method for bioimaging applications. They reported that blue luminescence was observed inside the HeLa cells, indicating that the QDs had penetrated the cell and mainly localized in the cytoplasm region (**Figure 9B**). In addition, they stated that molybdenum disulfide quantum dots were not toxic on cells and the morphology of cells remained normal [59]. Shi and co-worker introduced a Bottom-up hydrothermal approach for the preparation of MoS2 QDs with Na2MoO4 2H2O as molybdenum source and GSH as sulfur source. The in vitro result showed that blue fluorescence was observed in the cytoplasm of SW480 cells, which indicating that GSH-MoS2 QDs has successfully entered the cell. In the in vivo study, the GSH-MoS2 QDs were injected into the mice with

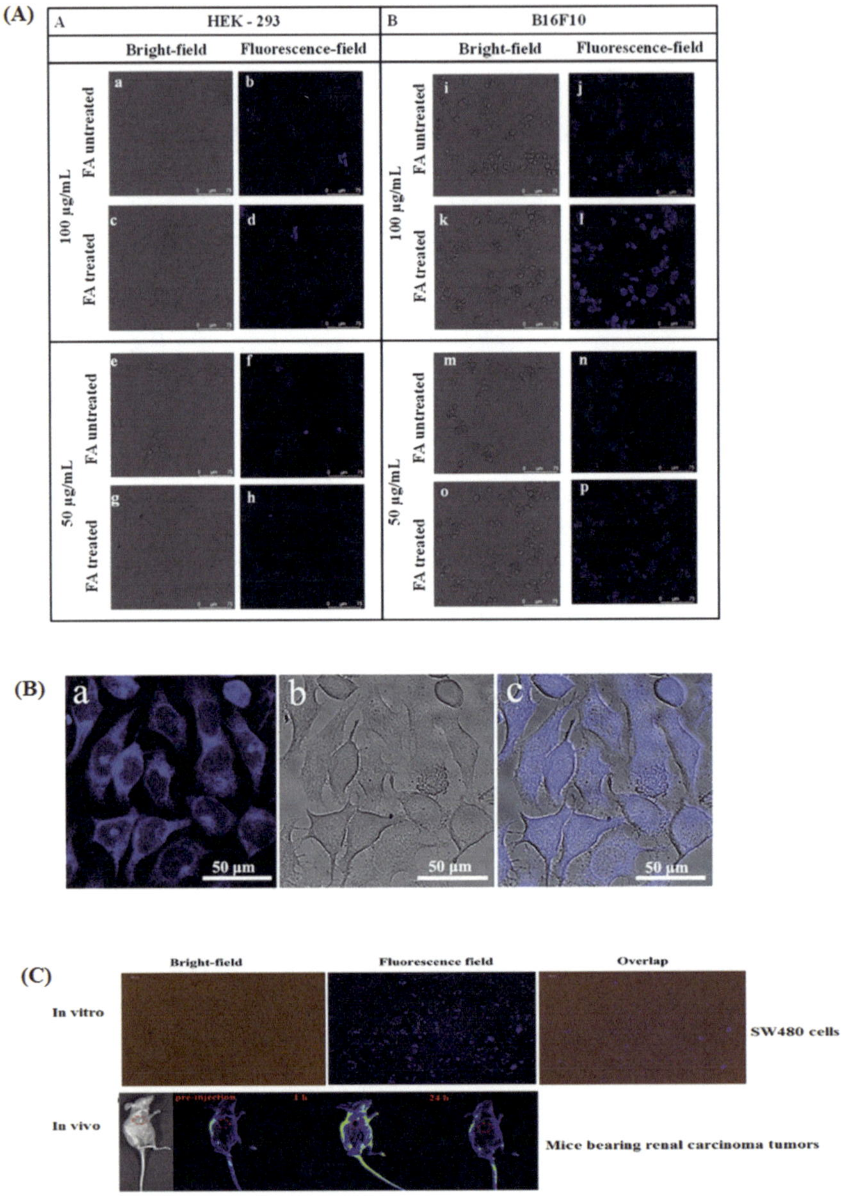

Figure 9.
Application of MoS2 QDs in cancer cell imaging. (A) Confocal microscopic images of HEK-293 and B16F10 cells after treating MoS2 QDs in FA-pretreated and untreated cells. Reprinted with permission from [51]. (B) Confocal fluorescence microphotograph of HeLa cells incubated with MoS2 QDs. Reprinted with permission from [52]. (C) In vitro fluorescence imaging of SW480 tumor cells with MoS2QDs; in vivo fluorescence images of mice bearing renal carcinoma tumors pre- and post-injection (1 h and 24 h) of MoS2QDs. The red circle in pictures indicated tumor region. Reprinted with permission from [53].

colon cancer via the tail vein. After the injection of QDs, the contrast of the tumor area increased and the environment around the tumor became blue (**Figure 9C**). In addition, the results of the biocompatibility test as well as histological analysis showed that GSH-MoS2 QDs do not have obvious toxicity on SW480 cells and no abnormality was observed in different tissues such as heart, kidney, and lung [60].

5. Black phosphorus QDs: synthesis methods, modification, cytotoxicity, and application in cancer cell imaging

Black phosphorus (BP) is another two-dimensional nanostructure that has attracted a lot of attention in less than a decade [52, 61]. like graphene, BP is composed of only one element (phosphorus) and is widely used in bioimaging and drug delivery due to its high biocompatibility, biological activity, biodegradability [62, 63]. Due to the high surface-to-volume ratio, BP nanosheets are highly efficient in loading anticancer drugs. In addition, with excellent photothermal conversion efficiency, BP nanosheets generate heat locally under laser irradiation in the NIR region, which can be used in photothermal therapy (PTT) of cancer [64, 65]. Also, due to its unique electronic structure, BP can be used as a photosensitizer to produce singlet oxygen and be effective in photodynamic therapy (PDT) of cancer [66, 67]. Moreover, BP nanosheets can also be used for photoacoustic imaging of tumors [68]. Black Phosphorus Quantum Dots (BPQDs) are a metal-free layered semiconductor derived from BP nanosheets that were first introduced in 2015 [69, 70]. There are many ways for synthesis of BPQDs, including ultrasonic exfoliation, electrochemical exfoliation, solvothermal, milling crash, blender breaking and pulsed laser irradiation, which classified into two general categories: top-down and bottom-up methods (**Figure 10**). Since BPQDs are not very stable in aqueous environment and may react

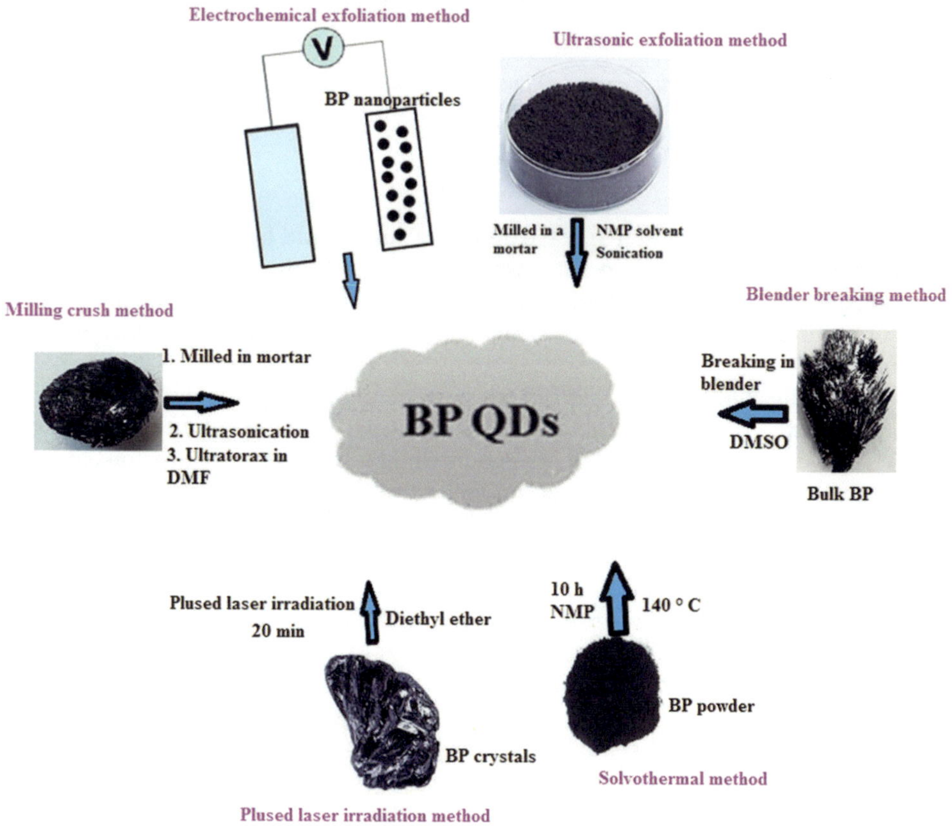

Figure 10.
The typical approaches for the synthesis of BPQDs [5].

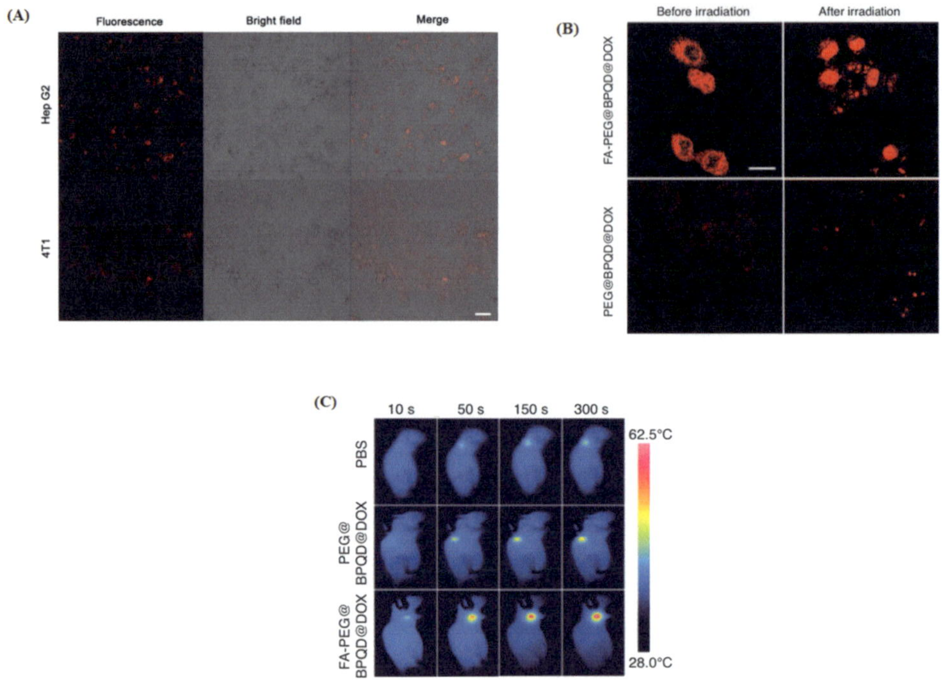

Figure 11.
Application of BPQDs in cancer cell imaging. (A) fluorescence images of Hep G2 and 4 T1 cells incubated with RdB/PEG-BPQDs. reprinted with permission from [57]. (B) In vitro imaging of 293 T cells treated with FA-PEG@BPQD@DOX and PEG@BPQD@DOX. (C) In vivo thermal images of mice with 293 T tumors irradiated at 808 nm. reprinted with permission from [58].

with oxygen and oxidize, their surface modification should be done. Polymer modified black phosphorus quantum dot, assembly off quantum dot in device, quantum dot molecule complex, quantum dot doping film and quantum dot nanosheet hybrid are among the common modifications that can be performed on black phosphorus quantum dot [5].

Several studies have shown that BPQDs have negligible cytotoxicity. However Gui et al. reported that BPQDs at a concentration up to 200 mg/mL had significant apoptotic effects on HeLa cells [69]. This type of QDs has many applications in fluorescence, photoacoustic, and thermal imaging of cancer cells [5]. In one study Li et al. reported that in the presence of RdB/PEG-BPQDs, the distinct fluorescence signals were observed in Hep G2 cells and 4 T1 cells (**Figure 11A**). Their observations also showed that the QDs were located almost exclusively in the cytoplasm, and no visible fluorescence was observed in cancer cells without the nanoprobes [71]. Wang and colleagues synthesized BPQDs modified with poly ethylene glycol and folic acid for cancer imaging. There in vitro experiment showed that the fluorescence was initially observed in the cytoplasm and was observed in the nucleus (**Figure 11B**). In addition, the thermal image indicated that after 4 h intravenous injection of FA-PEG@BPQD@DOX, the temperature of tumor site was reached to 56.8°C, which is sufficient for killing tumor cells (**Figure 11C**) [72].

6. A brief comparison between carbon-based QDs and BPQDs

Carbon-based QDs and BPQDs both consist of only one type of element (carbon and phosphorus) and comparing their efficiency in cancer cell imaging can be very important. The results of various studies have shown that both types of quantum dots have a high ability in cancer cells imaging [73, 74]. However, concentrations of BPQDs used for cancer cell imaging are lower than carbon-based QDs [74, 75]. In this context, two hypothesis are raised: First, it is likely that BPQDs increase the intensity of the PL signal to a greater extent than carbon-based QDs. Second, the cytotoxicity of BPQDs is higher than that of carbon-based QDs, and they should be used in low concentrations. However, due to the few studies, especially on BPQDs, these two hypotheses cannot be decisive, and the need for more studies in this field will help to clarify the issue more.

7. Conclusion

In this chapter, various QDs were introduced and in addition to expressing methods of synthesis, modification and cytotoxicity, the bioimaging applications (*in vitro and in vivo*) were discussed. SQDs due to having heavy metal cannot be used for clinical application. But, nontoxic $CuInS_2$ QDs are suitable alternatives for SQDs in biological imaging. Beside these QDs, carbon-based QDs, MoS2 QDs, and BP QDs have low toxicity and fluorescence emission in the NIR region of the light spectrum. The easy synthesis and the possibility of modification are other advantages of these QDs compared to SQDs. However, since studies have shown that QDs are not toxic at low concentrations, more research is needed to evaluate their blood circulation, cytotoxicity, etc. for bioimaging application. In addition, different methods of synthesis and modifications may affect the optical properties as well as the biocompatibility of QDs, and for this reason, more investigations should be done in this field. In the end, it should be noted that the use of safe QDs in clinical imaging techniques such as CT and MRI seems very likely in the near future.

Conflicts of interest

There are no conflicts to declare.

Author details

Salar Khaledian[1,2,3], Mohadese Abdoli[1,2], Reza Fatahian[3,4*] and Saleh Salehi Zahabi[4,5*]

1 Faculty of Innovative Science and Technology, Department of Nanobiotechnology, Razi University, Kermanshah, Iran

2 Nano Drug Delivery Research Center, Health Technology Institute, Kermanshah University of Medical Sciences, Kermanshah, Iran

3 Clinical Research Development Center, Taleghani and Imam Ali Hospitals, Kermanshah University of Medical Sciences, Kermanshah, Iran

4 Department of Neurosurgery, School of Medicine, Kermanshah University of Medical Sciences, Kermanshah, Iran

5 Radiology and Nuclear Medicine Department, Kermanshah University of Medical Sciences, Kermanshah, Iran

*Address all correspondence to: rfatahian@gmail.com and sszahabi@gmail.com

References

[1] Fitzmaurice C et al. Global, regional, and national cancer incidence, mortality, years of life lost, years lived with disability, and disability-adjusted life-years for 32 cancer groups, 1990 to 2015: A systematic analysis for the global burden of disease study. JAMA Oncology. 2017;**3**(4):524-548

[2] McHugh KJ et al. Biocompatible semiconductor quantum dots as cancer imaging agents. Advanced Materials. 2018;**30**(18):1706356

[3] Alshehri S et al. Progress of cancer nanotechnology as diagnostics, therapeutics, and theranostics nanomedicine: Preclinical promise and translational challenges. Pharmaceutics. 2020;**13**(1):24

[4] Smith RA et al. Cancer screening in the United States, 2018: A review of current American Cancer Society guidelines and current issues in cancer screening. A Cancer Journal for Clinicians. 2018;**68**(4):297-316

[5] Khaledian S et al. Applications of novel quantum dots derived from layered materials in cancer cell imaging. FlatChem. 2021;**27**:100246

[6] Kubota SI et al. Whole-body profiling of cancer metastasis with single-cell resolution. Cell Reports. 2017;**20**(1):236-250

[7] James ML, Gambhir SS. A molecular imaging primer: Modalities, imaging agents, and applications. Physiological Reviews. 2012;**92**(2):897-965

[8] Alivisatos AP. Semiconductor clusters, nanocrystals, and quantum dots. Science. 1996;**271**(5251):933-937

[9] Chen H et al. Quantum dots-enhanced chemiluminescence: Mechanism and application. Coordination Chemistry Reviews. 2014;**263**:86-100

[10] Mason EA, Lopez R, Mason RP. Wavelength shifting of chemiluminescence using quantum dots to enhance tissue light penetration. Optical Materials Express. 2016;**6**(4):1384-1392

[11] Nurunnabi M et al. In vivo biodistribution and toxicology of carboxylated graphene quantum dots. ACS Nano. 2013;**7**(8):6858-6867

[12] Shao L, Gao Y, Yan F. Semiconductor quantum dots for biomedicial applications. Sensors. 2011;**11**(12): 11736-11751

[13] Chung S, Revia RA, Zhang M. Graphene quantum dots and their applications in bioimaging, biosensing, and therapy. Advanced Materials. 2021;**33**(22):1904362

[14] Fan H y, et al. Graphene quantum dots (GQDs)-based nanomaterials for improving photodynamic therapy in cancer treatment. European Journal of Medicinal Chemistry. 2019; **182**:111620

[15] Pandey S, Bodas D. High-quality quantum dots for multiplexed bioimaging: A critical review. Advances in Colloid and Interface Science. 2020;**278**:102137

[16] Nair A et al. Natural carbon-based quantum dots and their applications in drug delivery: A review. Biomedicine & Pharmacotherapy. 2020;**132**:110834

[17] Madani SY et al. Conjugation of quantum dots on carbon nanotubes for medical diagnosis and treatment. International Journal of Nanomedicine. 2013;**8**:941

[18] Zhou W, Coleman JJ. Semiconductor quantum dots. Current Opinion in Solid State and Materials Science. 2016;**20**(6):352-360

[19] García de Arquer FP et al. Semiconductor quantum dots: Technological progress and future challenges. Science. 2021;**373**(6555):eaaz8541

[20] Gidwani B et al. Quantum dots: Prospectives, toxicity, advances and applications. Journal of Drug Delivery Science and Technology. 2021;**61**:102308

[21] Parani S, Lakshmi BS, Pandian K. Biopolymer encapsulation of CdTe quantum dot for In vitro controlled drug delivery release of 6-mercaptopurine. In: Advanced Materials Research. Trans Tech Publications Ltd. 2012;**584**:258-262

[22] Kang B et al. Synthesis of green CdSe/chitosan quantum dots using a polymer-assisted γ-radiation route. Radiation Physics and Chemistry. 2008;**77**(7):859-863

[23] Chen X et al. Microwave-assisted synthesis of glutathione-capped CdTe/CdSe near-infrared quantum dots for cell imaging. International Journal of Molecular Sciences. 2015;**16**(5):11500-11508

[24] Wang L et al. Semiconducting quantum dots: Modification and applications in biomedical science. Science China Materials. 2020;**63**(9):1631-1650

[25] Xu Q et al. Quantum dots in cell imaging and their safety issues. Journal of Materials Chemistry B. 2021;**9**(29):5765-5779

[26] Gao X et al. In vivo cancer targeting and imaging with semiconductor quantum dots. Nature Biotechnology. 2004;**22**(8):969-976

[27] Parak WJ, Pellegrino T, Plank C. Labelling of cells with quantum dots. Nanotechnology. 2005;**16**(2):R9

[28] Liu Z et al. Facile construction of near infrared fluorescence nanoprobe with amphiphilic protein-polymer bioconjugate for targeted cell imaging. ACS Applied Materials & Interfaces. 2015;**7**(34):18997-19005

[29] Kim E-M et al. Facile synthesis of near-infrared CuInS2/ZnS quantum dots and glycol-chitosan coating for in vivo imaging. Journal of Nanoparticle Research. 2017;**19**(7):1-12

[30] Zhao C et al. Small GSH-capped CuInS2 quantum dots: MPA-assisted aqueous phase transfer and bioimaging applications. ACS Applied Materials & Interfaces. 2015;7(32):17623-17629

[31] Kays JC et al. Shell-free copper indium sulfide quantum dots induce toxicity in vitro and in vivo. Nano Letters. 2020;**20**(3):1980-1991

[32] Xu X et al. Electrophoretic analysis and purification of fluorescent single-walled carbon nanotube fragments. Journal of the American Chemical Society. 2004;**126**(40):12736-12737

[33] Li H et al. Water-soluble fluorescent carbon quantum dots and photocatalyst design. Angewandte Chemie International Edition. 2010;**49**(26):4430-4434

[34] Zaini MS et al. Quantum confinement effect and photoenhancement of photoluminescence of PbS and PbS/MnS quantum dots. Applied Sciences. 2020;**10**(18):6282

[35] Kim S et al. Anomalous behaviors of visible luminescence from graphene quantum dots: Interplay between size and shape. ACS Nano. 2012;**6**(9):8203-8208

[36] Zhu S et al. Photoluminescence mechanism in graphene quantum dots: Quantum confinement effect and surface/edge state. Nano Today. 2017;**13**:10-14

[37] Yang S et al. Large-scale fabrication of heavy doped carbon quantum dots with tunable-photoluminescence and sensitive fluorescence detection. Journal of Materials Chemistry A. 2014;**2**(23):8660-8667

[38] Khaledian S et al. Rapid detection of diazinon as an organophosphorus poison in real samples using fluorescence carbon dots. Inorganic Chemistry Communications. 2021;**130**:108676

[39] Zhu S et al. Highly photoluminescent carbon dots for multicolor patterning, sensors, and bioimaging. Angewandte Chemie. 2013;**125**(14):4045-4049

[40] Tajik S et al. Carbon and graphene quantum dots: A review on syntheses, characterization, biological and sensing applications for neurotransmitter determination. RSC Advances. 2020;**10**(26):15406-15429

[41] Wu X et al. Fabrication of highly fluorescent graphene quantum dots using L-glutamic acid for in vitro/in vivo imaging and sensing. Journal of Materials Chemistry C. 2013;**1**(31):4676-4684

[42] Molaei MJ. Carbon quantum dots and their biomedical and therapeutic applications: A review. RSC Advances. 2019;**9**(12):6460-6481

[43] Farshbaf M et al. Carbon quantum dots: Recent progresses on synthesis, surface modification and applications. Artificial Cells, Nanomedicine, and Biotechnology. 2018;**46**(7):1331-1348

[44] Zou W-S et al. Insecticide as a precursor to prepare highly bright carbon dots for patterns printing and bioimaging: A new pathway for making poison profitable. Chemical Engineering Journal. 2016;**294**:323-332

[45] Li K et al. Recent advances in the cancer bioimaging with graphene quantum dots. Current Medicinal Chemistry. 2018;**25**(25):2876-2893

[46] Jiang D et al. Synthesis of luminescent graphene quantum dots with high quantum yield and their toxicity study. PLoS One. 2015;**10**(12):e0144906

[47] Wang Y et al. Direct solvent-derived polymer-coated nitrogen-doped carbon nanodots with high water solubility for targeted fluorescence imaging of glioma. Small. 2015;**11**(29):3575-3581

[48] Li S et al. Targeted tumour theranostics in mice via carbon quantum dots structurally mimicking large amino acids. Nature Biomedical Engineering. 2020;**4**(7):704-716

[49] Lu H et al. Graphene quantum dots for optical bioimaging. Small. 2019;**15**(36):1902136

[50] Wang X et al. Multifunctional graphene quantum dots for simultaneous targeted cellular imaging and drug delivery. Colloids and Surfaces B: Biointerfaces. 2014;**122**:638-644

[51] Abdoli M et al. Polyvinyl alcohol/gum tragacanth/graphene oxide composite nanofiber for antibiotic delivery. Journal of Drug Delivery Science and Technology. 2020;**60**:102044

[52] Khaledian S et al. Two-dimensional nanostructure colloids in novel nano drug delivery systems. Colloids and Surfaces A: Physicochemical and Engineering Aspects. 2020;**585**:124077

[53] Khaledian S et al. Electrospun nanofiber patch based on gum tragacanth/polyvinyl alcohol/

molybdenum disulfide composite for tetracycline delivery and their inhibitory effect on Gram+ and Gram- bacteria. Journal of Molecular Liquids. 2021;**334**:115989

[54] Liu J et al. Multifunctional MoS2 composite nanomaterials for drug delivery and synergistic photothermal therapy in cancer treatment. Ceramics International. 1 Aug 2022;**48**(15): 22419-22427

[55] Zhang W et al. Tumor acidity and near-infrared light responsive drug delivery MoS2-based nanoparticles for chemo-photothermal therapy. Photodiagnosis and Photodynamic Therapy. 2022;**38**:102716

[56] Peng Y et al. Determination of folic acid via its quenching effect on the fluorescence of MoS2 quantum dots. Microchimica Acta. 2019;**186**(9):1-8

[57] Li L et al. Facile synthesis of MoS 2 quantum dots as fluorescent probes for sensing of hydroquinone and bioimaging. Analytical Methods. 2019;**11**(26):3307-3313

[58] Roy S et al. Targeted bioimaging of cancer cells using free folic acid-sensitive molybdenum disulfide quantum dots through fluorescence "turn-off". ACS Applied Bio Materials. 2021;**4**(3):2839-2849

[59] Liu Q, Hu C, Wang X. A facile one-step method to produce MoS 2 quantum dots as promising bio-imaging materials. RSC Advances. 2016;**6**(30):25605-25610

[60] Shi M et al. "Bottom-up" preparation of MoS2 quantum dots for tumor imaging and their in vivo behavior study. Biochemical and Biophysical Research Communications. 2019;**516**(4):1090-1096

[61] Zou B et al. Photothermal-healing, and record thermal stability and fire safety black phosphorus–boron hybrid nanocomposites: Mechanism of phosphorus fixation effects and charring inspired by cell walls. Journal of Materials Chemistry A. 2022;**10**(27):14423-14434

[62] Tao W et al. Black phosphorus nanosheets as a robust delivery platform for cancer theranostics. Advanced Materials. 2017;**29**(1):1603276

[63] Qiu M et al. Biocompatible and biodegradable inorganic nanostructures for nanomedicine: Silicon and black phosphorus. Nano Today. 2019;**25**:135-155

[64] Shao J et al. Biodegradable black phosphorus-based nanospheres for in vivo photothermal cancer therapy. Nature Communications. 2016;**7**(1):1-13

[65] Zhang W et al. Phycocyanin-functionalized black phosphorus quantum dots enhance PDT/PTT therapy by inducing ROS and irreparable DNA damage. Biomaterials Science. 2021;**9**(15):5302-5318

[66] Liu J et al. Dual-triggered oxygen self-supply black phosphorus nanosystem for enhanced photodynamic therapy. Biomaterials. 2018;**172**:83-91

[67] Qi F et al. Photosynthetic cyanobacteria-hybridized black phosphorus nanosheets for enhanced tumor photodynamic therapy. Small. 2021;**17**(42):2102113

[68] Chen H et al. Nanocomposite of Au and black phosphorus quantum dots as versatile probes for amphibious SERS spectroscopy, 3D photoacoustic imaging and cancer therapy. Giant. 2021;**8**:100073

[69] Gui R et al. Black phosphorus quantum dots: Synthesis, properties,

functionalized modification and applications. Chemical Society Reviews. 2018;**47**(17):6795-6823

[70] Ding H et al. Black phosphorus quantum dots as multifunctional nanozymes for tumor photothermal/catalytic synergistic therapy. Nano Research. 2022;**15**(2):1554-1563

[71] Li Y et al. Multifunctional nanoplatform based on black phosphorus quantum dots for bioimaging and photodynamic/photothermal synergistic cancer therapy. ACS Applied Materials & Interfaces. 2017;**9**(30):25098-25106

[72] Wang J et al. PEGylated-folic acid–modified black phosphorus quantum dots as near-infrared agents for dual-modality imaging-guided selective cancer cell destruction. Nano. 2020;**9**(8):2425-2435

[73] Chen H et al. Ultra-stable tellurium-doped carbon quantum dots for cell protection and near-infrared photodynamic application. Science Bulletin. 2020;**65**(18):1580-1586

[74] Li Z et al. Polydopamine-functionalized black phosphorus quantum dots for cancer theranostics. Applied Materials Today. 2019;**15**:297-304

[75] Bouzas-Ramos D et al. Carbon quantum dots codoped with nitrogen and lanthanides for multimodal imaging. Advanced Functional Materials. 2019;**29**(38):1903884

Chapter 10

Applications of Quantum Dots in the Food Industry

Shabnam Sistani and Hajar Shekarchizadeh

Abstract

Quantum dots (QDs) are spherical particles with a size of <10 nm and, due to their unique properties, have good potential for use in the food industry. Among the various QDs, food industry researchers have highly regarded carbon quantum dots (CQDs) due to their nontoxicity and environmental friendliness. Food analysis is essential for quality assessment as well as safety control. In this regard, QDs-based fluorescence sensors can provide faster, more accurate, more sensitive, and cheaper analysis methods. The use of QDs to detect food additives, pathogens, heavy metals, nutrients, antibiotics, and insecticide residues is investigated in this chapter. QDs in packaging materials, due to their antioxidant, antimicrobial, and inhibitory properties, increase product shelf life, reduce the growth of microorganisms, improve mechanical properties, prevent gases and UV light, and reduce food waste. Their application in improved, active, intelligent, and bio-packaging will also be described. Then, their application in water treatment will be discussed. QDs, due to properties such as high aspect ratio, reactivity, electrostatic, hydrophilic, and hydrophobic interactions, have good potential for use in various water treatment methods, including membranes in filtration, adsorbents, and photocatalysts. Finally, their use to track protein will be investigated.

Keywords: carbon quantum dots, water purification, smart packaging, fluorescence, UV barrier, molecularly imprinted polymer

1. Introduction

Quantum dots (QDs), which are spherical particles with a size of <10 nm, are classified based on the materials used into semiconductor quantum dots and carbon-based quantum dots [1]. Semiconductor quantum dots are three-dimensional nanoparticles composed of inorganic elements like heavy metals and non-heavy materials (such as cadmium, selenide, and zinc sulfide). Semiconductor quantum dots are prepared by a bottom-up approach. They have a spherical shape, crystal structure, and size of fewer than 6 nm, which have quantum mechanical properties due to their small size. Size-dependent photoluminescence, excitation-independent photoluminescence, long life, and good photostability are specific features of QDs. Quantum dots are highly stable against photobleaching and can be considered an alternative to fluorescent index dyes and proteins. Semiconductor quantum dots are one of the most popular nanoparticles of choice for fluorescent labels, detection, labeling, and tracking [1]. One of the disadvantages of semiconductor quantum dots based on heavy metals is their toxicity, which leads to environmental and health concerns. Graphene quantum dots (GQDs) have been considered in various research fields, including the food industry, due to their lower toxicity and environmental friendliness. The small size of GQDs and their high oxygen content lead to reduced toxicity. The GQDs refer to a 0D graphene plate with a size of less than 10 nm. GQDs have good potential for practical applications due to their specific characteristics, such as excellent luminescence, functional groups, and surface defects. They are widely used as fluorescent probes for several analytes because of their significant properties such as photobleach resistance, low cytotoxicity, good biocompatibility, stable photoluminescence, and high water solubility. Photoluminescence is the most interesting feature of GQD, which can be easily changed by managing its size, surface performance, and chemical doping. In addition, GQDs contain carboxyl, carbonyl, epoxy, and hydroxyl groups that can bind to various biological molecules such as proteins, antibodies, enzymes, and so on [2].

Carbon quantum dots (CQDs) were introduced as environmentally safe alternative nanomaterials that can provide satisfactory optical properties, good biocompatibility, and nontoxicity [3]. CQDs were first discovered in 2004 by Zhu et al. during the purification of single-walled carbon nanotubes. The size of these nanoparticles is less than 10 nm. Like common semiconductor quantum dots, their major advantage over organic dyes is their high optical stability. Also, unlike other QD, CQDs usually have low toxicity and good biocompatibility [4]. In terms of chemical structure, carbon atoms (sp^2 and sp^3) are located as nuclei in amorphous shells consisting of functional groups. Many hydroxyl and carboxyl functional groups are present on the surface of carbon dots, which cause good solubility in water and ease of functionalization by different species. The amount of oxygen in CQDs varies from 5 to 50 wt%, depending on the synthesis method [5].

There is a lot of research on the use of quantum dots in various fields, but few articles report the use of quantum dots in food science. The purpose of this chapter is to review the application of quantum dots in the food industry.

2. QDs in food analysis

The development of modern agriculture and the food processing industry can lead to chemical and biological food contamination. Food analysis is essential for quality assessment as well as safety control. Common methods of analysis include spectroscopy, immunoassay, culture and colony counting, chromatography, nuclear magnetic resonance, and electrochemical methods. The advantages of these methods include reproducibility, sensitivity, and high selectivity. However, these methods require expensive equipment, trained personnel, and long separations [6]. So, finding faster, more accurate, more sensitive, and cheaper methods of analysis has always been the focus of researchers and food manufacturers. In this regard, fluorescence sensors based on QDs can play a key role in ensuring food quality and safety [7]. Various strategies have been developed to improve the performance and selectivity of sensors based on QDs. The following are the most common techniques for food analysis:

a. **QD/CQD@MIP:** New and effective analytical methods in nanomaterial-based food samples include QDs and CQD. One effective approach to increase the selectivity of these nanomaterials is their combination with a molecularly imprinted polymer (MIP). Combining the unique optical properties of QD/CQD with a high selection of MIPs is a significant advantage of QD/CQD@MIP. In most cases, the adsorption process in MIP proportionally reduces the intensity of QD/CQD fluorescence, making it valuable for quantitative analysis, so it has promising potential to build sensors with high stability, sensitivity, and selectivity response. Preparation of QD/CQD@MIP conjugates consists of two main steps. In the first step, fluorescent QD/CQD are synthesized as nuclei. Then, the prepared fluorescent nucleus is used directly for QD/CQD@MIP synthesis. In some cases, the surface modification process is performed during or after the synthesis of fluorescent dots. The QD/CQD modification process provides appropriate functional groups or compatibility with synthetic medium for the better embedding of fluorescent nanoparticles in the MIP. Methods used to prepare the nucleus for the synthesis of QD/CQD@MIPs include silylation, treatment with mercaptoperboxylic acids, acrylic acids, oleic acid, polyethylene glycol, polyethylene imine, and L-cysteine. The second step is the MIP printing process, which often uses three polymerization techniques, including the sol-gel method, free radical polymerization, and reverse microemulsion method [3].

b. **MNPs@QDs:** Food pollutants are one of the main factors jeopardizing food safety, and with biological accumulation, they are considered as a serious threat to human health. Currently, there are several instrumental analytical techniques for identifying different food contaminants, which, due to having disadvantages such as the need for trained people, complex pretreatment process, and long time, make them unsuitable for quick and on-site detection. MNPs@QDs nanocomposites have good application potential for the analysis of food parameters. Magnetic nanoparticles (MNPs) have a strong magnetic response and are quickly recovered under an external magnetic field, so they have good potential for easy control of adsorption and release of target analytes at low concentrations in complex food samples. MNPs are environmentally friendly because they can be recycled multiple times. In addition to taking advantage of the magnetic properties of MNP, magnetic fluorescent QDs nanocomposites (MNPs@QDs) have QDs-induced fluorescence properties that can rapidly enrich target analytes and separate complex food matrices under a magnetic field. These nanocomposites are also capable of quantitative analysis of analytes, thus greatly simplifying the pretreatment process, which prevents analyte wastage, reduces detection time, and increases detection efficiency [8].

2.1 Detection of food additives

Additives are widely used in food processing to improve flavor and shelf life and enhance nutritional value. Additives have always been abused and are one of the potential causes of food safety problems. It is necessary to develop a rapid and reliable diagnostic method to strengthen the monitoring of food additives [9]. **Table 1** provides an overview of recent studies on using QDs to detect food additives.

2.2 Detection of pathogen

One of the main causes of death in the world is infectious diseases caused by food. Infection caused by food pathogens causes great financial losses to the industry due to delays in product distribution and market recall, in addition to threatening consumer

QDs	Analytes	LOD	Linear range	Samples	Reference
Sulfur. Quantum dots	Tartrazine	39 nM	0.1–20 μM	Vitamin water, orange juice	[10]
GQDs	Sunset Yellow	7.6 nM	2.5 nm to 25 μM	—	[11]
CQDs	Nitrite	0.15 μmol/L	0.50–50 μmol/L	Corn sausage, ham sausage, pickle, and hot dog	[12]
CQDs	Nitrite	0.47 μM	5–80 μM	—	[13]
CQDs	Borax	8.0 $\mu mol{\cdot}L^{-1}$	0.025–10 $mmol{\cdot}L^{-1}$	Flour, bread, instant noodle	[14]
MoS_2 quantum dots	Allura red	1.7×10^{-6} M	?	?	[15]
Perovskite quantum dots	Rhodamine 6G	0.01 μg/mL	0–10 μg/mL	Tap water, food sample	[16]
Au@CQDs	Melamine	3.6 nM	1 μM to 10 μM	Milk	[17]
Fe_3O_4@SiO_2@CdSe QDs	Bisphenol A	0.34 nM	10^{-9}–10^{-4} M	Water	[18]

Table 1.
Application of QDs in the detection of food additives.

health. A common method for detecting pathogens is culturing a specific species and then examining its biochemical and immunological properties, which is very time consuming [19]. For this reason, fluorescence nanosensors based on QDs, which are faster in detecting pathogens, can be effective. The research done in this field is presented in **Table 2**.

2.3 Detection of heavy metals

The presence of heavy metal ions such as lead, mercury, cadmium, and arsenic in food is attributed to water and soil pollution. Heavy metals cause irreversible changes in protein structure and negatively affect cell function. Excessive consumption can cause side effects such as neurological disorders, kidney damage, and bone damage [27].

In one study, fluorescent nanoprobes were used to determine the multiplicity of Hg^{2+}, Cu^{2+}, and Ag^{+} ions. Nanoprobes (CQDs-QDx) were designed by mixing CQDs and multicolor CdTe QDs. CQDs were not sensitive to heavy metal ions. At the same time, CdTe QDs showed a size-dependent fluorescence response to different heavy metal ions, thus creating a ratiometric detection scheme by measuring the fluorescence intensity ratios of CQDs-QDx systems. By evaluating the detection performance, CQDs-QDx (x = 570, 650, and 702) were successfully used to differentiate and quantify Hg^{2+}, Cu^{2+}, and Ag^{+} ions. In addition, they also detected heavy metal ions in actual samples with acceptable results [28]. **Table 3** lists other studies for the detection of heavy metals.

2.4 Detection of insecticide and antibiotic residues

Insecticides are widely used in agriculture to control weeds and pests and also to improve food production. Excessive use of insecticides can lead to contamination of

QDs	Analytes	LOD (CFU/mL)	Linear range (CFU/mL)	Samples	Reference
Graphene oxide quantum dots	*P. aeruginosa*	100	1.28×10^3–2.00×10^7	Drinking water, orange juice, and popsicle	[20]
CQDs@MNPs	*E. coli*	487	500–106	Milk	[21]
Fe_3O_4@CS@CQDs	*S. typhi*	1.38×10^2	10^3–10^6	Lettuce	[22]
Fe_3O_4@SiO_2@QDs	*E. coli*	2.39×10^2	2.5×10^2–5×10^5	Milk	[23]
Fe_3O_4@ CdZnTe QDs	*L. monocytogenes*	1	1–10^9	Tap water,pasteurized milk	[24]
MnO_2@QDs	*S. typhi*	43	1.0×10^2–1.0×10^7	Chicken, meats	[25]
CQDs@ gold nanoparticles	*S. aureus*	10	10^8–10^1	Milk, orange juice	[26]

Table 2.
Application of QDs in the detection of pathogen.

the target environment and, consequently, food contamination. Negative effects of insecticides on humans include neurotoxicity, endocrine disorders, mutagenicity, and carcinogenicity. Therefore, it is necessary to have a sensitive and selective method for analyzing these compounds [36]. A ratiometric fluorescent sensor was synthesized using CQDs in a study to detect acephate, an organophosphate pesticide. The fluorescent quenching mechanism is related to the inner filter effect (IFE). The limit of detection of this method was 0.052 ppb, which is much lower than the standards for acephate from the EU. The ratiometric fluorescent sensor was successfully evaluated for the detection of OPs in tap water and pear samples [37].

Antibiotics are widely used in agriculture, animal husbandry, aquaculture, and pharmaceutical industries. If left unattended, they can harm human health through the food chain and the environmental cycle and can threaten food quality as a potential hazard. Penicillin, for example, causes allergic skin reactions. Streptomycin damages the kidneys and auditory nerve, and tetracycline causes liver damage and yellowing of the teeth. Therefore, it is necessary to develop antibiotic detection methods that have a lower limit of detection than the legal limit [38]. Tetracycline is an antibacterial compound widely used in food-producing animals to treat various bacterial infections due to its low cost and high practicality. Residues of this antibiotic in food products enter the human body through the food chain and cause allergic reactions, digestive disorders, dizziness, muscle pain, and headaches. On the other hand, excessive use of this antibiotic causes the formation of antibiotic-resistant strains. The common methods of determining the residues of tetracyclines include high-performance liquid chromatography and enzyme-linked immunosorbent assays (ELISA). The disadvantages of these methods include a long time and low sensitivity. Therefore, it seems necessary to develop an extremely sensitive method for determining the content of tetracycline. Wang et al. synthesized tungsten oxide quantum dot for tetracycline detection. The fluorescence of the synthesized QDs is quenched after

QDs	Analytes	LOD	Linear range	Samples	Reference
Nitrogen-doped carbon dots	Hg^{2+}	0.24 μM	0.3–2.0 μM	Beverage	[29]
CQDs	Fe^{3+}	9.55 μM	30–600 μM	Aqueous Solution	[30]
GQDs	Fe^{3+}	0.28 μM	0–100 μM	Tap and drinking water	[31]
Si, N-CQDs	Cr(VI)	0.995 μM	0–200 μM	—	[32]
Fe_3O_4@CQDs	Hg^{2+}	0.3 nM	0.003–0.01 μM	Lake/tap water and drinks	[33]
$CaCO_3$-Fe_3O_4 -$AgInS_2$/ZnS QDs	Co^{2+}, Ni^{2+}, Pb^{2+}	10, 100, 100 nM	No data	Water	[34]
Fe_3O_4@SiO_2-NH_2/ CQDs	Cu^{2+}	0.16 μM	0–80 μM	Water	[35]

Table 3.
Application of QDs in the detection of heavy metals.

the addition of tetracycline due to synergism of IFE, Förster resonance energy transfer (FRET), and photoinduced electron transfer (PET). They successfully tested the sensor to detect tetracycline with a high recovery rate in milk and milk powder [39]. **Table 4** lists some recent research to detect insecticide and antibiotic residues in food.

2.5 Detection of nutritional components

Numerous studies have been performed to use fluorescence sensors to detect nutrients such as ascorbic acid in fruits and vegetables [49], vitamin B_{12} [50], retinoic acid [51], and others. One of the common mechanisms in the detection of nutrient components is the fluorescence of QDs covered by metal ions and then the fluorescence recovery by the analyte. In other words, electrostatic interactions between metal ions and the surface groups of QDs cause electron transfer and fluorescence coating. By adding an analyte, fluorescence is restored due to analyte chelation and metal ions. This mechanism has been used to detect glutathione and thiamine in food [9].

3. QDs in food packaging

Food packaging is an essential component of the food supply cycle that acts as a shield or barrier against contamination, the external environment, and mechanical damage during transport, and its main purpose is to ensure the quality, health, integrity, and safety of the product. In addition to maintaining product quality, packaging systems also help reduce waste [52]. The use of nanoparticles in food packaging is a new technology that has received much attention due to its many benefits. Nanoparticles as fillers in packaging materials increase product shelf life, reduce the growth of microorganisms, improve mechanical properties, and block gases and UV light. These packages are a good alternative to nonbiodegradable plastic packaging materials due to their high-performance nanostructures and low weight.

QDs	Analytes	LOD	Linear range	Samples	Reference
Fe_3O_4@SiO_2@CQDs	4-nitrophenol	23.45 nM	0.08–10 μM	Water/fish	[40]
CQDs @ MIPs	Lambda-cyhalothrin	0.5 μg kg^{-1}	1–150 μg kg^{-1}	Vegetables and tea	[41]
Sulfur-doped graphene quantum dot	Omethoate	0.001 ppm	—	—	[42]
QDs-doped COFs@ MIP	Nereistoxin	1.60 μg L^{-1}	5–100 μg L^{-1}	Tap water	[43]
Cerium-nitrogen co-doped carbon quantum dots	Tetracycline	0.25 μM	0.4–35 μM	Milk and pork	[44]
Nitrogen-doped graphene quantum dots	Tetracycline	9.735×10^{-13} M	2.5×10^{-10}–5×10^{-6} M	Whole milk, skim milk, and honey	[45]
Graphene quantum dots (GQDs) and palladium nanoparticles (Pd NPs	Tetracyclines	45 ng.mL^{-1}	100–500 ng.mL^{-1}	Raw milk	[46]
CQDs	Tetracycline	0.36 μM	0.5–30 μM and 30–90 μM	Milk, egg, and chicken	[47]
PGr/CdTe QDs/Fe_3O_4@SiO_2/MIPs	Cefoperazone	0.09 ng/mL	0.1–25 ng/mL	Milk	[48]

Table 4.
Application of QDs in the detection of insecticide and antibiotic residues in food.

Food packaging is classified into four groups based on the application of nanoparticles: improved, active, smart, and bio-based packaging [53].

3.1 Improved packaging

The main purpose of this type of packaging is to use nanomaterials applied inside polymeric materials to improve the mechanical properties of the packaging, UV barrier, and the effect on permeability to water vapor and oxygen [54].

3.1.1 Improving the mechanical properties of packaging

Nanoparticles are the potential candidates for the synthesis of polymer nanocomposites due to their small size, high surface-to-volume ratio, high strength, and extremely high surface activity. Polymers reinforced with nanoparticles are very suitable for various applications such as food packaging due to high strength, low weight, and low cost [55]. In one study, CQDs were used to modify the polyvinyl alcohol matrix and cellulose nanofibers. The results showed that the addition of carbon dots to the matrix increases the tensile modulus and tensile strength of the composite due to the formation of hydrogen bonds, van der Waals force, and other chemical bonds between carbon surfactants, polyvinyl alcohol, and cellulose nanofibers. Electron microscopy images also showed that the carbon-modified film had a denser refractive index than the control film, indicating better composition and bonding in the film with carbon dots [56]. Yu and colleagues synthesized a film based on carbon dots and carboxymethylcellulose (CMC). The tensile strength of the synthesized film was 55% higher than that of CMC film, which indicates that the addition of carbon dots improves the mechanical properties of the film, which is attributed to the strong interaction between the carbon dots and the film [57].

3.1.2 UV barrier

Many nanomaterials can block UV light due to the combination of scattering and absorption effects, so they are used as a light blocker in packaging materials and protect light-sensitive foods. Research has shown that carbon dots are very effective in absorbing UV light. Although the mechanism of blocking UV light at carbon dots is not yet fully understood, some attribute this feature to functional groups on the surface of carbon dots. Transition electrons are transferred from n to π^* or π to π^*, which are mainly supplied by different functional groups, which in turn depend on the precursor used and the synthesis method. In films with carbon dots, intramolecular proton transfer in the excited state by O ... H ... O, O ... H ... N tunnels and abundant conjugate structures provide optical stability and film-blocking properties against UV light [58]. Polyvinyl alcohol is a semicrystalline polymer that is widely used in packaging due to its advantages, such as linear and strong structure, nontoxicity, biocompatibility, and thermal stability. One of its disadvantages is the passage of light, which limits its use in the packaging of light-sensitive products. CQDs were synthesized in one study and combined with the polyvinyl alcohol polymer drying method. The results showed that the ratio of UV light transmission is inversely proportional to the amount of CQDs, so it can be concluded that CQDs improve the inhibitory properties of the polymer against UV light [56].

3.1.3 Impact on gas and water vapor permeability

One of the important features of food packaging materials is permeability to oxygen and water vapor because they affect chemical reactions and microbial growth and thus food safety and quality. The addition of carbon dots to polymer films affects the permeability to oxygen and water vapor, the extent of which depends on the molecular and physicochemical properties of the carbon dots and polymer. Amphiphilic carbon dots are suitable for controlling the wettability and permeability of the polymer layer. Doping carbon dots and radical polymerization can be used to prevent phase separation of carbon dots in polymer networks and thus improve film permeability. The permeability of canola protein films increases after the addition of carbon dots with hydrophilic groups on their surfaces. Also, chitosan films with improved water absorption properties have been made using different types and amounts of carbon dots [58].

On the other hand, CQDs can increase water resistance, which is attributed to the reduction of layer porosity and the abundance of carboxyl groups at the surface of carbon dots, which result in a reduction in free radicals due to the reaction between hydroxyl and carboxyl. The hydroxyl functional group leads to a decrease in water absorption of the film. In one study, CQDs were used to improve the properties of polyvinyl alcohol/nanocellulose layers. By increasing the amount of carbon dots from 0.1 to 4 ml, the water absorption of the film decreased from 119.6 to 42%, which indicates the improvement of the film barrier property compared with moisture using CQDs [59].

3.2 Active packaging

Food spoilage causes environmental pollution, devastating impact on human health, economic losses, and increased treatment costs; therefore, technology development is needed to reduce food waste and improve food security. One possible way to reduce food waste and spoilage is to develop active ingredients for active packaging to increase product shelf life. Active packaging includes coatings with antimicrobial and antioxidant properties. These agents can be included in conventional nondegradable packaging or used in combination with biodegradable components [54].

3.2.1 Antimicrobial effect

Increasing resistance of microorganisms to antibiotics and other disinfectants has led food industry researchers to look for alternative antimicrobial approaches [51 m]. In one study, sulfur quantum dots (SQDs) were used to prepare food packaging film. The films showed strong antioxidant and antibacterial activity against bacterial food pathogens (*Escherichia. coli* and *Listeria monocytogenes*) and fungi (*Aspergillus niger* and *Penicillium chrysogenum*). When the film was used as a bread wrapping test, the film prevented mold growth for 14 days [60].

CQDs have shown great potential in killing and inhibiting bacteria, fungi, viruses, and drug-resistant species. The mechanisms of this property include the adhesion of CQDs to the bacterial surface, destruction of the bacterial membrane or cell wall, induction of oxidative stress through RNA/DNA degradation, and oxidative induction of proteins and other intracellular biological molecules. The most important killer effect of CQDs on microorganisms is due to the production of reactive oxygen species, which leads to the production of hydroxyl free radicals and single oxygen, which destroys vital biomolecules in the cell, and inactivates intracellular proteins, mitochondrial dysfunction, and lipid peroxidation, and eventually, the cell dies [61]. Ezzati et al. used sulfur-modified CQDs to modify the gelatin/pectin film in food packaging. The resulting film showed strong antimicrobial properties against foodborne pathogens such as *E. coli* and *L. monocytogenes* [62]. In another study, nitrogen-doped CQDs were synthesized and used to modify cellulose nanofibers. The obtained film showed high antibacterial and antifungal activity due to the production of reactive oxygen species. The prepared film was used for packaging tangerine and strawberry fruits. The results showed that the growth of fungi on the fruit surface was effectively inhibited, and their shelf life was increased by 10 and 2 days, respectively, indicating its good potential for use in active packaging [63].

Fan et al. used a combination of CQDs and chitosan to inhibit microorganisms in the packaging of freshly cut cucumbers. CQDs established a strong hydrogen bond with chitosan. The results showed that the diameter of the inhibitory zone against *E. coli* and *Staphylococcus aureus* improved with the increasing concentration of CQDs. In addition, the total number of bacteria, molds, and yeasts decreased during storage [64].

3.2.2 Antioxidant properties

ROSs, including single oxygen and free radicals (hydrogen peroxide, hydroxyl, and superoxide), play an important role in food spoilage, chemical degradation, polymers, and the destruction of biological structures. Therefore, the presence of antioxidant compounds or free radical scavengers in food and food packaging is essential for good health. Recently, a range of carbon nanoparticles such as nitrogen-doped carbon dots, selenium-doped carbon dots, and nitrogen/sulfur-doped carbon dots have been turned into antioxidants to remove ROS. The antioxidant properties of carbon dots are attributed to electron transfer, unpaired electrons due to surface defects, hydrogen donor behavior, doping element, and surface functional groups. The carbon dots synthesized in the NaOH electrolysis medium have more active oxygen-containing groups such as carbonyl and hydroxyl, which can be used as hydrogen donors to scavenge free radicals, resulting in higher antioxidant capacity. In addition, recent studies have shown that carboxyl and amine groups directly or indirectly through the hydrogen atom transfer (HAT) mechanism cause the removal of ROS and thus increase the antioxidant activity of CQDs. In this mechanism, hydrogen or electron is transferred from carboxyl and amine groups to active species such as ROS and deactivates them [65].

3.3 Smart packaging

The purpose of this type of packaging is to monitor the condition of the food or the surrounding environment. In this technology, a visual indicator gives the supplier or customer information about information such as product freshness, packaging leakage, storage temperature throughout the production chain, and corruption. Nanoparticles are used as reactive particles in packaging materials to inform the status of the package. In other words, they are nanosensors that respond to environmental changes (such as temperature, humidity, oxygen level, and microbial contamination). When nanosensors are integrated into food packaging, they can indicate specific chemical compositions and product freshness [66].

3.3.1 Product freshness indicator

This is done using intelligent detection technology and labels that can measure changes in environmental conditions inside the package. Among the many methods, smart labels have received much attention due to their advantages, such as accurate results, high sensitivity, and ease of use. The color change of the smart label attached to the packaging container indicates the freshness of the product. By monitoring the color change of the indicator, it can be seen that the food is unsuitable for consumption. For example, in marking the freshness of meat packaging, the basic principle is that the marker on the label is sensitive to volatile nitrogen compounds and amines in the packaging environment or changes in ambient pH caused by such components. When such sensitive changes are detected, they can be converted to response values, which are usually color changes that can be detected by the naked eye, so that the freshness of the meat is detected in real time [65]. Cauchy and colleagues used soy protein isolate carbon dots in anthocyanin-based smart films to monitor the freshness of packaged pork. When the meat samples were stored at 25°C, the activity of microorganisms increased the amount of volatile nitrogen inside the package, followed by a change in pH. As the shelf life increased, the fresh pork deteriorated, and the color of the detector changed from purple to green [67].

3.3.2 Oxygen indicator

At first, the need for oxygen markers in food packaging may not seem necessary until we realize that the main cause of further food spoilage is oxygen. Much of this spoilage is indirect because oxygen allows a myriad of food spoilage aerobic micro-organisms to grow. Oxygen can also cause direct spoilage of many foods through enzyme-catalyzed reactions, including browning fruits and vegetables, ascorbic acid degradation, oxidation of a wide range of flavors, and nonenzymatic reactions such as lipid oxidation. During storage, changes in oxygen levels in the packaging are an important indicator of food freshness and product respiration rate. Given the key role of oxygen in food spoilage, it is not surprising that most foods are packaged in a predominantly oxygen-free environment under a modified atmosphere. The main problem with the modified atmosphere is the lack of a simple, inexpensive oxygen indicator that assures the consumer that the package is safe and that oxygen input is insignificant. At present, this level of quality assurance is not possible in modified atmospheric foods that rely solely on a routine sampling of the packaging line. Typically, one packet out of every 300–400 packets is taken out of line for testing by a technician and checked using an expensive analytical system such as FT-IR or GC, and if sufficient packages are found that are not sealed enough, all 300–400 previous food packages are considered unsafe and are destroyed or repackaged. This unfavorable situation has led to special attention to cheap, reliable, and simple oxygen markers for food packaging [68].

One of the common methods for detecting oxygen in food packaging is the use of optical sensors, which use fluorophore nanoparticles in a solid matrix. Zhu et al. used a combination of silane-activated carbon dots as oxygen-insensitive fluorophore and ruthenium dihydrochloride as oxygen-sensitive fluorophore to prepare the fibrous membrane. The fibrous membrane prepared by photoluminescence ratiometric imaging is well able to evaluate the amount of oxygen in the packaging of fruits so that the fluorescence of CQDs is at a wavelength of 400 nm and its intensity does not change with changing oxygen concentration, but the intensity of fluorescence emitted by ruthenium at 600 nm depends on the oxygen concentration. They also used a digital camera instead of an expensive optical spectrometer to monitor oxygen concentration. Then, a fiber membrane was used to wrap the grapes, which initially emitted bright blue fluorescence but changed to purple on the third day of storage due to fluorescent respiration. As the respiration process increased, the fluorescent dye turned bright red on days 6 and 7 of storage. Therefore, it can be said that these fibrous membranes have a good potential for assessing oxygen in the packaging of agricultural products [69].

3.4 Biomaterial-based packaging

Common packaging materials include plastics such as polyvinyl chloride, polystyrene, polyethylene, and polypropylene, which are widely used due to their strength, heat, and moisture resistance. The use of these materials is a threat to the environment due to their nondegradable nature. Therefore, it can be said that biodegradable materials are the best choice for packaging. These coatings and films are monolayers, bilayers, or multilayers of polysaccharides, lipids, and proteins that, due to their sensitive nature, poor mechanical properties, and poor sealing properties, cannot be completely substituted for other materials [70]. Xylan is one of the hemicelluloses and the second most abundant polysaccharide in plants, which usually has poor mechanical performance due to its relatively low molecular weight. Modifying them with carbon dots improves their filmmaking and mechanical performance. Yang et al. used carbon dots to modify the carboxymethylzylan matrix. The results showed that carbon dots, in addition to creating excellent optical properties, improve the thermal stability and mechanical strength of nanocomposite films, so that by adding 92.1 wt% of carbon dots, due to creating a chemical bond between xylan and carbon dots, the tensile strength and modulus of elasticity increase by 114.3 and 90.7%, respectively. Also, this film can effectively absorb UV light and convert it into blue light. Due to the mentioned advantages, this film has a good capability for food packaging [71].

Schmitz and colleagues synthesized a nanocomposite by combining zein and QDs. The prepared QDs were used as nanofillers to obtain zein-based nanocomposite films that showed good visual appearance, homogeneity, and clarity. The presence of QDs increased hydrophobicity and reduced the water absorption of composite films up to three times compared with pure zein. The presence of QDs in the films led to extensive UV blocking in the absorption spectrum. Antimicrobial assays showed that zinc oxide nanoparticles loaded into zein films were promising antibacterial agents, as growth inhibition of *S. aureus* reached (96.5 ± 4.9) % to 44.8 wt% of ZnO nanoparticles [72]. Grzebieniarz and colleagues synthesized films based on natural resources such as potato starch and chitosan by combining QDs and cadmium sulfide. They performed a storage experiment using poultry meat coated with films produced to assess microbiological quality. The results showed that composites restrict the growth of selected microorganisms in poultry meat [73]. Priyadarshi et al. used sulfur quantum dots (SQDs) as functional fillers for biopolymer films in food packaging. SQDs showed antimicrobial activity with almost negligible toxicity. The addition of SQD to gelatin/agar composite films also increased the UV barrier property without compromising color and clarity. The film showed excellent antioxidant activity, moderate antibacterial effect on *L. monocytogenes*, and inhibitory effect on *E. coli* [74]. Zhang et al. made functional films based on carboxymethylcellulose (CMC) by adding different amounts of CQD. The CQDs were evenly distributed in the polymer matrix to form a very clear UV-blocking film. The addition of CQD increased the tensile strength (up to 27.6%) and elastic modulus (up to 61.5%). The films showed excellent antioxidant and antimicrobial activity. Lemon fruits were coated with film solutions to test the performance of the film. Lemons coated with CMC/CQD film showed an excellent appearance without mold growth even after 21 days of storage [75].

4. Water purification

The development of industry and agriculture leads to the release of various pollutants and their mixing with underground and surface water, which is a threat to human health and the environment. On the other hand, due to the growth of the population in the world, the demand for water will increase in the coming years. Therefore, it seems necessary to purify polluted water for various uses, including drinking, agriculture, and industrial use [76]. Water pollutants include heavy metals, inorganic and organic pollutants, especially dyes, polycyclic aromatic hydrocarbons, pesticides, and pharmaceuticals. Most of the mentioned pollutants exist in very low concentrations, but their risks to humans and living organisms are very high because they are very persistent in normal environments and water treatment systems. Organic compounds are important pollutants that come out of manufacturing plants. These organic pollutants lead to the reduction of dissolved oxygen and endanger human health. Heavy metals such as lead, nickel, cadmium, essential oil, copper, manganese, chromium, and cobalt are very harmful to human health even in low concentrations, and their frequent consumption can damage the liver and brain and cause cancer [77]. The most important water purification methods are filtration, crystallization, sedimentation, gravity separation, flotation, coagulation, ionic oxidation, solvent extraction, evaporation, distillation, reverse osmosis, ion exchange, electrodialysis, electrolysis, absorption, centrifugal and membrane fluid separation, neutralization and remineralization, reduction, and oxidation. None of the mentioned methods are completely effective for providing safe drinking water. Nanotechnology has provided new solutions for water purification. Nanomaterials are suitable for adsorption, catalysis, and sensing applications due to properties such as high aspect ratio, reactivity, adjustable pore volume, electrostatic, hydrophilic, and hydrophobic interactions. Nanomaterials are widely used in various water treatment methods, including membranes in filtration, adsorbents, and photocatalysts for pollutant degradation and detection. Recently, much research has been conducted on the potential application of QDs in water treatment due to their unique properties. Of course, it should be noted that nanomaterials used for drinking water purification must also be environmentally friendly and nontoxic, because unsafe particles in contact with the human body cause severe damage to vital organs and their dimensional properties aggravate biological damage slowly [78]. There is a lot of research on the use of CQDs in water treatment. In general, the use of CQDs in water purification can be investigated from two aspects.

4.1 CQDs as membranes

Nanofiltration membranes have good potential for water purification. One disadvantage of the traditional water filtration methodology is that soluble salts and a few soluble minerals and organic substances cannot be removed. Nanotechnology provides new solutions for water purification. This is done using nanoporous polymers, nanomembranes, etc., which usually have pore sizes between 1 and 50 nm and separate most bacterium and harmful substances. Desalinization also belongs to the current methodology. The general structure of the membrane consists of many layers. In a simple membrane structure, CQDs are uniformly distributed between the dense upper layer and the porous substrate. The addition of CQDs to thin-film nanocomposite (TFN) membranes will increase membrane efficiency, water flux, power density, and water purity. Also, because of the electrostatic repulsions between the deposits and also the membrane surface, a major improvement in antifouling properties happens within the designated layer. Efficient and large surfaces and large intermediate spaces with several functional groups are the most reasons for increasing fluxes. The presence of specific hydrophilic groups in CQDs reduces the nonspecific absorption and therefore will increase the selectivity within the needed absorption of pollutants. These membranes also are utilized in reverse osmosis because of enhanced water permeability, high permeation flux, and antifouling capability. By incorporating CQDs on the membrane surfaces, all processes like desalination performance, porosity, permeability, hydrophilicity, selectivity are increased [79].

4.1.1 Thin-film nanocomposite (TFN) membranes

In recent years, many studies are conducted on modifying selected skins by adding CQDs. The CQDs disperse within the aqueous phase and then participate within the surface polymerization process to create TFN membranes. CQD-modified TFN membranes will perform higher than virgin membranes, even with a small addition of CQD in the water phase. The resulting membranes show surface hydrophilicity and better permeability while maintaining solute selectivity, excellent stability, and improved antifouling features. The performance of TFN membranes can be maximized by properly functionalizing CQDs and optimizing their value [80]. In one study, CQDs were used to make new thin-film nanocomposite (TFN) membranes. First, the amino carbon quantum dots (ACQDs) were synthesized through a simple single-pot hydrothermal method and then used in surface polymerization to fabricate the ACQD-TFN membrane. Using ACQDs in membrane led to a water flux increase of 23.2 $kg{\cdot}m^{-2}{\cdot}h^{-1}$ at 70°C during the treatment of a 10 wt% NaCl solution, which was 44% higher performance than thin-film composite membrane without modification [81].

4.1.2 CQD/polymer composite membranes

In this method, CQDs are added to polymeric dopes to form homogeneous solutions and to form mixed matrices membranes through various spinning methods. The small size of CQDs leads to better dispersion of particles in doped solutions and the formation of membranes with a uniform structure without loss of mechanical strength. One of the challenges of this method is finding a suitable solvent for CQD and polymer. Another challenge is how to control the distribution of CQD in membranes. Therefore, in addition to ensuring the long-term stability of CQDs within the membrane, practical tools for the formation of chemical bonds between CQDs and the polymer matrix must be identified [80].

Colbum's team used an ionic liquid (1-ethyl-3-methylimidazolium acetate) as the common solvent for both CQDs and cellulose to create uniform membrane properties. CQDs are bound with the cellulose domain through hydrogen bond networks, and a stable composite membrane was formed. The presence of CQDs on the surface can also make the membrane negatively charged and more hydrophilic. Also, Field Emission Scanning Electron Microscopy images of the cross section of the membrane indicate that CQDs act as pore formers and help to form membranes with higher permeability [80].

4.1.3 Membranes with CQDs on top of substrates

In this method, CQDs are coated above the membrane surface with the help of various coating agents such as polydopamine (PDA) and trimethoxy silane (3-aminopropy) (APTMS). One way to stabilize CQDs is to create a covalent bond between the oxygen-containing groups in CQDs and the amine-containing agents on the membrane surface [28]. In research, CQDs were fabricated from citric acid *via* a simple method. Subsequently, they are immobilized onto the polydopamine (PDA) layer grafted on. The carboxylic groups of CQD react with the $-NH_2$ group of polydopamine, resulting in increased resistance of the modified membrane to deposition. The modified membranes possess much enhanced antibacterial activity and anti-biofouling propensity. The continuous PRO operations at 15 bar also confirm that the CQD-modified membranes exhibit a much higher power density (11.0 vs. 8.8 W/m^2) and water recovery after backwash (94 vs. 89%) than the unmodified ones [82].

4.2 CQDs for removal of pollutants

One of the appropriate strategies for water treatment is photocatalysis. Both sunlight and UV light are employed to destroy pollutants. It is an oxidation process, which is stable and environmentally friendly for water purification. CQDs show excellent photoluminescence. They are excellent fluorescent materials and efficient photocatalysts in UV light. CQDs also are sensitive to visible light, which results in enhanced charge carriers and photocatalytic performance. CQDs transfer electrons at different positions and can reduce the recombination of light-generated charges. All the mentioned properties have made them an appropriate candidate for water purification and removal of organic and inorganic pollutants from water [79].

CQDs can be used to take away organic and inorganic pollutants Cd^{2+} and Pb^{2+} ions from wastewater by absorption treatment. N-CQDs were with success incorporated into this treatment. Another adsorbent, polyethyleneimine-functionalized CQDs onto the magnetic materials ($MnFe_2O_4$) to provide a nanocomposite (PECQDs/$MnFe_2O_4$), is applied for the removal of uranium [79].

5. Protein tracking by QDs

Proteins are the main components of food and play an important role in nutrition, formation, and maintenance of food structure. Today, the function of protein in food matrices is well known, but it is necessary to study the role of protein in different food matrices. For example, in bread dough, gluten causes the dough to be elastic and viscous and forms a matrix to hold gases in the bread. On the other hand, knowing the function and amount of gluten in the dough is not enough. In the past, gluten function was studied by extracting gluten proteins and studying the behavior of the extracted proteins, but the behavior of the extracted gluten due to its interaction with other proteins may not be similar to that found in the food matrix. Knowing the distribution of proteins leads to a better understanding of their function. One of the effective strategies in this field is labeling proteins with QDs, which leads to a better understanding of their function [83].

Gluten and zein, the most common and consumed proteins, are cereals. The main storage protein of wheat grain is gluten. Gluten is a complex mixture of hundreds of protein components, mainly gliadin and glutenin. Based on solubility in alcohol-water solutions, gluten proteins are divided into soluble gliadin and insoluble glutenins. Both parts play an important role in the rheological properties of dough, but their functions are different. Gliadins contribute to the viscosity and expandability of the dough system, while glutenin is responsible for the strength and elasticity of the dough. In terms of amino acid composition, gluten has high amounts of glutamine and proline and low content of amino acids with charged side groups [84]. The most important storage protein in corn is zein. Zein contains four components (alpha, beta, gamma, and delta) with different peptide chains, molecular sizes, and solubilities. Zein is rich in glutamic acid, leucine, proline, and alanine, but it is exceptional among vegetable proteins in terms of the lack of tryptophan. This protein is considered hydrophobic due to significant amounts of nonpolar amino acids. Also, the high proportion of nonpolar amino acids and the lack of basic and acidic amino acids lead to a decrease in the solubility of zein [85].

Suzer et al. were the first to use QDs to image and label food structures. The distribution of gluten and zein in the cereal matrix strongly affects the structure and texture. Thus, knowledge of their distribution offers new insights into how distribution affects food structure. In this study, they used nucleus/shell CdSe/ZnS QDs to image the gluten network in flatbread and zein in corn extrudates. A confocal laser scanning microscope (CLSM) was used to observe the structure of QD-labeled cereal proteins and visualize their location in the food matrix. The results showed that QDs could be covalently conjugated with gluten and zein [86]. In one study, QDs were conjugated to gliadin antibodies and used as fluorescent probes to detect gliadin proteins in dough and baked bread samples. CLSM was used to investigate QDs-gliadin antibody conjugates and obtain 3D images of the gliadin distribution in the dough and flatbread matrix. CLSM images showed significant changes in the fluorescence intensity distribution generated by the gliadin-QD conjugate with cooking time. Based on the results obtained from the dough and flatbread samples, they stated that the distribution of gliadin in different layers (top, center, and bottom) is nonuniform and the baking time and location of the layers play an important role in the distribution of flatbread gliadin protein. From the successful binding of QDs to gliadin antibodies, it can be concluded that QDs have good potential as a probe to target protein subunits in food matrices [87].

In the mentioned studies, the antibody-antigen method was used to label the proteins. When the antibodies were conjugated with the QDs, to break the disulfide bonds in the antibodies, and then crosslink the QDs to the new free-SH groups, diethritol (DTT) was used in the antibody. The antigen detection region is strong in antibodies with disulfide bonds. Therefore, the use of DTT may lead to antibody damage. A new method has been proposed for the conjugation of QDs to antibodies in which disulfide bonds are not broken performed with acetylglucosamine. The next step is the integration of azides from modified N-acetylgalactosamine monosaccharides into antibody glycans. The third step is the catalyst-free click conjugation of desferrioxamine-modified dibenzocyclooctynes to the azide-bearing sugars. And the last step is the radiolabeling of modified chelator antibodies with zirconium. In this method, there is no risk of reducing the effectiveness of the antibody protein, because the antibody binding site is not manipulated in any way [83].

6. Conclusion

Undoubtedly, quantum dot-based sensors have promising prospects for food analysis. Quantum dots are superior to fluorescent dyes due to their relatively long luminescence lifetimes, photobleaching resistance, and high quantum yield. Because food samples have different biological and chemical components, identifying a specific target without the involvement of other sample components is a challenge. This highlights the importance of using connectors that can detect a specific target, one of the most important of which is the integration of QDs with MIP, which provides sensors with a low detection limit. QDs offer many benefits for food packaging, including improved mechanical properties, better thermal stability, increased water resistance, UV barrier properties, and antimicrobial and antioxidant activity. In this review, we reviewed recent research using quantum dots in water treatment as nanomembranes, nanosensors, and absorbent photocatalysts. Despite recent advances in this area, more fundamental research on the antifouling mechanism, biofilm growth, and industrial development is needed. Although CQDs are more secure than other QDs, are within the appropriate concentration range, and are not toxic to human and animal cells, research into the toxicity, migration, and degradation of CQDs is still in its infancy, and further studies are carried out to examine toxicity, behavior. Possible migration of CQDs from sensor films or packaging to food systems and their effect on the gastrointestinal tract is required.

Author details

Shabnam Sistani and Hajar Shekarchizadeh*
Department of Food Science and Technology, College of Agriculture, Isfahan University of Technology, Isfahan, Iran

*Address all correspondence to: shekarchizadeh@iut.ac.ir

References

[1] Gidwani B, Sahu V, Shukla SS, Pandey R, Joshi V, Jain VK, et al. Quantum dots: Prospectives, toxicity, advances and applications. Journal of Drug Delivery Science and Technology. 2021;**61**:102308. DOI: 10.1016/j.jddst.2020.102308

[2] Kumar P, Dhand C, Dwivedi N, Singh S, Khan R, Verma S, et al. Graphene quantum dots: A contemporary perspective on scope, opportunities, and sustainability. Renewable and Sustainable Energy Reviews. 2022;**157**:111993. DOI: 10.1016/j.rser.2021.111993

[3] Sobiech M, Luliński P, Wieczorek PP, Marć M. Quantum and carbon dots conjugated molecularly imprinted polymers as advanced nanomaterials for selective recognition of analytes in environmental, food and biomedical applications. TrAC Trends in Analytical Chemistry. 2021;**142**:116306. DOI: 10.1016/j.trac.2021.116306

[4] Dong Y, Cai J, Chi Y. Carbon Based Dots and their Luminescent Properties and Analytical Applications. In: Yang N, Jiang X, Pang DW, editors. Carbon Nanoparticles and Nanostructures. Cham, Switzerland: Springer; 2016. pp. 161-238. DOI: 10.1007/978-3-319-28782-9_6

[5] Molaei MJ. Principles, mechanisms, and application of carbon quantum dots in sensors: A review. Analytical Methods. 2020;**12**(10):1266-1287. DOI: 10.1039/C9AY02696G

[6] Luo X, Han Y, Chen X, Tang W, Yue T, Li Z. Carbon dots derived fluorescent nanosensors as versatile tools for food quality and safety assessment: A review. Trends in Food Science & Technology. 2020;**95**:149-161. DOI: 10.1016/j.tifs.2019.11.017

[7] Ragavan K, Neethirajan S. Nanoparticles as biosensors for food quality and safety assessment. In: Lopez Rubio A, Fabra Rovira MJ, Martinez Sanz M, Gomez Gomez-Mascaraque L, editors. Nanomaterials for Food Applications. Netherlands: Elsevier; 2019. p. 147-202. DOI: 10.1016/B978-0-12-814130-4.00007-5

[8] Xiong J, Zhang H, Qin L, Zhang S, Cao J, Jiang H. Magnetic fluorescent quantum dots nanocomposites in food contaminants analysis: Current challenges and opportunities. International Journal of Molecular Sciences. 2022;**23**(8):4088. DOI: 10.3390/ijms23084088

[9] Song-Ling Y, Huang J-J, Lin L, Hui-Jun F, Yuan-Ming S, Yu-Dong S, et al. Preparation of carbon dots and their application in food analysis as signal probe. Chinese Journal of Analytical Chemistry. 2017;**45**(10):1571-1581. DOI: 10.1016/S1872-2040(17)61045-4

[10] Peng X, Wang Y, Wang Q, Tang J, Zhang M, Yang X. Selective and sensitive detection of tartrazine in beverages by sulfur quantum dots with high fluorescence quantum yield. Spectrochimica Acta Part A: Molecular and Biomolecular Spectroscopy. 2022;**279**:121454. DOI: 10.1016/j.saa.2022.121454

[11] Niu H, Yang X, Wang Y, Li M, Zhang G, Pan P, et al. Electrochemiluminescence detection of sunset yellow by graphene quantum dots. Frontiers in Chemistry. 2020;**8**:505. DOI: 10.3389/fchem.2020.00505

[12] Zhang Q, Tian F, Zhou Q, Zhang C, Tang S, Jiang L, et al. Targeted ginkgo kernel biomass precursor using eco-friendly synthesis of efficient carbon quantum dots for detection

of trace nitrite ions and cell imaging. Inorganic Chemistry Communications. 2022;**140**:109442. DOI: 10.1016/j.inoche.2022.109442

[13] Deng Y, Zhou Y, Li Q, Qian J. One-step hydrothermal synthesis of nitrogen-doped carbon dots for high-sensitivity visual detection of nitrite and ascorbic acid. Analytical Methods. 2021;**13**(33):3685-3692. DOI: 10.1039/D1AY00885D

[14] Zhou C, Xiong Z-L, Zhao L. Preparation of carbon quantum dots from tobacco leaves and its application for fluorescence detection of borax in flour and its products. SSRN. 2022. In Press. DOI: 10.2139/ssrn.4004733

[15] Coloma A, Del Pozo M, Martínez-Moro R, Blanco E, Atienzar P, Sánchez L, et al. MoS2 quantum dots for on-line fluorescence determination of the food additive allura red. Food Chemistry. 2021;**345**:128628. DOI: 10.1016/j.foodchem.2020.128628

[16] Chan KK, Yap SHK, Giovanni D, Sum TC, Yong K-T. Water-stable perovskite quantum dots-based FRET nanosensor for the detection of rhodamine 6G in water, food, and biological samples. Microchemical Journal. 2022;180:107624. DOI: 10.1016/j.microc.2022.107624

[17] Hu X, Shi J, Shi Y, Zou X, Arslan M, Zhang W, et al. Use of a smartphone for visual detection of melamine in milk based on Au@ carbon quantum dots nanocomposites. Food Chemistry. 2019;**272**:58-65. DOI: 10.1016/j.foodchem.2018.08.021

[18] Shi J, Zhang X, Zhang Q, Yang P. Ultrasensitive and highly selective detection of bisphenol a using core-shell magnetic molecularly imprinted quantum dots electrochemiluminescent probe. Bulletin of Environmental Contamination and Toxicology. 2022;**108**(2):379-385. DOI: 10.1007/s00128-021-03351-z

[19] Huang C-C, Hung Y-S, Weng Y-M, Chen W, Lai Y-S. Sustainable development of carbon nanodots technology: Natural products as a carbon source and applications to food safety. Trends in Food Science & Technology. 2019;**86**:144-152. DOI: 10.1016/j.tifs.2019.02.016

[20] Gao R, Zhong Z, Gao X, Jia L. Graphene oxide quantum dots assisted construction of fluorescent aptasensor for rapid detection of *Pseudomonas aeruginosa* in food samples. Journal of Agricultural and Food Chemistry. 2018;**66**(41):10898-10905. DOI: 10.1021/acs.jafc.8b02164

[21] Hu X, Li Y, Xu Y, Gan Z, Zou X, Shi J, et al. Green one-step synthesis of carbon quantum dots from orange peel for fluorescent detection of Escherichia coli in milk. Food Chemistry. 2021;**339**:127775. DOI: 10.1016/j.foodchem.2020.127775

[22] Guo Z, Huang X, Li Z, Shi J, Zhai X, Hu X, et al. Rapid and highly sensitive detection of Salmonella typhimurium in lettuce by using magnetic fluorescent nanoparticles. Analytical Methods. 2020;**12**(48):5861-5868. DOI: 10.1039/D0AY01744B

[23] Huang Z, Peng J, Han J, Zhang G, Huang Y, Duan M, et al. A novel method based on fluorescent magnetic nanobeads for rapid detection of Escherichia coli O157: H7. Food Chemistry. 2019;**276**:333-341. DOI: 10.1016/j.foodchem.2018.09.164

[24] Liang S, Ji L, Zhong Y, Wang T, Yang H, Li Q-L, et al. Fluorescence immunoassay for the targeted determination of trace Listeria

monocytogenes based on immunomagnetic separation and CdZnTe quantum dot indication. Analytical Methods. 2022;**14**(11):1124-1133. DOI: 10.1039/D1AY02106K

[25] Hao L, Xue L, Huang F, Cai G, Qi W, Zhang M, et al. A microfluidic biosensor based on magnetic nanoparticle separation, quantum dots labeling and mno2 nanoflower amplification for rapid and sensitive detection of salmonella typhimurium. Micromachines. 2020;**11**(3):281. DOI: 10.3390/mi11030281

[26] Pebdeni AB, Hosseini M, Ganjali MR. Fluorescent turn-on aptasensor of Staphylococcus aureus based on the FRET between green carbon quantum dot and gold nanoparticle. Food Analytical Methods. 2020;**13**(11):2070-2079. DOI: 10.1007/s12161-020-01821-4

[27] Yang T, Huang H, Zhu F, Lin Q, Zhang L, Liu J. Recent progresses in nanobiosensing for food safety analysis. Sensors. 2016;**16**(7):1118. DOI: 10.3390/s16071118

[28] He Y, Wang Y, Mao G, Liang C, Fan M. Ratiometric fluorescent nanoprobes based on carbon dots and multicolor CdTe quantum dots for multiplexed determination of heavy metal ions. Analytica Chimica Acta. 2022;**1191**:339251. DOI: 10.1016/j.aca.2021.339251

[29] Fu W-J, Peng Z-X, Dai Y, Yang Y-F, Song J-Y, Sun W, et al. Highly fluorescent N doped C-dots as sensor for selective detection of Hg2+ in beverages. Spectrochimica Acta Part A: Molecular and Biomolecular Spectroscopy. 2022;**265**:120392. DOI: 10.1016/j.saa.2021.120392

[30] Zulfajri M, Gedda G, Chang C-J, Chang Y-P, Huang GG. Cranberry beans derived carbon dots as a potential fluorescence sensor for selective detection of Fe3+ ions in aqueous solution. ACS Omega. 2019;**4**(13):15382-15392. DOI: 10.1021/acsomega.9b01333

[31] Rodwihok C, Tam TV, Choi WM, Suwannakaew M, Woo SW, Wongratanaphisan D, et al. Preparation and characterization of photoluminescent graphene quantum dots from watermelon rind waste for the detection of ferric ions and cellular bio-imaging applications. Nanomaterials. 2022;**12**(4):702. DOI: 10.3390/nano12040702

[32] Zhang J, Jing C, Wang B. A label-free fluorescent sensor based on Si, N-codoped carbon quantum dots with enhanced sensitivity for the determination of Cr (VI). Materials. 2022;**15**(5):1733. DOI: 10.3390/ma15051733

[33] Xie R, Qu Y, Tang M, Zhao J, Chua S, Li T, et al. Carbon dots-magnetic nanocomposites for the detection and removal of Hg^{2+}. Food Chemistry. 2021;**364**:130366. DOI: 10.1016/j.foodchem.2021.130366

[34] Kurshanov DA, Khavlyuk PD, Baranov MA, Dubavik A, Rybin AV, Fedorov AV, et al. Magneto-fluorescent hybrid sensor $CaCO_3$-Fe_3O_4-$AgInS_2$/ZnS for the detection of heavy metal ions in aqueous media. Materials. 2020;**13**(19):4373. DOI: 10.3390/ma13194373

[35] Dong S, Wang S, Wang X, Zhai L. Superparamagnetic nanocomposite Fe_3O_4@ SiO_2-NH_2/CQDs as fluorescent probe for copper (II) detection. Materials Letters. 2020;**278**:128404. DOI: 10.1016/j.matlet.2020.128404

[36] Nsibande S, Forbes P. Fluorescence detection of pesticides using quantum dot materials–A review. Analytica Chimica Acta. 2016;**945**:9-22. DOI: 10.1016/j.aca.2016.10.002

[37] Li H, Deng R, Tavakoli H, Li X, Li X. Ultrasensitive detection of Acephate based on carbon quantum dots-mediated fluorescence inner filter effects (Ife). SSRN. 2022. In Press. DOI: 10.2139/ssrn.4104463

[38] Wang Q, Zhao W-M. Optical methods of antibiotic residues detections: A comprehensive review. Sensors and Actuators B: Chemical. 2018;**269**:238-256. DOI: 10.1016/j.snb.2018.04.097

[39] Wang X, Li L, Jiang H, Zhangsun H, Wang Q, Sun X, et al. Highly selective and sensitive fluorescence detection of tetracyclines based on novel tungsten oxide quantum dots. Food Chemistry. 2022;**374**:131774. DOI: 10.1016/j.foodchem.2021.131774

[40] Zhu W, Zhou Y, Liu S, Luo M, Du J, Fan J, et al. A novel magnetic fluorescent molecularly imprinted sensor for highly selective and sensitive detection of 4-nitrophenol in food samples through a dual-recognition mechanism. Food Chemistry. 2021;**348**:129126. DOI: 10.1016/j.foodchem.2021.129126

[41] Zhang D, Tang J, Liu H. Rapid determination of lambda-cyhalothrin using a fluorescent probe based on ionic-liquid-sensitized carbon dots coated with molecularly imprinted polymers. Analytical and Bioanalytical Chemistry. 2019;**411**(20):5309-5316. DOI: 10.1007/s00216-019-01912-0

[42] Nair RV, Chandran PR, Mohamed AP, Pillai S. Sulphur-doped graphene quantum dot based fluorescent turn-on aptasensor for selective and ultrasensitive detection of omethoate. Analytica Chimica Acta. 2021;**1181**:338893. DOI: 10.1016/j.aca.2021.338893

[43] Zhang Y, Yuan X, Jiang W, Liu H. Determination of nereistoxin-related insecticide via quantum-dots-doped covalent organic frameworks in a molecularly imprinted network. Microchimica Acta. 2020;**187**(8):1-9. DOI: 10.1007/s00604-020-04435-z

[44] Zhang Z, Fan Z. Application of cerium–nitrogen co-doped carbon quantum dots to the detection of tetracyclines residues and bioimaging. Microchemical Journal. 2021;**165**:106139. DOI: 10.1016/j.microc.2021.106139

[45] Xie H, Lu Y, You R, Qian W, Lin S. Green synthetic nitrogen-doped graphene quantum dot fluorescent probe for the highly sensitive and selective detection of tetracycline in food samples. RSC Advances. 2022;**12**(13):8160-8171. DOI: 10.1039/D2RA00337F

[46] Ahmed SR, Kumar S, Ortega GA, Srinivasan S, Rajabzadeh AR. Target specific aptamer-induced self-assembly of fluorescent graphene quantum dots on palladium nanoparticles for sensitive detection of tetracycline in raw milk. Food Chemistry. 2021;**346**:128893. DOI: 10.1016/j.foodchem.2020.128893

[47] Cao Y, Wang X, Bai H, Jia P, Zhao Y, Liu Y, et al. Fluorescent detection of tetracycline in foods based on carbon dots derived from natural red beet pigment. LWT. 2022;**157**:113100. DOI: 10.1016/j.lwt.2022.113100

[48] Chaitong N, Chansud N, Orachorn N, Limbut W, Bunkoed O. A magnetic nanocomposite optosensing probe based on porous graphene, selective polymer and quantum dots for the detection of cefoperazone in milk. Microchemical Journal. 2021;**171**:106838. DOI: 10.1016/j.microc.2021.106838

[49] Liu J, Chen Y, Wang W, Feng J, Liang M, Ma S, et al. "Switch-on" fluorescent sensing of ascorbic acid in food samples based on carbon

quantum dots–MnO_2 probe. Journal of Agricultural and Food Chemistry. 2016;**64**(1):371-380. DOI: 10.1021/acs.jafc.5b05726

[50] Wang J, Wei J, Su S, Qiu J. Novel fluorescence resonance energy transfer optical sensors for vitamin B 12 detection using thermally reduced carbon dots. New Journal of Chemistry. 2015;**39**(1):501-507. DOI: 10.1039/C4NJ00538D

[51] Majumdar S, Bhattacharjee T, Thakur D, Chowdhury D. Carbon dot based fluorescence sensor for retinoic acid. ChemistrySelect. 2018;**3**(2):673-677. DOI: 10.1002/slct.201702458

[52] Kalpana S, Priyadarshini S, Leena MM, Moses J, Anandharamakrishnan C. Intelligent packaging: Trends and applications in food systems. Trends in Food Science & Technology. 2019;**93**:145-157. DOI: 10.1016/j.tifs.2019.09.008

[53] Emamhadi MA, Sarafraz M, Akbari M, Fakhri Y, Linh NTT, Khaneghah AM. Nanomaterials for food packaging applications: A systematic review. Food and Chemical Toxicology. 2020;**146**:111825. DOI: 10.1016/j.fct.2020.111825

[54] Primožič M, Knez Ž, Leitgeb M. (Bio) nanotechnology in food science—Food packaging. Nanomaterials. 2021;**11**(2):292. DOI: 10.3390/nano11020292

[55] Thippeswamy BH, Maligi AS, Hegde G. Roadmap of effects of biowaste-synthesized carbon nanomaterials on carbon nano-reinforced composites. Catalysts. 2021;**11**(12):1485. DOI: 10.3390/catal11121485

[56] Xu L, Zhang Y, Pan H, Xu N, Mei C, Mao H, et al. Preparation and performance of radiata-pine-derived polyvinyl alcohol/carbon quantum dots fluorescent films. Materials. 2019;**13**(1):67. DOI: 10.3390/ma13010067

[57] You Y, Zhang H, Liu Y, Lei B. Transparent sunlight conversion film based on carboxymethyl cellulose and carbon dots. Carbohydrate Polymers. 2016;**151**:245-250. DOI: 10.1016/j.carbpol.2016.05.063

[58] Moradi M, Molaei R, Kousheh SA, T. Guimarães J, McClements DJ. Carbon dots synthesized from microorganisms and food by-products: Active and smart food packaging applications. Critical Reviews in Food Science and Nutrition. 2021;**13**:1-17. DOI: 10.1080/10408398.2021.2015283

[59] Xu L, Li Y, Gao S, Niu Y, Liu H, Mei C, et al. Preparation and properties of cyanobacteria-based carbon quantum dots/polyvinyl alcohol/nanocellulose composite. Polymers. 2020;**12**(5):1143. DOI: 10.3390/polym12051143

[60] Riahi Z, Priyadarshi R, Rhim J-W, Lotfali E, Bagheri R, Pircheraghi G. Alginate-based multifunctional films incorporated with sulfur quantum dots for active packaging applications. Colloids and Surfaces B: Biointerfaces. 2022;**215**:112519. DOI: 10.1016/j.colsurfb.2022.112519

[61] Dong X, Liang W, Meziani MJ, Sun Y-P, Yang L. Carbon dots as potent antimicrobial agents. Theranostics. 2020;**10**(2):671. DOI: 10.7150/thno.39863

[62] Ezati P, Roy S, Rhim J-W. Pectin/gelatin-based bioactive composite films reinforced with sulfur functionalized carbon dots. Colloids and Surfaces A: Physicochemical and Engineering Aspects. 2022;**636**:128123. DOI: 10.1016/j.colsurfa.2021.128123

[63] Ezati P, Rhim J-W, Molaei R, Priyadarshi R, Han S. Cellulose nanofiber-based coating film integrated with nitrogen-functionalized carbon dots for active packaging applications of fresh fruit. Postharvest Biology and Technology. 2022;**186**:111845. DOI: 10.1016/j.postharvbio.2022.111845

[64] Fan K, Zhang M, Fan D, Jiang F. Effect of carbon dots with chitosan coating on microorganisms and storage quality of modified-atmosphere-packaged fresh-cut cucumber. Journal of the Science of Food and Agriculture. 2019;**99**(13):6032-6041. DOI: 10.1002/jsfa.9879

[65] Zhao L, Zhang M, Mujumdar AS, Wang H. Application of carbon dots in food preservation: A critical review for packaging enhancers and food preservatives. Critical Reviews in Food Science and Nutrition. 2022;**17**:1-19. DOI: 10.1080/10408398.2022.2039896

[66] Silvestre C, Duraccio D, Cimmino S. Food packaging based on polymer nanomaterials. Progress in Polymer Science. 2011;**36**(12):1766-1782. DOI: 10.1016/j.progpolymsci.2011.02.003

[67] Amin U, Khan MU, Majeed Y, Rebezov M, Khayrullin M, Bobkova E, et al. Potentials of polysaccharides, lipids and proteins in biodegradable food packaging applications. International Journal of Biological Macromolecules. 2021;**183**:2184-2198. DOI: 10.1016/j.ijbiomac.2021.05.182

[68] Mills A. Oxygen indicators in food packaging. In: Baraton MI, editors. Sensors for Environment, Health and Security. Germany: Springer; 2009. p. 371-388. DOI: 10.1007/978-1-4020-9009-7_24

[69] Xu Y, Yang D, Huo S, Ren J, Gao N, Chen Z, et al. Carbon dots and ruthenium doped oxygen sensitive nanofibrous membranes for monitoring the respiration of agricultural products. Polymer Testing. 2021;**93**:106957. DOI: 10.1016/j.polymertesting.2020.106957

[70] Chausali N, Saxena J, Prasad R. Recent trends in nanotechnology applications of bio-based packaging. Journal of Agriculture and Food Research. 2022;**7**:100257. DOI: 10.1016/j.jafr.2021.100257

[71] Yang Y, Zhao Y, Hu Y, Peng X, Zhong L. Xylan-derived light conversion nanocomposite film. Polymers. 2020;**12**(8):1779. DOI: 10.3390/polym12081779

[72] Schmitz F, de Albuquerque MBS, Alberton MD, Riegel-Vidotti IC, Zimmermann LM. Zein films with ZnO and ZnO: Mg quantum dots as functional nanofillers: New nanocomposites for food package with UV-blocker and antimicrobial properties. Polymer Testing. 2020;**91**:106709. DOI: 10.1016/j.polymertesting.2020.106709

[73] Grzebieniarz W, Nowak N, Khachatryan G, Krzan M, Krystyjan M, Kosiński J, et al. The preparation and characterization of quantum dots in polysaccharide carriers (starch/chitosan) as elements of smart packaging and their impact on the growth of microorganisms in food. Materials. 2021;**14**(24):7732. DOI: 10.3390/ma14247732

[74] Priyadarshi R, Riahi Z, Rhim J-W, Han S, Lee S-G. Sulfur quantum dots as fillers in gelatin/agar-based functional food packaging films. ACS Applied Nano Materials. 2021;**4**(12):14292-14302. DOI: 10.1021/acsanm.1c03925

[75] Riahi Z, Rhim J-W, Bagheri R, Pircheraghi G, Lotfali E. Carboxymethyl cellulose-based functional film integrated with chitosan-based

carbon quantum dots for active food packaging applications. Progress in Organic Coatings. 2022;**166**:106794. DOI: 10.1016/j.porgcoat.2022.106794

[76] Subala AS, Anand K, Sibi MSA. Quantum Dots in Waste Water Treatment-Review Article. 2020:**5**:1449-1455

[77] Tewari C, Kumar S, Pandey N, Pandey S, Sahoo NG, Campus D. Quantum dots based materials for water treatment. Quantum Dots: Properties and Applications. 2021;**96**:280-304. DOI: 10.21741/9781644901250-1

[78] Kunduru KR, Nazarkovsky M, Farah S, Pawar RP, Basu A, Domb AJ. Nanotechnology for water purification: Applications of nanotechnology methods in wastewater treatment. In: Mihai GA, editors. Water Purification. United States of America: Elsevier; 2017. p. 33-74. DOI: 10.1016/B978-0-12-804300-4.00002-2

[79] Sridharan K, Ilango V, Samuel RS. Water purification by carbon quantum dots. In: Lichtfouse E, Senthilkannan Muthu S, Khadir A, editors. Inorganic-Organic Composites for Water and Wastewater Treatment. Singapore: Springer; 2022. p. 113-160. DOI: 10.1007/978-981-16-5928-7_4

[80] Zhao DL, Chung T-S. Applications of carbon quantum dots (CQDs) in membrane technologies: A review. Water Research. 2018;**147**:43-49. DOI: 10.1016/j.watres.2018.09.040

[81] Sun J, Jia W, Guo J, Khanzada NK, Jin P, Wong PW, et al. Amino-embedded carbon quantum dots incorporated thin-film nanocomposite membrane for desalination by pervaporation. Desalination. 2022;**533**:115742. DOI: 10.1016/j.desal.2022.115742

[82] Zhao DL, Das S, Chung T-S. Carbon quantum dots grafted antifouling membranes for osmotic power generation via pressure-retarded osmosis process. Environmental Science & Technology. 2017;**51**(23):14016-14023. DOI: 10.1021/acs.est.7b04190

[83] Bonilla JC, Bozkurt F, Ansari S, Sozer N, Kokini JL. Applications of quantum dots in food science and biology. Trends in Food Science & Technology. 2016;**53**:75-89. DOI: 10.1016/j.tifs.2016.04.006

[84] Biesiekierski JR. What is gluten? Journal of Gastroenterology and Hepatology. 2017;**32**:78-81. DOI: 10.1111/jgh.13703

[85] Kasaai MR. Zein and zein-based nano-materials for food and nutrition applications: A review. Trends in Food Science & Technology. 2018;**79**:184-197. DOI: 10.1016/j.tifs.2018.07.015

[86] Sozer N, Kokini JL. Use of quantum nanodot crystals as imaging probes for cereal proteins. Food Research International. 2014;**57**:142-151. DOI: 10.1016/j.foodres.2013.12.031

[87] Ansari S, Bozkurt F, Yazar G, Ryan V, Bhunia A, Kokini J. Probing the distribution of gliadin proteins in dough and baked bread using conjugated quantum dots as a labeling tool. Journal of Cereal Science. 2015;**63**:41-48. DOI: /10.1016/j.jcs.2014.12.001

Section 3

Quantum Dots in Other Important Applications

Chapter 11

Quantum Dot Scattering in Monolayer Molybdenum Disulfide

Rachid Houça, Abdelhadi Belouad, Abdellatif Kamal, El Bouâzzaoui Choubabi and Mohammed El Bouziani

Abstract

This chapter looks at how electrons propagate in a circular quantum dot (QD) of monolayer molybdenum disulfide (MoS_2) that is exposed to an electric potential. Mathematical formulas for the eigenstates, scattering coefficients, scattering efficiency, and radial component of the reflected current and electron density are presented using the continuum model. As a function of physical characteristics such as incident electronic energy, potential barrier, and quantum dot radius, we discover two scattering regimes. We demonstrate the presence of scattering resonances for low-energy incoming electrons. We should also point out that the far-field dispersed current has unique favored scattering directions.

Keywords: scattering, monolayer molybdenum disulfide, quantum dot, electric potential, electron density

1. Introduction

The hunt for novel two-dimensional materials has been sparked by graphene research [1, 2]. Transition metal dichalcogenides (TMDs) are among them. TMD (MX_2 such as M = Mo, W; X = S, S_e, and T) monolayers have recently emerged as promising nanostructures for optics, electronics, and spintronic applications. For many years, molybdenum disulfide (MoS_2) has been a strong material that has gotten a lot of attention because of its intriguing electrical and optical characteristics [3–5]. It has a straight bandgap in the visible frequency band [6–10] and high carrier mobility at ambient temperature [11–15]. As a result, it is a strong contender for future electrical and optoelectronic technologies.

A strong spin-orbit interaction characterizes the MoS_2 monolayer with its honeycomb atomic structure. As a result, totally new electron spin characteristics will emerge. On the one hand, conduction or valence electrons should be significantly less susceptible to the ultrafast spin relaxation effects observed in two-dimensional semiconductor structures such as GaAs quantum wells. The absorption of circularly polarized light, on the other hand, might result in a population of spin-polarized electrons (i.e. an imbalance between the number of *spin-up* and *spin-down* electrons). Moreover, the circularly polarized excitation allows control of the distribution of these electrons in one of two valleys of reciprocal space, according to [16]. We call this *valley polarization*, and we are now working to better understand it so that we can use it in information storage and processing uses. The borders of the valence and conduction bands are positioned at the two corners of the Brillouin zone, i.e. the k and k' points, making the MoS_2 monolayer a semiconductor. This provides an additional degree of freedom for electrons and holes, which may be employed for encoding information and subsequent processing [16–19].

Monolayer MoS_2 quantum dots (QDs), which feature different physical and chemical characteristics, strong quantum confinement properties, edge effects [20], and a direct bandgap, are key MoS_2 related nanostructures that have gotten a lot of interest in recent years. Some many approaches for preparing MoS_2 QD have been suggested to date, including solvothermal treatment synthesizing [21], hydrothermal synthesis [22], grinding exfoliation [23], liquid exfoliation in organic solvents [24], electrochemical etching [25], and reaction processing [26].

The contours of the valence and conduction bands determine edge states, which are confined on the boundaries and have energies in the bandgap. The electrical behavior of MoS_2 quantum dots was studied using the tight-binding model [22]. According to the orbital asymmetry [22], it was demonstrated that quantum dots with the same form but distinct electrical characteristics may be created.

In this work, we investigate electron propagation in the presence of a potential barrier in a circular electrostatically defined quantum dot monolayer molybdenium disulfide MoS_2. Different scattering ranges are identified based on the quantum dot radius, potential barrier, electron spin, and electron energy.

The following is the structure of the present study. We give a theoretical investigation of electron propagation wave plane in a circular quantum dot of monolayer molybdenium disulfide MoS_2 in Section 2. The spinors of the Dirac equation solutions corresponding to each area of varied scattering parameters are given. The scattering coefficients are calculated using the continuity criterion. We evaluate the scattering efficiency, square modulus of the scattering coefficients, radial component of the far-field scattered current, and electron density in Section 3, and we describe our findings using various plots. The basic findings of the study are presented in Section 4.

2. Theoretical model

As shown in **Figure 1**, we investigate a quantum dot of radius ρ in the presence of an electric potential U. In the proximity of the valleys k and k' by generating the wave functions via the base of conduction and valence bands, the Dirac-Weyl Hamiltonian for low-energy charge carriers in monolayer molybdenum disulfide (MoS_2) yields [27, 28]

$$\mathcal{H} = \mathcal{H}_0 + \frac{\kappa}{2}\sigma_z + \frac{\Delta_{so}}{2}\tau s_z(1-\sigma_z) + U(r) \tag{1}$$

so the $\mathcal{H}_0$ is supplied by:

$$\mathcal{H}_0 = v_F \boldsymbol{\sigma} \cdot \boldsymbol{p} \tag{2}$$

where v_F, $\boldsymbol{p} = \left(p_x, p_y\right)$ and $\boldsymbol{\sigma} = \left(\sigma_x, \sigma_y, \sigma_z\right)$ are, respectively, the Fermi velocity, the $2D$ momentum operator, and Pauli matrices acting on the atomic orbitals, $U(r)$ is

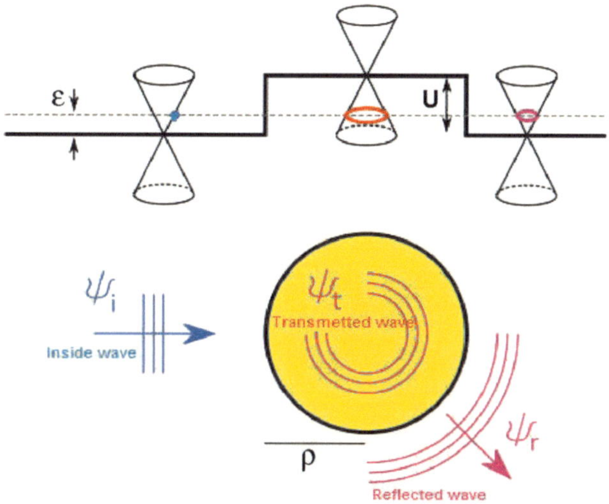

Figure 1.
The energy ε of Dirac electrons propagate in a circular quantum dot of monolayer molybdenum disulfide (MoS_2)*. The dot is defined by its radius ρ and the bias introduced U. The incoming plane wave with energy ε > U (blue) belongs to a conduction band state (upper cone). The reflected wave (purple) is now in the conduction band, but the transmitting wave (red) is in the valence band (lower cone).*

the potential barrier, and $\kappa = 166.10^{-3}$ eV [28] is related to the material bandgap energy, $s_z = \pm 1$ represents the electron *spin-up* and *spin-down* and $\tau = \pm 1$ denotes the k and k' valleys, $\Delta_{so} = 75.10^{-3}$ eV [28] is the splitting of the valence band owing to spin-orbit coupling, and ρ is the dot radius. For simplicity, the values $v_F = \hbar = 1$ shall be assumed.

We then investigate localized-state solutions in our system, which is characterized as a circularly symmetric quantum dot, utilizing the potential barrier $U(r)$ and the energy gap $\kappa(r)$ defined by:

$$U(r) = \begin{cases} 0, & r > \rho \\ U, & r \le \rho \end{cases}, \qquad \kappa(r) = \begin{cases} 0, & r > \rho \\ \kappa, & r \le \rho \end{cases}. \tag{3}$$

We conduct our research using the polar coordinates (r, ϖ), so that the Hamiltonian (1) has the format:

$$\mathcal{H} = \begin{pmatrix} U_+ & \pi^- \\ \pi^+ & U_- + \tau s_z \lambda_{so} \end{pmatrix} \tag{4}$$

in which the two potentials and two operators have been established

$$U_\pm = U \pm \frac{\kappa}{2}, \qquad \pi^\pm = e^{\pm i\varpi}\left(-i\frac{\partial}{\partial r} \pm \frac{1}{r}\frac{\partial}{\partial \varpi}\right). \tag{5}$$

The eigenvalue equation is used to determine the energy spectrum:

$$\mathcal{H}\psi_l(r,\varpi) = \varepsilon\psi_l(r,\varpi). \tag{6}$$

Because $J_z = -i\hbar\partial_\varphi + \hbar\sigma_z/2$ commutes with $\mathcal{H}$ by fulfilling $[\mathcal{H}, J_z] = 0$. The separability of ψ_l into the radial $\Phi^\pm(r)$ and angular $F^\pm(\varpi)$ parts is required for this commutation, and then we obtain [29, 30]

$$\psi_l(r,\varpi) = \Phi_l^+(r)F_l^+(\varpi) + \Phi_{l+1}^-(r)F_{l+1}^-(\varpi) \tag{7}$$

where the two angular components are

$$F_l^+(\varpi) = \frac{e^{il\varpi}}{\sqrt{2\pi}}\begin{pmatrix}1\\0\end{pmatrix}, \qquad F_{l+1}^-(\varpi) = \frac{e^{i(l+1)\varpi}}{\sqrt{2\pi}}\begin{pmatrix}0\\1\end{pmatrix} \tag{8}$$

and l is an integer, which denotes the total angular quantum number.

To obtain the energy spectrum solutions, we must first solve the eigenvalue equation:

$$\mathcal{H}\psi_l(r,\varpi) = \varepsilon\psi_l(r,\varpi) \tag{9}$$

by taking into account two areas, as shown in **Figure 1**, outside $(r > \rho)$ and within $(r \leq \rho)$ the quantum dot. As a result, we get an incident wave ψ_i propagating in the x-direction, an outgoing reflected wave ψ_r, and a transmitted wave ψ_t within the dot.

The radial components $\Phi_l^+(r)$ and $\Phi_{l+1}^-(r)$ satisfy two linked differential equations outside the dot $(r > \rho)$:

$$-i\frac{\partial}{\partial r}\Phi_l^+(r) + i\frac{l}{r}\Phi_l^+(r) = \varepsilon\Phi_{l+1}^-(r) \tag{10}$$

$$-i\frac{\partial}{\partial r}\Phi_{l+1}^-(r) - i\frac{l+1}{r}\Phi_{l+1}^-(r) = \varepsilon\Phi_l^+(r). \tag{11}$$

This may be solved by inserting (10) into (11) to get a second differential equation that $\Phi_l^+(r)$ can fulfill

$$\left(r^2\frac{\partial^2}{\partial^2 r} + r\frac{\partial}{\partial r} + r^2\varepsilon^2 - l^2\right)\Phi_l^+(r) = 0 \tag{12}$$

where the last equation's solutions are the Bessel functions $J_l(\varepsilon r)$. Furthermore, the incident electron's wave function moving along the x-direction $(x = r\cos\varpi)$ has the expression:

$$\psi_l^i(r,\varpi) = \frac{1}{\sqrt{2}}\sum_l i^l\left[J_l(kr)e^{il\varpi}\begin{pmatrix}1\\0\end{pmatrix} + iJ_{l+1}(kr)e^{i(l+1)\varpi}\begin{pmatrix}0\\1\end{pmatrix}\right] \tag{13}$$

as well as the reflected wave:

$$\psi_l^r(r,\varpi) = \frac{1}{\sqrt{2}}\sum_l i^l\alpha_l\left[H_l^{(1)}(kr)e^{il\varpi}\begin{pmatrix}1\\0\end{pmatrix} + iH_{l+1}^{(1)}(kr)e^{i(l+1)\varpi}\begin{pmatrix}0\\1\end{pmatrix}\right] \tag{14}$$

wherein $H_l^{(1)}(kr)$ is the first class of Hankel function [31], α_l is the scattering parameters, and $k = \varepsilon$ is the wave number.

The radial functions Φ_l^+ and Φ_{l+1}^- are represented by the following equations inside the dot ($r \leq \rho$):

$$i\left(\frac{\partial}{\partial r} - \frac{l}{r}\right)\Phi_l^+(r) + (\varepsilon - U_- - \tau s_z \Delta_{so})\Phi_{l+1}^-(r) = 0 \tag{15}$$

$$i\left(\frac{\partial}{\partial r} + \frac{l+1}{r}\right)\Phi_{l+1}^-(r) + (\varepsilon - U_+)\Phi_l^+(r) = 0 \tag{16}$$

Expressing (15) as

$$\Phi_{l+1}^-(r) = -\frac{i}{\varepsilon - U_- - \tau s_z \Delta_{so}}\left(\frac{\partial}{\partial r} - \frac{l}{r}\right)\Phi_l^+(r) \tag{17}$$

By substituting the Eq. (17) in (16), we find a differential equation for $\Phi_l^+(r)$:

$$\left(r^2 \frac{\partial^2}{\partial^2 r} + r\frac{\partial}{\partial r} + r^2\gamma^2 - l^2\right)\Phi_l^+(r) = 0 \tag{18}$$

where

$$\gamma^2 = (\varepsilon - U_+)(\varepsilon - U_- - \tau s_z \Delta_{so}). \tag{19}$$

The solution of (18) can be worked out to get the transmitted wave as:

$$\psi_l^t(r, \varpi) = \frac{1}{\sqrt{2}}\sum_l i^l \beta_l \left[J_l(\gamma r) e^{il\varpi}\begin{pmatrix}1\\0\end{pmatrix} + i\mu J_{l+1}(\gamma r) e^{i(l+1)\varpi}\begin{pmatrix}0\\1\end{pmatrix}\right] \tag{20}$$

where the β_l denote the transmission coefficients and

$$\mu = \sqrt{\frac{\varepsilon - U_+}{\varepsilon - U_- - \tau s_z \Delta_{so}}}. \tag{21}$$

Requiring the eigenspinors continuity at the boundary $r = \rho$ of the quantum dot,

$$\psi_l^i(\rho) + \psi_l^r(\rho) = \psi_l^t(\rho), \tag{22}$$

to obtain the conditions

$$J_l(k\rho) + \alpha_l H^{(1)}(k\rho) = \beta_l J_l(\gamma\rho), \tag{23}$$

$$J_{l+1}(k\rho) + \alpha_l H_{l+1}^{(1)}(k\rho) = \mu\beta_l J_{l+1}(\gamma\rho). \tag{24}$$

Solving these equations to get the scattering coefficients:

$$\alpha_l = -\frac{J_l(\gamma\rho)J_{l+1}(k\rho) - \mu J_{l+1}(\gamma\rho)J_l(k\rho)}{J_l(\gamma\rho)H_{l+1}^{(1)}(k\rho) - \mu J_{l+1}(\gamma\rho)H_l^{(1)}(k\rho)} \tag{25}$$

and the transmission coefficients by:

$$\beta_l = \frac{J_l(k\rho)H_{l+1}^{(1)}(k\rho) - J_{l+1}(k\rho)H_l^{(1)}(k\rho)}{J_l(\gamma\rho)H_{l+1}^{(1)}(k\rho) - \mu J_{l+1}(\gamma\rho)H_l^{(1)}(k\rho)} \tag{26}$$

The density is described as $\boldsymbol{j} = \psi^{\dagger}\boldsymbol{\sigma}\psi$, with $\psi = \psi_r + \psi_i$ outside the dot and $\psi = \psi_t$ inside the dot.

Angular scattering is described by the far-field radial component of the reflecting current, which giving by

$$j_r = \psi^{\dagger}\left(\cos(\varpi)\sigma_x + \sin(\varpi)\sigma_y\right)\psi = \psi^{\dagger}\begin{pmatrix} 0 & e^{-i\varpi} \\ e^{i\varpi} & 0 \end{pmatrix}\psi \tag{27}$$

the corresponding radial current can be written as:

$$j_r^r = \frac{1}{2}\sum_{l=0}^{\infty} A_l(kr) \times \begin{pmatrix} 0 & e^{-i\varpi} \\ e^{i\varpi} & 0 \end{pmatrix} \times \sum_{l=0}^{\infty} A_l^*(kr) \tag{28}$$

where

$$A_l(kr) = (-i)^l\left[H_l^{(1)*}(kr)\begin{pmatrix} \alpha_l^* e^{-il\varpi} \\ \alpha_{-l-1}^* e^{il\varpi} \end{pmatrix} - iH_{l+1}^{(1)*}(kr)\begin{pmatrix} \alpha_{-l-1}^* e^{i(l+1)\varpi} \\ \alpha_l^* e^{-(l+1)\varpi} \end{pmatrix}\right] \tag{29}$$

By injecting the asymptotic behavior of the Hankel function of the first kind for $kr \gg 1$

$$H_l(kr) \simeq \sqrt{\frac{2}{\pi kr}} e^{i\left(kr - \frac{l\pi}{2} - \frac{\pi}{4}\right)}, \tag{30}$$

into the Eq. (29), the density of the system (28) may be simplified to the following form:

$$j_r^r(\varpi) = \frac{4}{\pi kr}\sum_{l=0}^{\infty}[1 + \cos(2l+1)\varpi]\,|c_l|^2 \tag{31}$$

where

$$|c_l| = \frac{1}{\sqrt{2}}\left(\left[\left|\alpha_{-(l+1)}\right|^2 + |\alpha_l|^2\right]\right)^{12}. \tag{32}$$

The scattering cross section σ is defined by [32]:

$$\sigma = \frac{I_r^r}{I^i/A_u} \tag{33}$$

where I_r^r and $\left(I^i/A_u\right)$ denote, respectively, the total reflected flux across a concentric circle and the incident flux per unit area. Furthermore, I_r^r is defined by:

$$I_r^r = \int_0^{2\pi} r j_r^r(\varpi)\mathrm{d}\varpi = \frac{8}{k}\sum_{l=0}^{\infty}|c_l|^2. \tag{34}$$

For the incident wave (13), we note that the incident flux is equal to $I^i/A_u = 1$.

To go deeper into our investigation of the scattering problem for a plane Dirac electron at various sizes of the circular quantum dot, we define the scattering efficiency Q by splitting the scattering cross section by the geometric cross section. It is written by:

$$Q = \frac{\sigma}{2\rho} = \frac{4}{k\rho}\sum_{l=0}^{\infty}|c_l|^2. \tag{35}$$

3. Results and discussions

Figure 2a and **b** depict the scattering efficiency Q for low energies under barriers scattering (*n-p* junction) $\varepsilon = 0.01, 0.02, 0.04, 0.06 < U = 1$, in terms of quantum dot radius ρ for the *spin-up* state in the two valleys $k(\tau = 1, s_z = 1)$ and $k'(\tau = -1, s_z = 1)$ and the *spin-down* state in the two valleys $k(\tau = 1, s_z = -1)$ and $k'(\tau = -1, s_z = -1)$. We see that when $\rho \to 0$ implies that $Q \to 0$, and when ρ rises, Q raises to a maximum value $Q_{max} = 20, 25, 39.5, 68$ for $\varepsilon = 0.06, 0.04, 0.02, 0.01$ for the state (k', s_z) (**Figure 2a**) and for the state (k, -s_z) (**Figure 2b**), then the scattering efficiency Q has a highly damped oscillatory behavior with the appearance of net transverse resonant peaks, comparable to graphene quantum dots [32–34]. Moreover, the height of the peak decreases as the radius ρ rises, but its breadth increases, which shows the peculiarities of the energy dispersion. Furthermore, by comparing **Figure 2a** and **b**, we remark that Q's dependency on the *spin-up* and *spin-down* states in the two valleys is symmetrical, i.e. $Q(-\tau, s_z) = Q(\tau, -s_z)$.

Figure 2c and **d** present, respectively, the scattering efficiency Q for energies across the barrier (*n-n* junction) $\varepsilon = 1.16, 1.2, 1.3, 1.5 > U = 1$, in terms of quantum dot radius ρ for the *spin-up* state in the two valleys k and k' as well as the *spin-down* state in

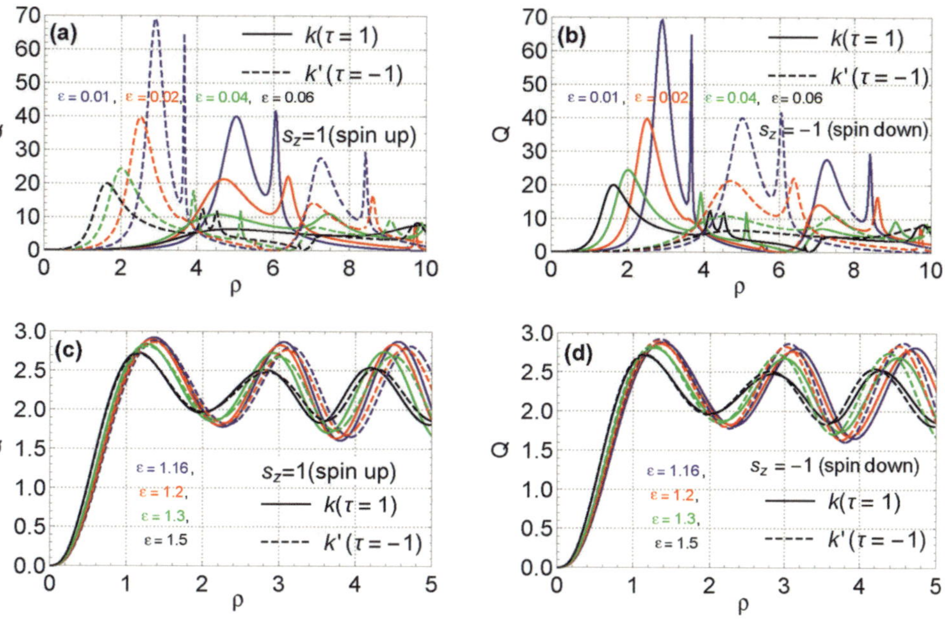

Figure 2.
Scattering efficiency Q, for the potential U = 1, in terms of the dot radius ρ for different values of ε for the spin-up *and* spin-down *states in two valleys k and k′.*

valleys k and k'. However, when ρ grows, the scattering efficiency Q roughly linearly until it reaches a maximum value $Q_{max} = 2.9$ relates to a certain value of ρ. Furthermore, by raising ρ and for the four energy levels, the four curves display oscillating attitude [35]. In this domain ($\varepsilon > U$), we find attitudes of symmetry Q ($Q(-\tau, s_z) = Q(\tau, -s_z)$) with regard to s_z and τ comparable with those of the previous domain ($\varepsilon < U$).

To study the scattering for $\varepsilon < U$ in more detail, we present in **Figure 3** the scattering efficiency as a function of the electron energy ε. In **Figure 3a** and **b**, we consider dots with small radius $\rho = 0.9, 01, 1.2, 1.3$, for the *spin-up* state in the two valleys $k(\tau = 1)$ and $k'(\tau = 1)$ and the *spin-down* state in two valleys $k(\tau = 1)$ and $k'(\tau = -1)$.

Figure 3a shows that for $0 \leq \varepsilon \leq 0.6$, Q exhibits a maximum for the *spin-up* state in the valley $k'(\tau = -1, s_z = 1)$ with the emergence of a single peak related to $\varepsilon = 0.55$ and a minima for the *spin-up* state in the valley $k(\tau, s_z = 1$ without any visible peaks. For $E > 0.6$, we observe that Q has a maximum for the *spin-up* state in the valley $k(\tau = 1, s_z = 1)$ with the emergence of a single peak appropriate for $\varepsilon = 0.75$ and a minima for the *spin-up* state in the valley $k'(\tau = -1, s_z = 1)$ without the emergence of peaks. For $E > 0.6$, we notice that Q has a maximum for the *spin-up* state in the valley $k(\tau = 1, s_z = 1)$ with the emergence of a single peak appropriate for $\varepsilon = 0.75$ and a minima for the *spin-up* state in the valley $k'(\tau = -1, s_z = 1)$ without the emergence of peaks.

Figure 3b indicates that for $0 \leq \varepsilon \leq 0.6$, Q exhibits a maximum for the *spin-down* state in the valley $k(\tau = 1, s_z = -1)$ with the emergence of a single peak relates to $\varepsilon = 0.55$ and a minima for the *spin-up* state in the valley $k'(\tau = 1, s_z = -1)$ without the emergence of peaks. We note that for $\varepsilon > 0.6$, Q shows a maximum for the *spin-down* state in the valley $k'(\tau = -1, s_z = -1)$ with the emergence of a single peak appropriate

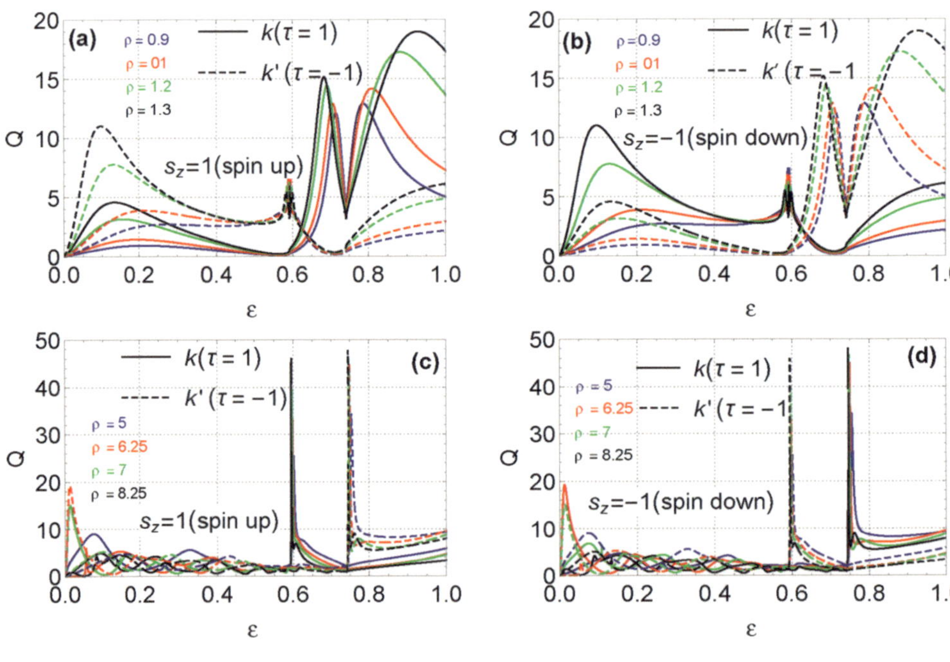

Figure 3.
Scattering efficiency Q, for the potential $U = 1$, in terms of the energy ε of the incident electron for distinct values of ρ for the spin-up *and* spin-down *states in two valleys k and k'.*

for $\varepsilon = 0.75$ and a minima for the *spin-down* state in the valley $k(\tau = 1, s_z = -1)$ without the emergence of peaks. The electron scattering efficiency, on the other hand, is invariant by the transformation $Q(\tau, s_z) \rightarrow Q(-\tau, -s_z)$.

Figure 3c and **d** represent Q in terms of ε for increasing values of $\rho = 5, 6.25, 7, 8.25$ for the *spin-up* state in the two valleys $k(\tau = 1, s_z = 1)$ and $k'(\tau = -1, s_z = 1)$ and the *spin-down* state in two valleys $k(\tau = 1, s_z = -1)$ and $k'(\tau = -1, s_z = -1)$, respectively. Additionally, we notice that Q has significant maxima for low energies as well. However, when ε rises, we see the emergence of a peak with damped oscillations for both *spin-up* and *spin-down* states. The resonant excitation of the quantum dot's normal modes causes these sharp peaks. As a result, the dependency of Q on s_z in the two valleys k and k' is symmetric with regard to $\pm s_z$, i.e $Q(-\tau, s_z) = Q(\tau, -s_z)$.

The square modulus of the scattering $|c_l|^2$ is presented in **Figure 4** for $l = 0, 1, 2, 3$ in terms of the energy ε, for the *spinup* and *spin-down* states in the two valleys $k(\tau = 1)$ and $k'(\tau = 1)$ for various size of the dot radius: **Figure 4a** and **b**: $\rho = 2$, **Figure 4c** and

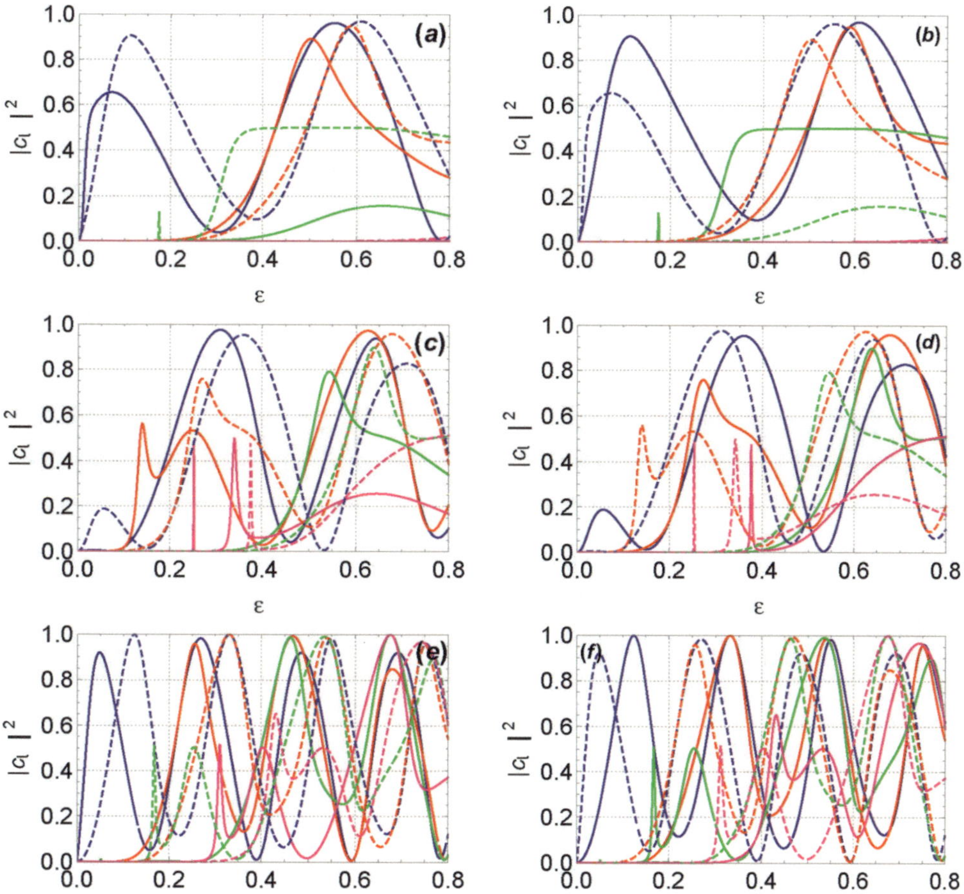

Figure 4.
Square modulus of the scattering coefficients $|c_l|^2$ in terms of the energy ε at $U = 1$ for $l = 0$ (blue curve), 1 (red curve), 2 (green curve), and 3 (magenta curve). (a): (k, k', spin-up, $\rho = 2$) states. (b): (k, k', spin-down, $\rho = 2$) states. (c): (k, k', spin-up, $\rho = 3$) states. (d): (k, k', spin-down, $\rho = 3$) states, (e): (k, k', spin-up, $\rho = 7.75$) states, and (f): (k, k', spin-down, $\rho = 7.75$) states. Solid curve relates to valley k and dashed curve relates to valley k'.

d: $\rho = 3$, **Figure 4e** and **f**: $\rho = 4$ within all panels $U = 1$. Except for the situation corresponding to $l = 0$, all scattering coefficients are zero for zero or near to zero energy. Furthermore, when the energy increases, the scattering coefficients $|c_l|^2$ exhibit oscillatory behavior [35]. We may observe that using the spin-orbit interaction results in an increase in the number of oscillations. Furthermore, we see that some energy values, $|c_l|^2$, have strong peaks. Additionally, the resonances of the dot's normal modes result in the high peaks previously seen for the scattering efficiency Q

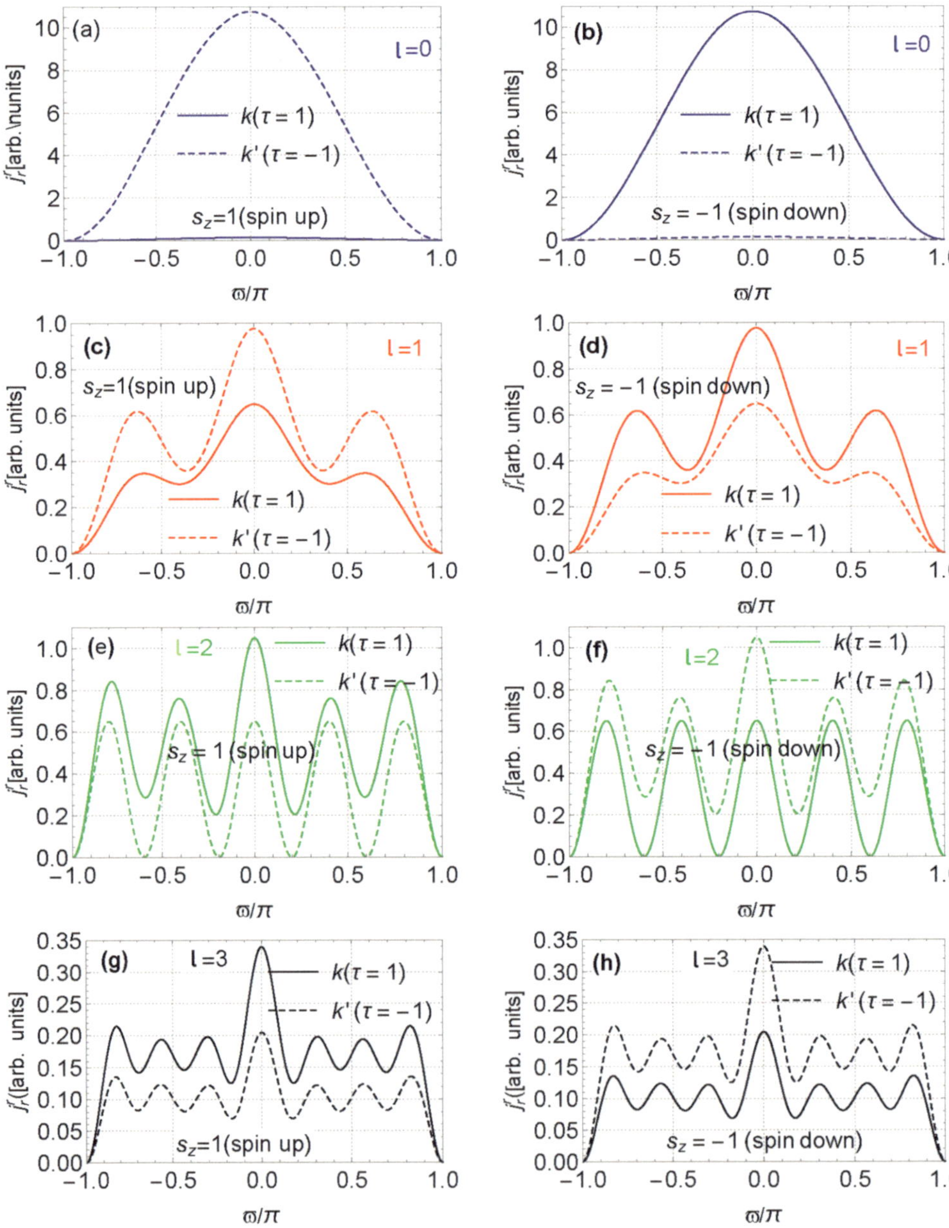

Figure 5.
Radial component of the far-field scattered current j_r^f in terms of angle ϖ for various l for the spin-up *and* spin-down *states in two valleys k and k' with fixed values $\rho = 7.75$, $U = 1$, and $\varepsilon = 0.0704$.*

in terms of energy (**Figure 3**). These findings indicate that the term spin-orbit interaction in Eq. (1) needs symmetry $|c_l|^2(-\tau, s_z) = |c_l|^2(\tau, -s_z)$.

For the *spin-up* and *spin-down* states, we graph the angular characteristic of the reflected radial component j_r^r in terms of *varpi* as shown in **Figure 5**. We note that j_r^r has a maximum for $\varpi = 0$ and a minimum for $\varpi = \pm\pi$. Furthermore, only forward scattering is preferred for the mode c_0 (**Figure 5a** and **b**). More favored scattering directions appear for higher modes. As a result, there are three preferred scattering directions for $l = 1$ (**Figure 5c** and **d**). Additionally, there are five favored scattering directions for $l = 2$ (**Figure 5e** and **f**) and seven preferred scattering directions for $l = 3$ (**Figure 5g** and **h**). Generally, each mode has a $2l + 1$ favored scattering direction that is apparent but with a different amplitude [30], although the mode ($l = 0$) has a larger amplitude than the higher modes ($l > 0$). The electron density profile at the dot reflects resonant scattering by only one of the normal modes. As a result, in both the up and down states, the dependency of j_r^r on τ is symmetric with respect to $\pm\tau$ and $\pm s_z$, i.e. $j_r^r(-\tau, s_z) = j_r^r(\tau, -s_z)$.

The radial component of the far-field scattered current j_r^r as a function of incident energy is shown in **Figure 6** for the states (a): ($k(\tau = 1), k'(\tau = -1)$, *spin-up*), and (b): ($k(\tau = 1), k'(\tau = -1)$, *spin-down*) for fixed values $U = 1$ and $\rho = 4$. In all $\varpi = 0$ (red curve) and $\varpi = 2\pi/3$ (blue curve) [30]. When $\varepsilon \to 0$, j_r^r for the two values of ϖ, when

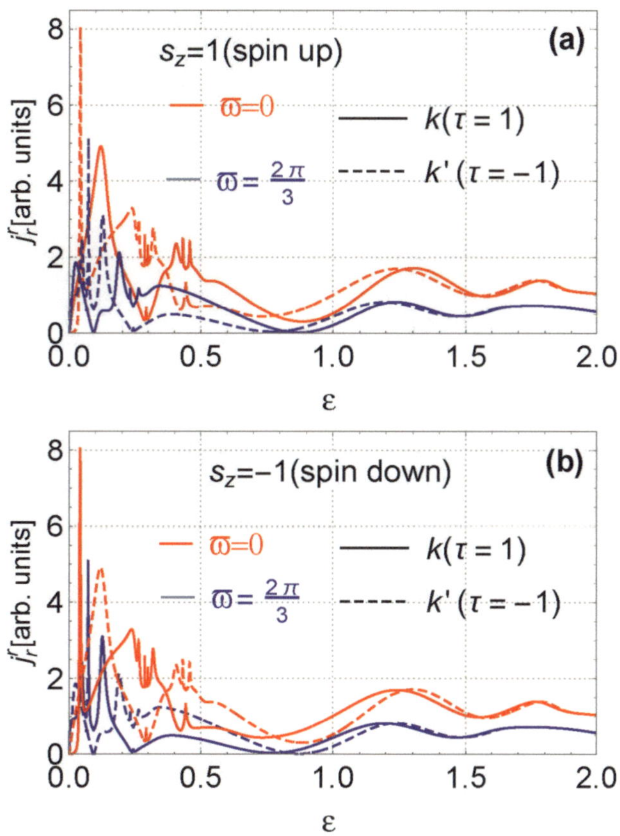

Figure 6.
The radial component of the far-field scattered current j_r^r in terms of the incident energy ε for various ϖ with fixed values of $U = 1$ and $\rho = 2$ for the spin-up *and* spin-down *states in two valleys k and k'.*

ε grows to the value 0.5, we see the emergence of peaks of resonances with a maximum peak for $(\varpi, \tau, s_z) = (0, -1, 1)$, as shown in **Figure 6a**). Although j_r^r presents oscillatory behavior in the regime $0, 5 < \varepsilon < 1.5$, in the regime $\varepsilon \geq 1.5$, j_r^r exhibits damped oscillatory behavior with the symmetry $j_r^r(\varpi, \tau, s_z) = j_r^r(\varpi, -\tau, s_z)$. **Figure 6b** indicates that when $\tau \to -\tau$ and $s_z \to -s_z$, in an analogous method to write $j_r^r(\varpi, -\tau, s_z) = j_r^r(\varpi, \tau, -s_z)$, the behavior of j_r^r is identical to that of **Figure 6a**.

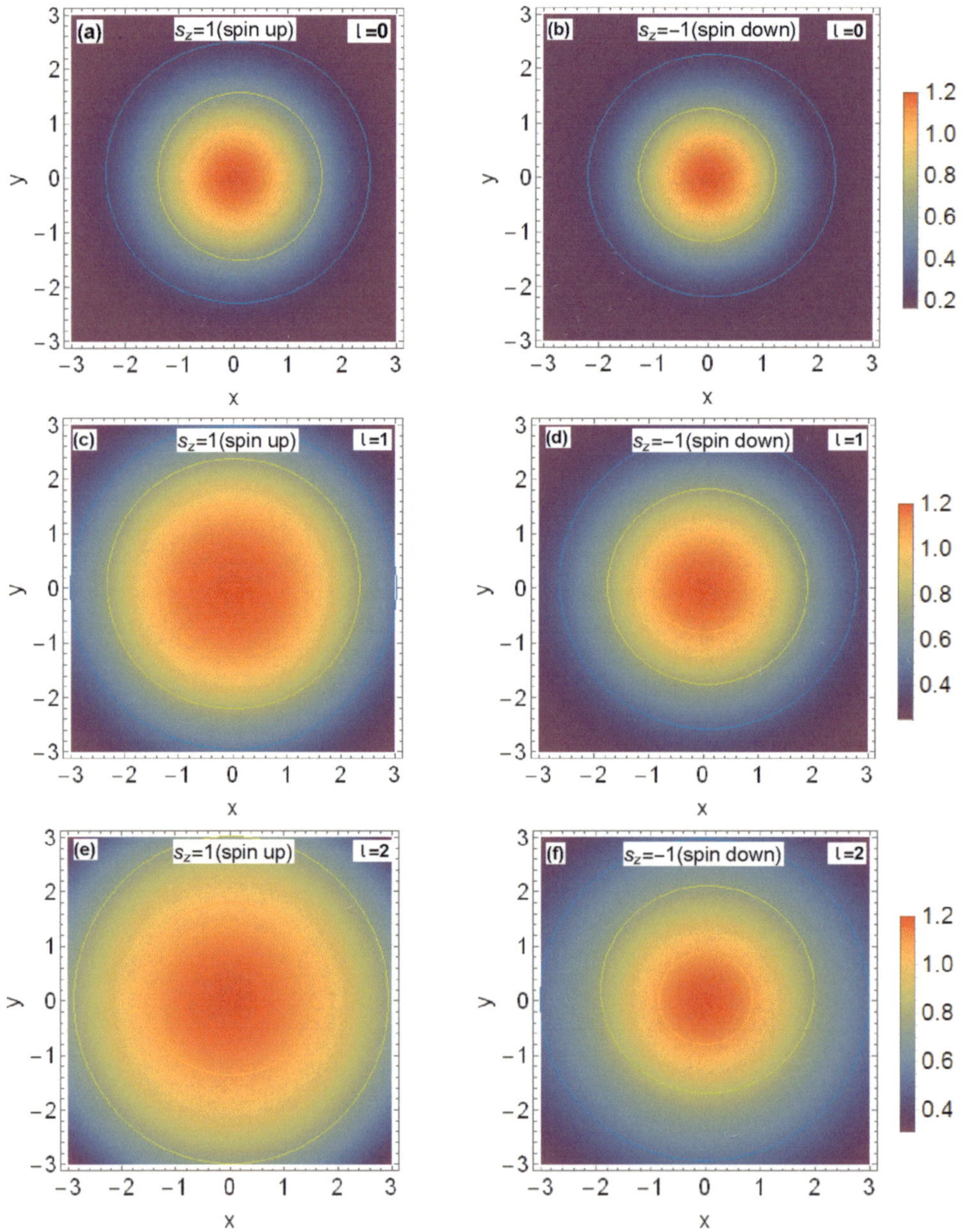

Figure 7.
Spatial density profile $\psi_t^\dagger \psi_t$ in the vicinity of the quantum dot for various l for the spin-up *and* spin-down *states in two valleys k and k′ for fixed values of $\rho = 3$, $\varepsilon = 0.078$, and $U = 1$.*

The electron density profile around the dot reflects the resonant scattering of a single mode. The density inside the quantum dot is described by:

$$\psi_t^\dagger \psi_t = |\beta_l|^2 \left[\left(j_l(\gamma r)^2 + \left(\mu j_{l+1}(\gamma r) \right)^2 \right] . \tag{36}$$

For the modes β_0, β_1, β_2, and β_3, we represent the spatial density $\psi_t^\dagger \psi_t$ in the quantum dot, as shown in **Figure 7**. For both *spin-up* and *spin-down* states, the modes β_l have a maximum electron density at the center of the quantum dot, and as the size of the quantum dot rises, the electron density drops. We also show that the electron density is becoming significant as the angular monument l is steadily increased. Furthermore, the *spin-up* and *spin-down* electron densities are not comparable; more interestingly, the electron density inside the dot has considerably risen, indicating momentary particle entrapment at scattering resonances.

4. Summary

The scattering of a planar Dirac electron wave on a circular quantum dot defined electrostatically in the MoS_2 monolayer has been investigated. The scattering parameters α_l and β_l, which characterize the features of our systems, were calculated using the continuity equation at the quantum dot's borders. The radial component of thev current density, as well as the scattering efficiency and square modulus of the scattering coefficient, were computed. In two energy regimes of the incoming electron, $\varepsilon < U$ and $\varepsilon \geq U$, the scattering of a planar Dirac electron wave has been examined.

In the regime $\varepsilon < U$, where the incoming electron has a low energy, Q exhibits a damped oscillatory behavior with emergent peaks owing to the excitation of the dot's normal modes; tiny values of ε correlate with high amplitudes of Q. We have proven that Q exhibits an oscillatory behavior in the other regime $\varepsilon \geq U$. On the other hand, we discovered that $Q(-\tau, s_z) = Q(\tau, -s_z)$ has a remarkable valley symmetry.

To find the resonances, we looked at the energy dependence of the square module of the scattering parameters $|c_l|^2$. We discovered that only the lowest scattering coefficient is non-null near $\varepsilon \to 0$, but as ε increased, the remaining coefficients began to notice significant contributions. The consecutive emergence of modes is interspersed with sudden and sharp peaks of various ε, $|c_l|^2$, but for a not large ε, we have shown that the sequential appearance of modes is interspersed with abrupt and sharp peaks of distinct $|c_l|^2$. We discovered that each mode has $(2l + 1)$ preferred paths of scattering visible with distinct amplitudes when it comes to the angular feature of the reflected radial component. Furthermore, we have demonstrated that the density of electrons inside the quantum dot has grown significantly, indicating that electrons have been temporarily trapped during the scattering resonances.

Author details

Rachid Houça[1,2]*, Abdelhadi Belouad[1], Abdellatif Kamal[1,3], El Bouâzzaoui Choubabi[1] and Mohammed El Bouziani[1]

1 Faculty of Sciences, L.P.M.C. Laboratory, Theoretical Physics Group, Chouaïb Doukkali University, El Jadida, Morocco

2 Faculty of Sciences, L.P.T.H.E. Laboratory, Theoretical Physics and High Energy, Ibn Zohr University, Agadir, Morocco

3 ISPS2I Laboratory, National Higher School of Arts and Crafts (ENSAM), Hassan II University, Casablanca, Morocco

*Address all correspondence to: houca.rachid@gmail.com

References

[1] Novoselov KS, Geim AK, Morozov SV, Jiang D, Katsnelson MI, Grigorieva IV, et al. Two-dimensional gas of massless Dirac fermions in graphene. Nature. 2005;**438**:197

[2] Geim AK, Novoselov KS. The rise of grapheme. Nature Materials. 2007;**6**:183

[3] Mak KF, He K, Shan J, Heinz TF. Control of valley polarization in monolayer MoS_2 by optical helicity. Nature Nanotechnology. 2012;**7**:494

[4] Jones AM, Yu H, Ghimire NJ, Wu S, Aivazian G, Ross JS, et al. Optical generation of excitonic valley coherence in monolayer WSe_2. Nature Nanotechnology. 2013;**8**:634

[5] Zaumseil J. Electronic control of circularly polarized light emission. Science. 2014;**344**:702

[6] Mak KF, Lee C, Hone J, Shan J, Heinz TF. Atomically Thin MoS_2: A New Direct-Gap Semiconductor. Physical Review Letters. 2010;**105**: 136805

[7] Splendiani A, Sun L, Zhang Y, Li T, Kim J, Chim CY, et al. Emerging Photoluminescence in Monolayer MoS_2. Nano Letters. 2010;**10**:1271

[8] Tongay S, Zhou J, Ataca C, Lo K, Matthews TS, Li J, et al. Thermally Driven Crossover from Indirect toward Direct Bandgap in 2D Semiconductors: MoSe2 versus MoS_2. Nano Letters. 2012; **12**:5576

[9] Ross JS, Wu S, Yu H, Ghimire N, Jones A, Aivazian G, et al. Electrical control of neutral and charged excitons in a monolayer semiconductor. Nature Communications. 2013;**4**:1474

[10] Zeng H, Liu GB, Dai J, Yan Y, Zhu B, He R, et al. Optical signature of symmetry variations and spin-valley coupling in atomically thin tungsten dichalcogenides. Scientific Reports. 2013;**3**:1608

[11] Radisavljevic B, Radenovic A, Brivio J, Giacometti V, Kis A. Single-layer MoS_2 transistors. Nature Nanotechnology. 2011;**6**:147

[12] Lembke D, Kis A. Breakdown of High-Performance Monolayer MoS2 Transistors. ACS Nano. 2012;**6**:10070

[13] Lin MW, Liu L, Lan Q, Tan X, Dhindsa KS, Zeng P, et al. Mobility enhancement and highly efficient gating of monolayer MoS2 transistors with polymer electrolyte. Journal of Physics D. 2012;**45**:345102

[14] Bao W, Cai X, Kim D, Sridhara K, Fuhrer MS. High mobility ambipolar MoS_2 field-effect transistors: Substrate and dielectric effects. Applied Physics Letters. 2013;**102**:042104

[15] Larentis S, Fallahazad B, Tutuc E. Field-effect transistors and intrinsic mobility in ultra-thin $MoSe_2$ layers. Applied Physics Letters. 2012;**101**: 223104

[16] Sallen G, Bouet L, Marie X, Wang G, Zhu CR, Han WP, et al. Robust optical emission polarization in MoS_2 monolayers through selective valley excitation. Physical Review B. 2012;**86**: 081301

[17] Xiao D, Yao W, Niu Q. Valley-contrasting physics in graphene: Magnetic moment and topological transport. Physical Review Letters. 2007; **99**:236809

[18] Yao W, Xiao D, Niu Q. Valley-dependent optoelectronics from inversion symmetry breaking. Physical Review B. 2008;**77**:235406

[19] Gunawan O, Shkolnikov YP, Vakili K, Gokmen T, De Poortere EP, Shayegan M. Valley susceptibility of an interacting two-dimensional electron system. Physical Review Letters. 2006; **97**:186404

[20] Wilcoxon J, Samara G. Strong quantum-size effects in a layered semiconductor: MoS_2 nanoclusters. Physical Review B. 1995;**51**:7299

[21] Benson J, Li M, Wang S, Wang P, Papakonstantinou P. Microwave-Assisted Reactant-Protecting Strategy toward Efficient MoS2 Electrocatalysts in Hydrogen Evolution Reaction. ACS Applied Materials & Interfaces. 2015;**7**: 14113

[22] Ren X, Pang L, Zhang Y, Ren XD, Fan H, Liu SZ. One-step hydrothermal synthesis of monolayer MoS_2 quantum dots for highly efficient electrocatalytic hydrogen evolution. Journal of Materials Chemistry A. 2015;**3**:10693

[23] Xu S, Li D, Wu P. One-Pot, Facile, and Versatile Synthesis of Monolayer MoS_2/WS_2 Quantum Dots as Bioimaging Probes and Efficient Electrocatalysts for Hydrogen Evolution Reaction. Advanced Functional Materials. 2015;**25**: 1136

[24] Gopalakrishnan D, Damien D, Shaijumon M. MoS_2 Quantum Dot-Interspersed Exfoliated MoS_2 Nanosheets. ACS Nano. 2014;**8**:5297

[25] Gopalakrishnan D, Damien D, Li B, Gullappalli H, Pillai VK, Ajayan PM, et al. Electrochemical synthesis of luminescent MoS_2 quantum dots. Chemical Communications. 2015;**51**:6293

[26] Li BL, Chen LX, Zou HL, Lei JL, Luo HQ, Li NB. Electrochemically induced Fenton reaction of few-layer MoS_2 nanosheets: preparation of luminescent quantum dots via a transition of nanoporous morphology. Nanoscale. 2014;**6**:9831

[27] Oliveira D, Jiyong F, Villegas-Lelovsky L, Dias AC, Fanyao Q. Valley Zeeman energy in monolayer MoS_2 quantum rings: Aharonov-Bohm effect. Physical Review B. 2016;**93**: 205422

[28] Xiao D, Liu GB, Feng W, Xu X, Yao W. Coupled Spin and Valley Physics in Monolayers of MoS_2 and Other Group-VI Dichalcogenides. Physical Review Letters. 2012;**108**:196802

[29] Schnez S, Ensslin K, Sigrist M, Ihn T. Analytic model of the energy spectrum of a graphene quantum dot in a perpendicular magnetic field. Physical Review B. 2008;**78**:195427

[30] Heinisch RL, Bronold FX, Fehske H. Mie scattering analog in graphene: Lensing, particle confinement, and depletion of Klein tunneling. Physical Review B. 2013;**87**:155409

[31] Berry MV, Mondragon RJ. Neutrino billiards: time-reversal symmetry-breaking without magnetic fields. Proceedings of the Royal Society. London A. 1987;**412**:53

[32] Grujic M, Zarenia M, Chaves A, Tadie M, Farias GA, Peeters FM. Electronic and optical properties of a circular graphene quantum dot in a magnetic field: influence of the boundary conditions. Physical Review B. 2011;**84**:205441

[33] Schulz C, Heinisch RL, Fehske H. Scattering of two-dimensional Dirac fermions on gate-defined oscillating

quantum dots. Physical Review B. 2015; **91**:045130

[34] Belouad A, Zahidi Y, Jellal A, Bahlouli H. Electron scattering in gapped graphene quantum dots. EPL. 2018; **28002**:123

[35] Xue Wu Z, Long CY, Wu J, Z. Wen Long and T. Xu. Scattering in graphene quantum dots under generalized uncertainty principle. International Journal of Modern Physics A. 2019;**4**:1950212

Chapter 12

Determination of Qubit Entanglement in One-step Double Photoionization of Helium Atom

Minakshi Chakraborty and Sandip Sen

Abstract

Quantum entanglement is a unique phenomenon of quantum mechanics that explains how two subatomic particles are correlated even if they are separated by a vast distance. The phenomena of quantum entanglement are useful resources for quantum information. In this chapter, we will study the entanglement properties of bipartite states of two electronic qubits, without observing spin-orbit interaction (SOI), produced by single-step double photoionization in helium atom following the absorption of a single photon. In absence of SOI, Russell-Saunders coupling (L-S coupling) is applicable. We observe that the entanglement depends significantly on the direction of the ejection, as well as the spin quantization of photoelectrons.

Keywords: quantum entanglement, qubit, double photoionization, concurrence, density matrix

1. Introduction

Quantum entanglement is the foremost prediction of quantum mechanics and one of the resources needed in quantum computing [1, 2]. Nowadays, it is an approach to solving time and processing power-consuming problems using quantum computers. A qubit is a quantum bit, the counterpart in quantum computing to the binary digit or bit of classical computing [3–5]. An important distinguishing feature between a qubit and a classical bit is that qubits can exhibit quantum entanglement. Quantum entangle ment is a nonlocal property that allows a set of qubit to express a higher correlation that is not possible in classical systems. Qubits are employed in areas other than quantum computing, such as sensors. This helps in network and communication channels, improving the security of data stored in the quantum field and protecting it, as well as the speedier delivery of messages and encrypted networks for security-related information.

Einstein et al. began studying quantum entanglement in 1935 after presenting the EPR conundrum [6]. Bell [7] demonstrated the fault in EPR arguments by demonstrating that the notion of locality invoked in the EPR conundrum was incompatible with the hidden variables interpretation of the quantum theory. On the other hand, Bell's theorem offers one of the potential ways to determine if two specific particles form an EPR pair or entangled state. The strong correlations of entangled particles are used as resources for quantum cryptography [8], quantum teleportation [9], and quantum computation [10] because entanglement allows multiple states of qubits to be acted on simultaneously unlike classical bits that can have one value at a time.

The bipartite entangled states of photons have been generated [11–13]. Being an excellent carrier of information, a photon is not suitable for long-term storage as it is immediately destroyed as soon as one tries to detect it. Some theories have been developed for studying bipartite states of electronic qubits produced by photoionization [14]. One of the simplest processes for producing EPR pairs of particles with non-zero rest mass is the simultaneous ejection of two electrons following the absorption of a single photon in an atom or molecule [15, 16]. This process is known as double photoionization (DPI). Here, entanglement is produced due to the process taking place inside an atom. We discuss DPI of an atom for generating bipartite states of two electronic qubits (say e_1 and e_2). In the absence of spin-orbit interaction (SOI), these particles are entangled with regard to their spin-angular momenta. In addition, the electrons in DPI are released in all directions and with all kinetic energies (subject to the conservation of total energy). Additionally, any direction can be quantized for its spins. Utilizing the density matrix (DM) of DPI concurrence [17–19] allows us to measure the level of entanglement.

We first establish some pertinent conventions in Section 2 and then quickly discuss the density operator (DO) and states for an atom's DPI. This operator is appropriate when the target atom is in its ground state before DPI and the ionizing electromagnetic radiation is in a pure state of polarization. In the absence of SOI, this DO is used to develop an expression for the density matrix (DM) required to examine the quantum entanglement characteristics of the photoelectrons' and photons' spin states in Russell-Saunders coupling. In Section 3, we examine entanglement in a DPI system for qubits. Section 4 provides a quantitative application of the qubits for the DPI of the helium atom. The conclusion is given in Section 5, the last section.

2. Preliminaries

2.1 Density operator

Let's use e_1 and e_2 to symbolize the two freely moving electrons whose entanglement characteristics we are interested in. The $i(=1, 2)$th electron's propagation vector is $\vec{k}_i = (k_i, \theta_i, \phi_i)$, and its kinetic energy is provided by $\varepsilon_i = \hbar^2 k_i^2/2m$. These two electrons are a crucial component of atom A and are supposed to be simultaneously released from it after the absorption of a single photon. $\mu_i(= \pm 1/2)$ denotes the projection of the spin angular momentum of the ith electron along its spin quantization direction $\hat{u}_i = (\alpha_i, \beta_i)$. If A^{2+} denotes the residual dication, then our process can schematically be represented by

$$h\nu_r(|l_r| = 1, m_r) + A|0\rangle \rightarrow A^{2+}|f\rangle + e_1\left(\vec{k}_1; \mu_1\hat{u}_1\right) + e_2\left(\vec{k}_2; \mu_2\hat{u}_2\right). \quad (1)$$

Here, in the electric dipole (E_1) approximation, $E_r = h\nu_r$ and $|l_r| = 1$ represent, respectively, the energy and angular momentum of the photon absorbed by atom A. The parameter m_r stands for the photon's state of polarization, where $|0\rangle$ and $|f\rangle$, respectively, represent the bound electronic states of A with energy E_0 and the remaining doubly charged photoion A^{2+} with energy E_f. The direction of the electric vector of the linearly polarized ($m_r = 0$) radiation involved in the process (1) defines the quantization axis of our space (or photon) frame of reference; if the ionizing radiation is circularly polarized ($m_r = \pm 1$) or unpolarized, the direction of incidence then determines the polar axis of the photon frame.

Let us use $\rho_r = |m_r\rangle\langle m_r|$ and $\rho_0 = |0\rangle\langle 0|$ to represent the density operators of the ionizing radiation and the unpolarized atom A before DPI, respectively. Before the interaction between the two occurs, the incident photon and the atom are uncorrelated. This indicates that the direct product determines the density operator for the combined (atom + photon) system of Eq. (1).

$$\rho_i = \rho_0 \otimes \rho_r \quad (2)$$

Let us denote by F_p the photoionization operator in the E_1 approximation. Then the density operator of the combined (A^{2+} + e_1 + e_2) system in Eq. (1) after DPI becomes

$$\rho_f = K_p F_p \rho_i F_p^{+}. \quad (3)$$

The E_1 photoionization operator F_p has been defined in Appendix A. Here, $K_p = 3\pi\left(e^2/\alpha_0 E_r\right)^2$ with α_0 the dimensionless fine structure constant [20].

2.2 Definitions of E_1 photoionization operator and concurrence

In the present case, E_1 photoionization operator F_p for n_e-electron system can be defined as [20]:

$$F_p = \sqrt{\frac{4\pi\alpha^3 E_r^3 m_e}{4e^4\hbar^2}} \sum_{i=1}^{n_e} \hat{\xi}_{m_r} \cdot \vec{r}_i, \quad (4)$$

and

$$F_p = \sqrt{\frac{4\pi\alpha^3 E_r^3 a_0^2 m_e}{3\hbar^2}} \sum_{i=1}^{n_e} \hat{\xi}_{m_r} . \vec{\nabla}_i. \quad (5)$$

The operators of Eqs. (4) and (5) represent the interaction of the atomic electrons (their number being n_e) with the incident electromagnetic radiation in the E_1 length and velocity approximations, respectively. In Eq. (4), $\vec{r}_i$ is the position vector and in Eq. (5), $\vec{\nabla}_i \left[= (-1)^{1/2} \; \vec{p}_i/\hbar\right]$ is the linear momentum of ith bound atomic electron. Here $\hat{\xi}_{m_r}$ is the spherical unit vector [21] in the direction of polarization of the incident.

The *concurrence* (C) is a very successful and widely used measure for quantifying quantum entanglement between two qubits. It is an additive and operational measure of entanglement. Additivity is a very desirable property that can reduce calculational complexity of entanglement [4, 17, 22]. For any state $|\phi\rangle_{AB}$ in a $d \otimes d'$ $(d \leq d')$ quantized system, it can be written as:

$$C = \sqrt{2\left(1 - Trace\rho_A^2\right)}, \tag{6}$$

where ρ_A is the reduced density matrix defined as $\rho_A = trace_B\left(|\phi\rangle_{AB}\langle\phi|\right)$. When $0 \leq C \leq 1$ always; $C = 0$ for a separable (unentangled) state; $C > 0$ for a nonseparable (entangled) state.

3. Entanglement between two electronic qubits for DPI

Here, we calculate the DM for the angle- and spin-resolved DPI of an atom without considering SOI into account in either the bound electronic states of A and A^{2+} or the continua of the two photoelectrons (e_1, e_2) ejected in the process (1). Here LS-coupling is applicable. Therefore, the total orbital angular momenta $(\vec{L}_0, \vec{L}_f)$ and the spin angular momenta $(\vec{S}_0, \vec{S}_f)$ of A and A^{2+} are conserved quantities. If the orbital angular momentum of the photoelectron e_1 is $\vec{l}_1$ and that of e_2 is $\vec{l}_2$ with their respective spin angular momenta $\left(\frac{1}{2}\right)_1$ and $\left(\frac{1}{2}\right)_2$, we then have

$$\vec{L}_0 + \vec{l}_r = \vec{L}_f + \vec{l}\left(= \vec{l}_1 + \vec{l}_2\right) \tag{7}$$

and

$$\vec{S}_0 = \vec{S}_f + \vec{s}_t\left(= \left(\frac{1}{2}\right)_1 + \left(\frac{1}{2}\right)_2\right). \tag{8}$$

Here M_{L_0}, M_{S_0}, M_{L_f}, and M_{S_f} are the respective projections of $\vec{L}_0$, $\vec{S}_0$, $\vec{L}_f$, and $\vec{S}_f$ along the polar axis of the space frame, then, in Eq. (1), the bound electronic state of atom A is $|0\rangle \equiv |L_0 S_0 M_{L_0} M_{S_0}\rangle$ and that of the dication A^{2+} is $|f\rangle \equiv \left|L_f S_f M_{L_f} M_{S_f}\right\rangle$. The density operator (2) for the combined (atom + photon) system becomes

$$\rho_i = \frac{1}{(2L_0+1)(2S_0+1)} \sum_{M_{L_0} M_{S_0}} |0; 1m_r\rangle\langle 0; 1m_r|, \tag{9}$$

here we have defined $|0; 1m_r\rangle \equiv |0\rangle|1m_r\rangle$.

In order to calculate the DM for the $(A^{2+}+e_1+e_2)$ system in process (1), we now calculate the matrix elements ρ_i and ρ_f. Following the procedures given in Ref. [15], the matrix elements in the present case are given by

$$\left\langle f; \vec{k}_1, \mu_1\hat{u}_1; \vec{k}_2, \mu_2\hat{u}_2 \middle| \rho_f \middle| f; \vec{k}_1, \mu_1'\hat{u}_1; \vec{k}_2, \mu_2'\hat{u}_2 \right\rangle = \frac{K_p}{(2L_0+1)(2S_0+1)}$$
$$\sum_{M_{L_0}M_{S_0}} \left\langle f; \vec{k}_1, \mu_1\hat{u}_1; \vec{k}_2, \mu_2\hat{u}_2 \middle| F_p \middle| 0; 1\mathrm{m}_r \right\rangle \mathrm{x} \left\langle f; \vec{k}_1, \mu\prime_1\hat{u}_1; \vec{k}_2, \mu\prime_2\hat{u}_2 \middle| F_p \middle| 0; 1\mathrm{m}_r \right\rangle^*. \tag{10}$$

As the DM in Eq. (10) is Hermitian, we can write

$$\begin{aligned} &\left\langle f; \vec{k}_1, \mu_1\hat{u}_1; \vec{k}_2, \mu_2\hat{u}_2 \middle| \rho_f \middle| f; \vec{k}_1, \mu_1'\hat{u}_1; \vec{k}_2\mu_2'\hat{u}_2 \right\rangle \\ &= \left\langle f; \vec{k}_1, \mu\prime_1\hat{u}_1; \vec{k}_2, \mu_{\prime 2}\hat{u}_2 \middle| \rho_f \middle| f; \vec{k}_1, \mu_1\hat{u}_1; \vec{k}_2\mu_2\hat{u}_2 \right\rangle^* \end{aligned} \tag{11}$$

Next, we evaluate the matrix elements of *E1* photoionization operator F_p occurring on the right-hand side of Eq. (10). To this end, we first introduce the coupling suggested by Eqs. (7) and (8) in each of the bras and kets used to calculate the matrix elements of F_p. We therefore can write

$$|0, 1m_r\rangle = \sum_{LM_L} (-1)^{1-L_0-M_L} \sqrt{2L+1} \begin{pmatrix} L_0 & 1 & L \\ M_{L_0} & m_r & -M_L \end{pmatrix} |(L_0 1)LM_L; S_0 M_{S_0}\rangle \tag{12}$$

and

$$\begin{aligned} \left| f; \hat{k}_1, \mu_1\hat{u}_1; \hat{k}_2\mu_2\hat{u}_2 \right\rangle &= (-1)^{-L_f-S_f} \sum_{\substack{\ell_1\ell_2\ell \\ n_1n_2N}} \sum_{\substack{\nu_1\nu_2 \\ s_t\nu}} \sum_{\substack{L_TM_{L_T} \\ SM_S}} i^{\ell_1+\ell_2} e^{-i(\sigma_{\ell_1}+\sigma_{\ell_2})} (-1)^{-\ell_1+\ell_2+\ell+s_t-N-\nu-M_{L_T}-M_S} \\ &x\sqrt{(2\ell+1)(2s_t+1)(2L_T+1)(2S+1)} \begin{pmatrix} \ell_1 & \ell_2 & \ell \\ m_1 & m_2 & -m \end{pmatrix} \begin{pmatrix} \frac{1}{2} & \frac{1}{2} & s_t \\ \nu_1 & \nu_2 & -\nu \end{pmatrix} \begin{pmatrix} L_f & \ell & L_T \\ M_{L_f} & N & -M_{L_T} \end{pmatrix} \\ &x \begin{pmatrix} S_f & s_t & S \\ M_{S_f} & \nu & -M_S \end{pmatrix} Y_{\ell_1}^{n_1^*}\left(\hat{k}_1\right) Y_{\ell_2}^{n_2^*}\left(\hat{k}_1\right) D_{\mu_1\nu_1}^{\frac{1}{2}^*}(\omega_1) D_{\mu_2\nu_2}^{\frac{1}{2}^*}(\omega_2) \left| (L_f\ell)L_TM_{L_T}; (S_fs_t)SM_S \right\rangle. \end{aligned} \tag{13}$$

Here, σ_{ℓ_1} and σ_{ℓ_2} are the Coulomb phases for l_1th and l_2th partial waves of the photoelectrons, respectively; *Ds* are the well-known rotational harmonics [23] with $\omega_1(\alpha_1, \beta_1, 0)$ and $\omega_2(\alpha_2, \beta_2, 0)$, the Euler angles that rotate the axis of the space-frame into the spin-polarization directions $\hat{u}_1$ and $\hat{u}_2$ (**Figure 1**), respectively. Furthermore, in (13), the properly antisymmetrized and asymptotically normalized [24] ket $\left|(L_f\ell)L_TM_{L_T}; (S_fs_y)SM_S\right\rangle$ represents the photoion A^{2+} in its electronic state and the two photoelectrons with their total orbitaland spin angular momenta coupled according to the scheme expressed in Eqs. (7) and (8).

Now we substitute in Eq. (9) the normalized condition of reference [25]

$$\begin{aligned} &\langle (L_f\ell)L_TM_{L_T}; (S_fs_t)SM_S | F_p | (L_0 1)LM_L; S_0M_{S_0} \rangle \\ &= \delta_{LL_T}\delta_{M_LM_{L_T}}\delta_{S_0S}\delta_{M_{S_0}M_S} \langle (L_f\ell)L | F_p | (L_0 1)L \rangle. \end{aligned}$$

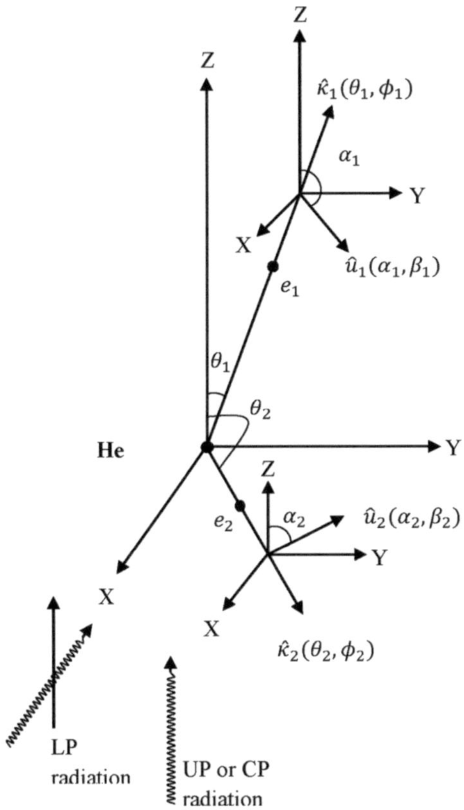

Figure 1.
Two electrons are emitted simultaneously from helium after photoabsorption.

This result arises from the conservation circumstances in (7) and (8).

By using Racah algebra to analytically evaluate as many of the sums that are contained there, the following subsequent equation is made simpler. It necessitates, for instance, the application of (a) the addition theorems (i.e., Eqs. (4.3.2) and (4.6.5) from [23] for rotational and spherical harmonics, (b) Eq. (6.2.5) [23] for transforming a single sum of the product of three 3-j symbols into a product of one 3-j and one 6-j symbols, (c) identity (5) given on page 453 in Ref. [26] for converting a product of two 3-j symbols and one 6-j symbol summed over two variables into a product of two 3-j and one 6-j symbols, (d) Eq. (14.42) from [27] that transforms a quadruple sum of the product of four 3-j symbols into a double sum containing two 3-j and one 9-j symbols, (e) Eq. (3.7.9) [23] to convert a phase factor into a 3-j symbol, (f) the orthogonality of 3-j symbols (3.7.7) [23], (g) relation (6.4.14) [23] to write a 9-j symbol in terms of a 6-j symbol, and (h) relation (6.4.14) [23] to turn the sum of the products of two 6-j symbols into one 6-j symbol. Due to these and other simplifications, the DM (10) is written as follows:

$$\left\langle f;\vec{k}_1,\mu_1\hat{u}_1;\vec{k}_2,\mu_2\hat{u}_2\middle|\rho_f\middle|f;\vec{k}_1,\mu_1'\hat{u}_1;\vec{k}_2\mu_2'\hat{u}_2\right\rangle =$$

$$\frac{K_p}{4\pi(2L_0+1)}\sum_{\substack{l_1l_2l\\ l_1'l_2'l'}}\sum_{\substack{L_1L_2M\\ LL'L_r}}\sum_{sQm_1m_2n}(-1)^{l_1+l_2+l'+Q-n+\mu_1+\mu_2+M_{S_f}}(2S_f+1)(2L_r+1)(2s+1)$$

$$\text{x}(2Q+1)\sqrt{(2L_1+1)(2L_2+1)}\begin{pmatrix}l_1 & l_1' & L_1\\ 0 & 0 & 0\end{pmatrix}\begin{pmatrix}l_2 & l_2' & L_2\\ 0 & 0 & 0\end{pmatrix}\begin{pmatrix}1 & 1 & L_r\\ m_r & -m_r & 0\end{pmatrix}\begin{pmatrix}L_1 & L_2 & L_r\\ M & -M & 0\end{pmatrix}$$

$$= \text{x}\begin{Bmatrix}1 & 1 & L_r\\ L' & L & L_0\end{Bmatrix}\begin{Bmatrix}l & l' & L_r\\ L' & L & L_f\end{Bmatrix}\begin{Bmatrix}l_1 & l_2 & l\\ l_1' & l_2' & l'\\ L_1 & L_2 & L_r\end{Bmatrix}\begin{pmatrix}\frac{1}{2} & \frac{1}{2} & s\\ \mu_1 & -\mu_1' & m_1\end{pmatrix}\begin{pmatrix}\frac{1}{2} & \frac{1}{2} & s\\ \mu_2 & -\mu_2' & m_2\end{pmatrix}\begin{pmatrix}S_f & S_f & Q\\ M_{S_f} & -M_{S_f}' & n\end{pmatrix}$$

$$\text{x}\begin{pmatrix}s & s & Q\\ m_1 & m_2 & n\end{pmatrix}\begin{Bmatrix}\frac{1}{2} & \frac{1}{2} & s\\ S_f & S_f & Q\end{Bmatrix}Y_{L_1}^{M}\left(\hat{k}_1\right)Y_{L_2}^{-M}\left(\hat{k}_2\right)D_{m_1n}^{s^*}(\omega_1)D_{m_2-n}^{s^*}(\omega_2)$$

$$\text{x}\langle L_fl|F(L)|L_01\rangle\langle L_fl'|F(L')|L_01\rangle^* \tag{14}$$

with

$$\langle L_fl|F(\mathrm{L})|L_01\rangle = (-i)^{l_1+l_2}e^{-i(\sigma_{l_1}+\sigma_{l_2})}(2L+1)\sqrt{(2l_1+1)(2l_2+1)(2l+1)}\langle(L_fl)\mathrm{L}|F_p|(L_01)L\rangle.$$

From Eq. (12), the DM for the angle- and spin-resolved DPI process (1) in the absence of SOI can be written in the following form:

$$\left\langle f;\vec{k}_1,\mu_1\hat{u}_1;\vec{k}_2,\mu_2\hat{u}_2\middle|\rho_f\middle|f;\vec{k}_1,\mu_1'\hat{u}_1;\vec{k}_2,\mu_2'\hat{u}_2\right\rangle = \frac{d^3\sigma(m_r)}{d\varepsilon_1d\hat{k}_1d\hat{k}_2}\sigma(S_0;S_f;\hat{u}_1,\hat{u}_2)_{M_{S_f}\mu_1\mu_2;M_{S_f}'\mu_1'\mu_2'}. \tag{15}$$

The first term, that is, the triple differential cross section (TDCS, i.e., $d^3\sigma(m_r)/d\varepsilon_1d\hat{k}_1d\hat{k}_2$) on the right-hand side of (15) depends upon the orbital angular momenta of A and A^{2+}; phase shifts, energies (ε_1, ε_2), and the directions ($\hat{k}_1$, $\hat{k}_2$) of the emitted electrons (e_1, e_2); the state of polarization (m_r) of the ionizing radiation; and the photoionization dynamics. It does not include spins of the photoelectrons or the target atom or the residual dication. Thus, $d^3\sigma(m_r)/d\varepsilon_1d\hat{k}_1d\hat{k}_2$ in the DM (15) describes purely the angular correlation between the photoelectrons in the L-S coupling scheme for the angular momenta of the particles involved in DPI (1). Its value is always positive. Here,

$$\frac{d^3\sigma(m_r)}{d\varepsilon_1 d\hat{k}_1 d\hat{k}_2} = (-1)^{m_r+L_0+L_f}\frac{K_p}{4\pi(2L_0+1)}\sum_{\substack{l_1l_2l\\ l'_1l'_2l'}}\sum_{\substack{L_1L_2M\\ LL'L_r}}(-1)^{l'_1+l'_2+l'}(2L_r+1)$$

$$\text{x}\sqrt{(2L_1+1)(2L_2+1)}\begin{pmatrix} l_1 & l'_1 & L_1\\ 0 & 0 & 0\end{pmatrix}\begin{pmatrix} l_2 & l'_2 & L_2\\ 0 & 0 & 0\end{pmatrix}\begin{pmatrix} 1 & 1 & L_r\\ m_r & -m_r & 0\end{pmatrix}\begin{pmatrix} L_1 & L_2 & L_r\\ M & -M & 0\end{pmatrix}\begin{Bmatrix} 1 & 1 & L_r\\ L' & L & L_0\end{Bmatrix}$$

$$\text{x}\begin{Bmatrix} l & l' & L_r\\ L' & L & L_f\end{Bmatrix}\begin{Bmatrix} l_1 & l_2 & l\\ l'_1 & l'_2 & l'\\ L_1 & L_2 & L_r\end{Bmatrix}Y^M_{L_1}(\hat{k}_1)Y^{-M}_{L_2}(\hat{k}_2)\langle L_f l|F(L)|L_0 1\rangle\langle L_f l'|F(L')|L_0 1\rangle^*. \tag{16}$$

The second term (i.e., $\sigma(S_0; S_f; \hat{u})_{M_{S_f}\mu; M'_{S_f}\mu'}$) present on the right-hand side of Eq. (15) is the spin-correlation density matrix (SCDM), which completely determines the entanglement properties among electronic qubit (e_p) and ionic qudit (M^+), can be written as:

$$\sigma_{e_1e_2} = (-1)^{\mu_1+\mu'_2}\sum_{ss_tm_1m_2n}(-1)^{s_t-n}(2s+1)$$

$$\text{x}\begin{pmatrix} \frac{1}{2} & \frac{1}{2} & s\\ \mu_1 & -\mu'_1 & m_1\end{pmatrix}\begin{pmatrix} \frac{1}{2} & \frac{1}{2} & s\\ \mu_2 & -\mu'_2 & m_2\end{pmatrix}\begin{Bmatrix} \frac{1}{2} & \frac{1}{2} & s\\ \frac{1}{2} & \frac{1}{2} & s_t\end{Bmatrix}D^{s^*}_{m_1n}(\omega_1)D^{s^*}_{m_2-n}(\omega_2)\ . \tag{17}$$

Considering the condition given by Eq. (8), we see that $|S_f - S_0|$ can take only two values: 0 and 1. Let us consider the entanglement between two electrons e_1 and e_2 in both cases. In order to calculate the SCDMs, we have to consider the real part of the outgoing wave function [28]. Before writing the DM, we first consider

$$\alpha = \alpha_1 - \alpha_2,\ c_1 = \cos\beta_1,\ s_1 = \sin\beta_1,\ c_2 = \cos\beta_2,\ s_2 = \sin\beta_2,\ \text{ and } c = \cos\alpha$$

i. Let us first consider the case $|S_f - S_0| = 0$. We obtain the following SCDM from (17):

$$\sigma_{e_1e_2} =$$

$\mu'_1\mu'_2$ / $\mu_1\mu_2$	$\frac{1}{2}\frac{1}{2}$	$\frac{1}{2}\ -\frac{1}{2}$	$-\frac{1}{2}\frac{1}{2}$	$-\frac{1}{2}-\frac{1}{2}$
$\frac{1}{2}\frac{1}{2}$	$0.25-\frac{1}{4}(c_1c_2+s_1s_2c)$	$\frac{1}{4}(c_1s_2-s_1c_2c)$	$\frac{1}{4}(s_1c_2-c_1s_2c)$	$\frac{1}{4}(c-s_1s_2-c_1c_2c)$
$\frac{1}{2}\ -\frac{1}{2}$	$\frac{1}{4}(c_1s_2-s_1c_2c)$	$0.25+\frac{1}{4}(c_1c_2+s_1s_2c)$	$-\frac{1}{4}(c+s_1s_2+c_1c_2c)$	$-\frac{1}{4}(s_1c_2-c_1s_2c)$
$-\frac{1}{2}\frac{1}{2}$	$\frac{1}{4}(s_1c_2-c_1s_2c)$	$-\frac{1}{4}(c+s_1s_2+c_1c_2c)$	$0.25+\frac{1}{4}(c_1c_2+s_1s_2c)$	$-\frac{1}{4}(c_1s_2-s_1c_2c)$
$-\frac{1}{2}-\frac{1}{2}$	$\frac{1}{4}(c-s_1s_2-c_1c_2c)$	$-\frac{1}{4}(s_1c_2-c_1s_2c)$	$-\frac{1}{4}(c_1s_2-s_1c_2c)$	$0.25-\frac{1}{4}(c_1c_2+s_1s_2c)$

Table 1.
SCDM of qubit-qubit system for the case $|S_f - S_0| = 0$.

We calculate from **Table 1** that the det $(\sigma_{e_1e_2}) = -\frac{1}{16}\cos^2\alpha$. In order to quantifying entanglement, we have obtained concurrence, using SCDM of **Table 1**, according to the definition given in Eq. (6), which yields

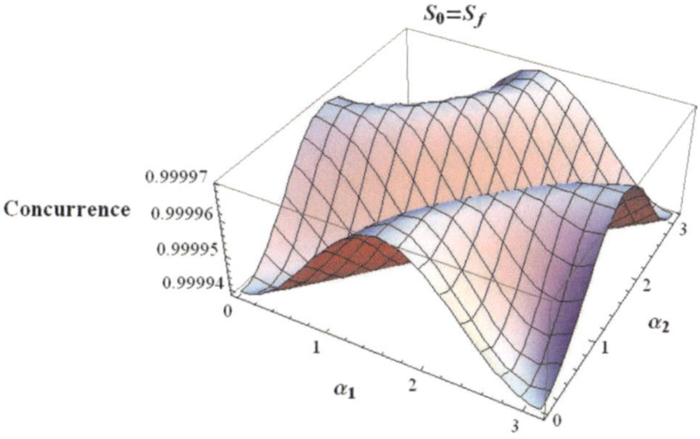

Figure 2.
Variation of concurrence for the case $|S_f - S_o| = 0$ *with respect to spin quantization directions* α_1 *and* α_2.

$$C = \sqrt{1 - \frac{1 + (\cos\alpha_1 \cos\alpha_2 + \cos\alpha \sin\alpha_1 \sin\alpha_2)^2}{16384}}. \tag{18}$$

Eq. (18) means [4, 17, 22] that the values of C are positive and so the spin state $|S_f - S_0| = 0$ is entangled, depending on α_1 and α_2. This variation of concurrence is shown in **Figure 2**.

We can see from **Figure 2** that the states are entangled (the values of concurrence are positive) and separable (for the zero value of concurrence) depending on spin quantization directions.

ii. Considering the case of $|S_f - S_0| = 1$, we procure the following SCDM from Eq. (17):

$$\sigma_{e_1 e_2} =$$

$\mu_1'\mu_2'$ / $\mu_1\mu_2$	$\frac{1}{2}\frac{1}{2}$	$\frac{1}{2}\ -\frac{1}{2}$	$-\frac{1}{2}\frac{1}{2}$	$-\frac{1}{2}-\frac{1}{2}$
$\frac{1}{2}\frac{1}{2}$	$0.25 + \frac{1}{12}(c_1c_2 + s_1s_2c)$	$-\frac{1}{12}(c_1s_2 - s_1c_2c)$	$-\frac{1}{12}(s_1c_2 - c_1s_2c)$	$-\frac{1}{12}(c - s_1s_2 - c_1c_2c)$
$\frac{1}{2}\ -\frac{1}{2}$	$-\frac{1}{12}(c_1s_2 - s_1c_2c)$	$0.25 - \frac{1}{12}(c_1c_2 + s_1s_2c)$	$\frac{1}{12}(c + s_1s_2 + c_1c_2c)$	$\frac{1}{12}(s_1c_2 - c_1s_2c)$
$-\frac{1}{2}\frac{1}{2}$	$-\frac{1}{12}(s_1c_2 - c_1s_2c)$	$\frac{1}{12}(c + s_1s_2 + c_1c_2c)$	$0.25 - \frac{1}{4}(c_1c_2 + s_1s_2c)$	$\frac{1}{12}(c_1s_2 - s_1c_2c)$
$-\frac{1}{2}-\frac{1}{2}$	$-\frac{1}{12}(c - s_1s_2 - c_1c_2c)$	$\frac{1}{12}(s_1c_2 - c_1s_2c)$	$\frac{1}{12}(c_1s_2 - s_1c_2c)$	$0.25 + \frac{1}{4}(c_1c_2 + s_1s_2c)$

Table 2.
SCDM of qubit-qubit system for the case $|S_f - S_o| = 1$.

According to **Table 2**, we thus obtain

$$\det\left(\sigma_{e_1e_2}\right) = \frac{1}{576}(140 + 8\cos 2\alpha_1 - 4\cos 4\alpha_1 + \cos(\alpha_1 - 5\alpha_2) - 8\cos 2(\alpha_1 - 2\alpha_2)$$
$$+8\cos(4\alpha_1 - 2\alpha_2) + \cos(5\alpha_1 - \alpha_2) - 8\cos\alpha - 120\cos 2\alpha - 7\cos 3\alpha - 4\cos 4\alpha - \cos 5\alpha$$
$$-8\cos 2\alpha_2 - 4\cos 4\alpha_2 - 8\cos 2(\alpha_1 + \alpha_2) + 7\cos(3\alpha_1 + \alpha_2) + 7\cos(\alpha_1 + 3\alpha_2)$$

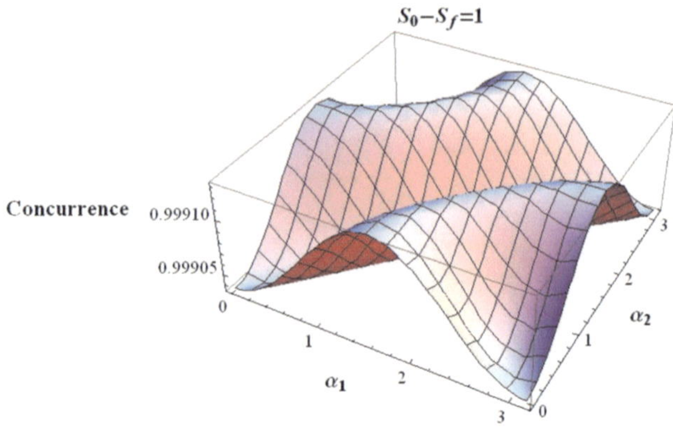

Figure 3.
Variation of concurrence for the case $|S_f - S_0| = 1$ *with respect to spin quantization directions* α_1 *and* α_2.

In order to quantifying entanglement, we obtain the concurrence, using SCDM of **Table 2**. According to the definition given in Eq. (6), which yields

$$C = \sqrt{1 - \frac{9 + (\cos\alpha_1 \cos\alpha_2 + \cos\alpha \sin\alpha_1 \sin\alpha_2)^2}{5184}}. \tag{19}$$

Eq. (19) indicates [4, 17, 22] that the values of C are positive and so the spin state $|S_f - S_0| = 1$ is entangled, depending on $\alpha_1$and α_2. This variation of concurrence is given in **Figure 3**.

We find from **Figure 3** that the states for $|S_f - S_0| = 1$ are entangled (the values of concurrence are positive) and separable at valley points (for the zero value of concurrence) depending on spin quantization directions.

4. Example for Entanglement in DPI for He

As an application qubit-qubit entanglement, we consider helium atom where DPI in its ground electronic state can be represented by

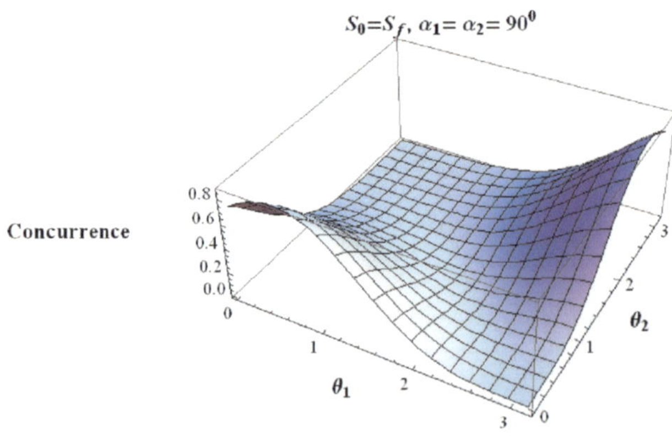

Figure 4.
Variation of concurrence for the case $|S_f - S_0| = 0$ *with respect to directions of ejection of photoelectrons.*

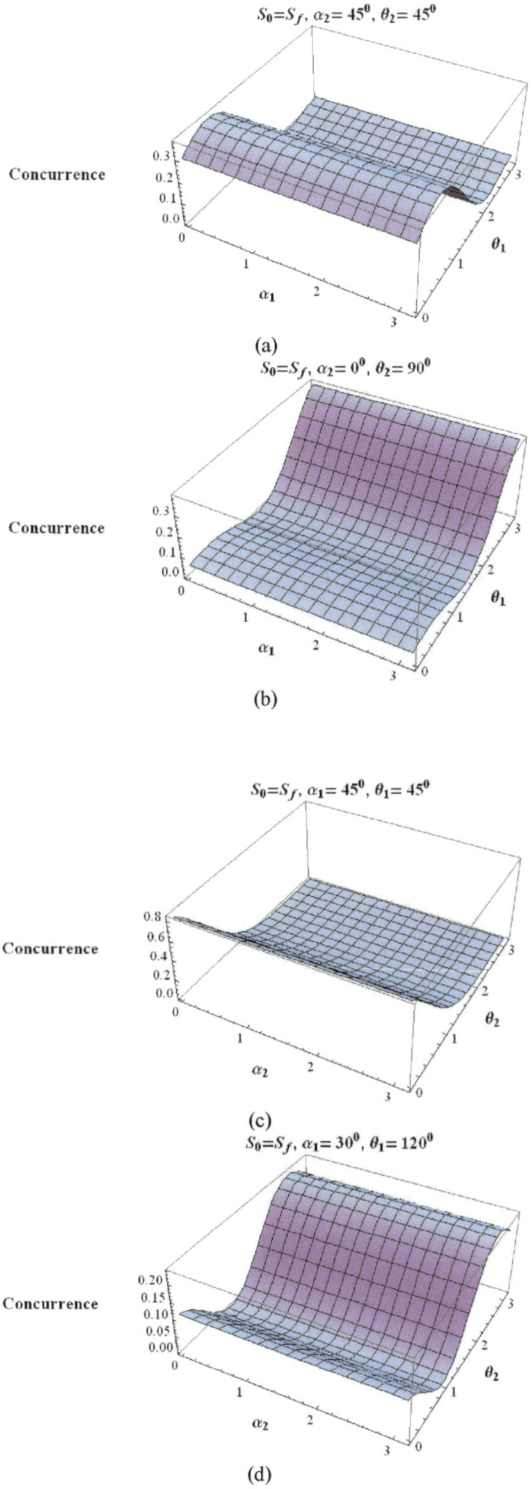

Figure 5.
Variation of concurrences for the case $|S_f - S_o| = 0$ with respect to spin quantization directions and directions of ejection of photoelectron.

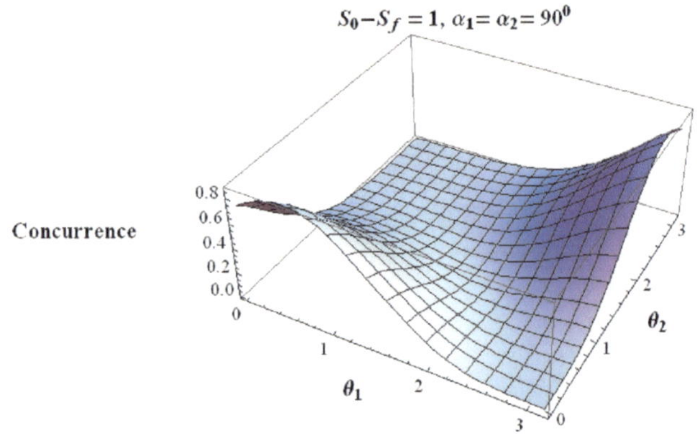

Figure 6.
Variation of concurrence for the case $|S_f - S_0| = 1$ with respect to directions of ejection of photoelectrons.

$$h\nu_r + He\left(1s^{21}S_0\right) \rightarrow He^{2+}\left(1s^{01}S_0\right) + e_1\left(\vec{k}_1; \mu_1\hat{u}_1\right) + e_2\left(\vec{k}_2; \mu_2\hat{u}_2\right) \quad . \tag{20}$$

We use the values of TDCS of helium given in Ref. [29] for photon energy 99 eV along with linear polarization of photon and for equal energy sharing between the two electrons. The TDCS is

$$\left.\frac{d^3\sigma}{d\varepsilon_1 d\hat{k}_1 d\hat{k}_2}\right|_{\varepsilon_1=\varepsilon_2} = a(\cos\theta_1 + \cos\theta_2)^2 \exp\left\{-\frac{1}{2}\left[(\theta_{12} - 180^0)/\gamma\right]^2\right\}, \tag{21}$$

where θ_{12} is the angle between two electrons, γ named Gaussian half-width of value 90.2 ± 2^0. The value of normalization factor $a = 107 \pm 16\ b\ eV^{-1}sr^{-2}$.

i. In case of $|S_f - S_0| = 0$, the variations of concurrence with respect to the direction of ejection and spin polarization of the photoelectrons are shown in **Figures 4** and 5.

From **Figures 4** and 5, we see that the values of concurrence are either zero or positive depending on the directions of ejection and spin quantization of e_1 and e_2. So we can conclude that depending on the values of θ_1, θ_2, α_1, and α_2, most of the states for $|S_f - S_0| = 0$ of helium atom are entangled.

ii. In case of $|S_f - S_0| = 1$, the variations of concurrence with respect to the direction of ejection and spin polarization of the photoelectrons are shown in **Figures 6** and 7.

From **Figures 6** and 7, we see that for the states for $|S_f - S_0| = 1$ of a helium atom, the nature of variations as well as magnitudes of concurrence (i.e., entanglement) depend on the directions of ejection of photoelectrons along with their spin polarization.

5. Conclusion

For concurrently creating two electrons in continuum in a single step, DPI is the most natural approach. It is the most obvious example of electron-electron correlation

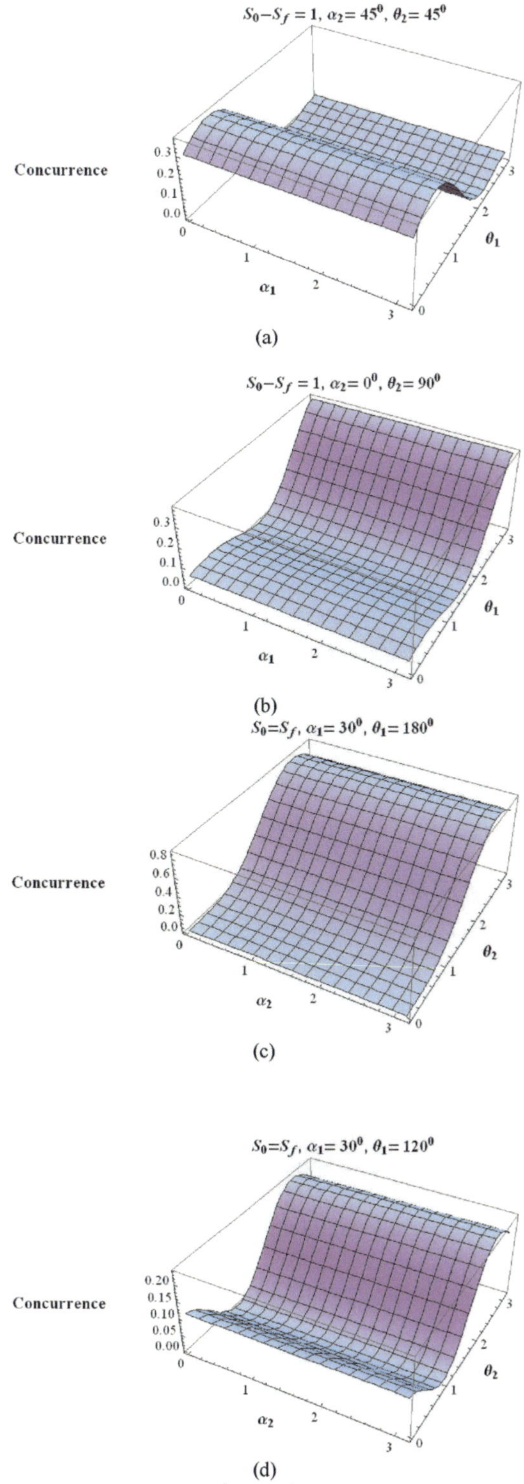

Figure 7.
Variation of concurrences for the case $|S_f - S_o| = 1$ *with respect to spin quantization directions and directions of ejection of photoelectron.*

in an atom because if the independent particle model were true, only one photon would have been absorbed before two electrons would have emerged at the same time. Thus, the presence of correlation effects between them could lead to the simultaneous ejection of two electrons from an atom following the absorption of a single photon. In this article, we have tried to demonstrate how effective the DPI method is for creating different types of entanglement between two qubits. A quantitative application for this case is studied for DPI in helium atom. For helium atom, we have studied the states for $|S_f - S_0| = 0$ and $|S_f - S_0| = 1$, we have shown that depending on the direction of ejection, as well as spin polarization of the ejected photoelectrons, the states are totally entangled, partially entangled, and separate.

Author details

Minakshi Chakraborty[1*] and Sandip Sen[2]

1 Department of Physics, Asansol Girls' College, Asansol, India

2 Department of Physics, Triveni devi Bhalotia College, Raniganj, India

*Address all correspondence to: bminakshi@yahoo.com

References

[1] Nielsen MA, Chuang IL. Quantum Computation and Quantum Information. 10th ed. Cambridge, England: Anniversary Edition Cambridge University Press; 2012. DOI: 10.1017/CBO9780511976667

[2] Alber G, Berth T, Horodecki M, Horodecki P, Horodecki R, Rötteler M, et al. Quantum Information: An Introduction to basic Theoretical Concepts and Experiments. Berlin: Springer; 2001

[3] Shumacher B. Quantum Coding. Physical Review A. 1995;**51**:2738-2747

[4] Horodecki M, Horodecki P, Horodecki R. Reviews of Modern Physics. 2009;**81**:865-942

[5] Rungta P, Munro W, Nemoto K, Deuar P, Milburn GJ, Caves CM. Qudit entanglement. In: Carmichael HJ, Glauber RJ, Scully MO, editors. Directions in Quantum Optics. Berlin: Springer; 2001. pp. 149-164

[6] Einstein A, Polodolsky B, Rosen N. Physiological Reviews. 1935;**47**:777-780

[7] Bell JS. Physics. 1965;**1**:195-200. DOI: 10.1103/PhysicsPhysiqueFizika.1.195

[8] Ekert A. Physical Review Letters. 1991;**67**:661-663. DOI: 10.1103/PhysRevLett.67.661

[9] Bannet CH, Brassard C, Jozsa R, Peres A, Wooters WK. Physical Review Letters. 1993;**70**:1895-1899

[10] Ekert A, Jozsa R. Reviews of Modern Physics. 1998;**68**:733-753

[11] Variri A, Pan JW, Jennewein T, Weins G, Zeilinger A. Physical Review Letters. 2003;**91**:227902. DOI: 10.1103/PhysRevLett.91.227902

[12] Zong-Cheng X, Mai-Lin L, Ya-Ting Z, Jian-Quan Y, Pramana. J. Physics. 2016;**86**:495-502

[13] Deng W and Deng Y, Pramana. J. Physics. 2018;**91**:45(1-5)

[14] Parida S, Chandra N. Physical Review A. 2012;**86**:062302

[15] Chandra N, Chakraborty M. Journal of Physics B: Atomic, Molecular and Optical Physics. 2002;**35**:2219-2238

[16] Chandra N, Ghosh R. Physical Review A. 2006;**74**:052329

[17] Wei TC, Nemoto K, Goldbart PM, Kwiat PG, Munro WJ, Verstraete F. Physical Review A. 2003;**67**:02110

[18] Benett CH, DiVincenzo DP, Smolin J, Wootters WK. Physical Review A. 1996;**54**:3824-3851

[19] Ishizaka S, Hiroshima T. Physical Review A. 2000;**62**:022310. DOI: 10.1103/PhysRevA.62.022310

[20] Chandra N. Chemical Physics. 1986; **108**:301-315

[21] Lee S, Chi DP, Oh SD, Kim J. Physical Review A. 2008;**41**:495301

[22] Wootters WK. Physical Review Letters. 1998;**80**:2245-2248

[23] Edmonds AR. Angular Momentum in Quantum Mechanics. Princeton: Princeton University Press; 1996

[24] Breit G, Bethe HA. Physics Review. 1954;**93**:888-890

[25] Manson ST, Starace AF. Reviews of Modern Physics. 1982;**54**:389-405

[26] Varshalovich DA, Moskalev AN, Khersonskii VK. Quantum Theory of Angular Momentum. Singapore: World Scientific; 1989

[27] de Shalit A, Talmi I. Nuclear Shell Theory. New York: Dover; 2004

[28] Martin F, Horner DA, Vanroose W, Rescigno TN, McCurdy CW. First Principles Calculations of the Double Photoionization of Atoms and Molecules using B-splines and Exterior Complex Scaling. Berkeley: Lawrence Berkeley National Laboratory; 2005

[29] Bräuning H, Dörner R, Cocke CL, Prior MH, Krässig B, Kheifets AS, et al. Journal of Physics B. 1998;**31**:5149-5160

Chapter 13

Pharmacodynamic Implications of Transcranial Photobiomodulation and Quantum Physics in Clinical Medicine

Kristin S. Williams

Abstract

Photobiomodulation (PBM) is the application of light therapy that utilizes photons to alter the activity of molecular and cellular processes in the tissue where the stimulation is applied. Because the photons associated with the therapeutic mechanisms of PBM affect processes associated with the mitochondria, it is hypothesized that PBM increases ATP synthesis. Alteration of the mitochondrial respiratory enzyme, cytochrome c oxidase (CCO), is hypothesized to induce healing to damaged tissues via regeneration. Utilization of PBM has been examined in clinical disorders which include but are not limited to Alzheimer's/dementia, epilepsy, and age-related macular degeneration. Transcranial PBM (tPBM) utilizes quantum dot light emitting diodes (QLEDs). QLEDs allow for narrow wavelength emissions from applications of PBM to alter electrophysiological activity and tissue regeneration. This chapter aims to evaluate the mechanisms of QLED applications of PBM and its applications as a photodynamic therapy in the medical sciences. Further, this chapter will examine the quantum mechanics of tPBM and its ability to affect electrophysiological activity according to the electroencephalogram (EEG) across the delta, theta, alpha, beta frequency bands.

Keywords: transcranial, stimulation, neurodegeneration, Alzheimer's, photobiomodulation, EEG, quantum, physics, photodynamic therapy

1. Introduction

Photobiomodulation (PBM) is the application of light therapy that utilizes photons to alter the activity of molecular and cellular processes in the tissue where the stimulation is applied [1]. Historical contexts of PBM therapy and its development indicate that this intervention has been synonymous with low level laser therapy (LLLT), intensive monochromatic light energy, and light emitting diode (LED) energy interventions [2–4]. The emission of monochromatic light within the range of 600 to 1000 nm from low level lasers and LEDs is thought to underlie alterations in cellular signaling pathways, as the photons from LLLT interventions can influence ATP synthesis, gene expression, and oxygen consumption [5]. This hypothesis is based upon the ability of cytochrome c oxidase (CCO), an enzyme within the electron transport chain, to alter mitochondrial membrane potential via exposure to photonic energy in the spectra of 360-860 nm [5]. Near infrared light (NIR) is defined as the emission of light within the 700 to 980 nm spectra whereas infrared light (IR) is energy that is emitted at 1000 nm or 300 GHz and above [6].

While PBM has shown efficacy for neurodegenerative disorders, this intervention has been applied to other neurological disorders which include, but are not limited to, retinal disorders, ischemic stroke, and epilepsy [7]. Neurodegenerative diseases are currently considered as incurable as pharmacological interventions have failed to slow the progression of neuronal necrosis [8–11]. Because many pharmacological interventions induce significant side effects and have largely proven ineffective in curtailing degenerative processes, determining new non-invasive therapies for these disorders is imperative [7]. As the mechanisms by which transcranial PBM (tPBM) are based allow for deep tissue penetration; exposure of subcortical and cortical structures to LLLT interventions is associated with increased perfusion, cellular oxygenation, and renormalization of functional electrophysiological networks [12].

Despite differences in the clinicopathogenesis of neurodegenerative disorders, tPBM is thought to directly influence mitochondrial dysfunction and oxidative inflammatory processes [13, 14]. Neuropathophysiological correlates of Alzheimer's dementia (AD) include, but are not limited to, neurofibrillary tangles; dystrophic neuritis; amyloid precursor protein deposits and increased phosphorylated tau concentrations [13–16]. The inability for β-amyloid concentrations to be adequately decreased results in the breakdown of microtubular assemblies due to hyperphosphorylated tau. Hyperphosphorylation is a biological process that mediates the regulation of mitosis. Because hyperphosphorylation is a signaling process that regulates cell division, abnormalities in microtubules can cause toxicity to cells. Disruptions of the polymerization dynamics of microtubules can result in synaptic failure as these cells are implicated in maintenance of cell structure and homeostatic regulation of cellular metabolic demand [13–16].

2. Applications of PBM to neurodegenerative disorders

Utilization of tPBM in AD populations aims to increase reactive oxygen species (ROS) [17]. Alterations of ROS are associated with improved adenosine triphosphate (ATP) synthesis as this transcription factor is implicated in mitochondrial electron transport, gene expression, and inactivation of pro-apoptotic proteins and nucleic acids [17, 18]. Because the clinicopathology of AD is associated with dysfunction of the mitochondria, clinical implications of hypometabolism and hypoxia are integral to the understanding the mechanisms that underlie tPBM interventions [17, 18]. As the photons from tPBM are absorbed by CCO and hemoproteins within the brain, this suggests that this modality may alter electrochemical reactions and intracellular signaling as a function of improvements in the mitochondrial electron transport chain [19].

Applications of monochromatic wavelengths across the NIR and IR spectra via tPBM interventions affect complexes within functional mitochondrial pathways differentially [20–22]. NIR is considered as light that is emitted at a frequency between 700 nm and 1 mm whereas IR is photic energy emitted at 1 mm or 300 GHz and above [6]. Examination of homeostatic processes that are mediated by CCO, or complex IV within the electron transport chain, suggests that this target of PBM interventions can be stimulated to increase ROS production, oxygen consumption, and ATP synthesis when activated by red light and NIR frequencies between 633 nm and 808 nm [20–22]. Evaluations of alterations within complexes I, II, or III were not significant when exposed to frequencies between 633 and 808 nm [20–22].

Examination of higher irradiances within the NIR spectra indicate that exposure to 980 nm induced significant alterations within complexes III and IV [20]. Similar results were examined for exposure to IR at 1064 nm despite the effects being robust across complexes I, III, and IV. These results suggest that complex II within the electron transport chain may require stimulation within a different spectrum that is not within the wavelength frequencies associated with NIR or IR irradiances to induce physiological modulation for processes related to homeostasis [20–22].

Dysregulation of mitochondrial signaling negatively affects homeostatic processes and ATP availability as these cells are significantly implicated in metabolic regulation [5, 20–22]. As the mitochondria influences cellular signaling, proliferation, DNA and RNA synthesis, and gene expression, implications of ATP availability and the mitochondrial membrane potential must be considered. The mitochondrial membrane potential can be modeled as $\Delta\Psi m$. As changes in mitochondrial signaling are derived as a function of $\Delta\Psi m$, which can be altered according to electrochemical changes in the photon gradient, this indicates that photic energy can alter the mitochondrial respiratory chain and ATP synthesis [5].

The utilization of animal models in PBM research has allowed for the elucidation that applications of laser light across the 450, 620–680, and 760–895 nm frequency bands may alter complexes within the mitochondrial respiratory chain [5]. These models have allowed for mechanistic evaluations of responsivity to PBM as the elucidation of photoacceptors and photoreceptors are hypothesized to underlie the cellular pathways that are modulated by LLLT therapies [20–24]. Photoreceptors are specialized neuroepithelial cells that respond to photic stimulation within the range of 660 to 1000 nm whereas photoacceptors are non-specialized cells which alter transcription signaling after the photons have been absorbed [20, 23–26].

Laws of photobiology indicate that photons must be absorbed within the tissue prior to transformation into chemical, heat, or kinetic energy [27, 28]. Chromophores are molecular compounds that can convert photons into sources of energy [29]. The activation of chromophores by a specific wavelength of light emission is associated with alterations of biological signaling within the tissue by which it is located [29].

3. Mechanisms of PBM

Technological advances have allowed quantum physics to be applied to the development of non-invasive clinical inventions that alter cellular signaling reactivity within mitochondrial pathways [28]. Elucidation of the mechanisms that underlie LLLT therapies may facilitate comprehensive understandings of applications of NIR and IR for energetic restorative medicine. As tPBM alters cellular responsivity via energy particles emitted from light, this suggests that oscillatory properties of cells within the body can be altered by endogenous or exogenous excitation [28].

PBM applications can be utilized as stimulatory or inhibitory clinical interventions depending upon specific irradiation parameters [28, 30, 31]. Irradiation accounts for the depth, wavelength frequency, and dosage. Stimulatory processes such as cell proliferation, tissue rejuvenation, and reduction of inflammatory processes have been reported within the range of 10–100 mW cm−2 whereas inhibitory processes require higher irradiance levels [28, 30, 31]. While the depth of the tissue for phototherapy applications must be considered, pharmacodynamic considerations such as the time course for light exposure must also be examined.

Applications of photonic therapy require that fluence rate and irradiance are differentiated despite utilizing mW cm−2 as a unit of measurement [28, 32]. Because transcranial applications of PBM are intended for a spherical structure within the body, the quantum mechanics of tPBM must be considered according to fluence rate. Fluence rate considers the geometric properties of tissues targeted by tPBM, scattering effects, and absorption rate as a time average. The time average is calculated according to the dispersion of photic energy in multiple directions within a spherical model. As the photons must be absorbed within a cross sectional area of the model by which it is applied, this calculation considers the average propagation of the emitted wavelength within the targeted tissue [28].

Calculations of fluence rate must use multifactorial analyses that account for divergent properties of biological tissues in tPBM applications. As tPBM interventions target subcortical structures and neurophysiological pathways, considerations of the radiation transfer equation (RTE) allow for derivation of photon transport and alterations in the mitochondrial membrane potential [6]. RTE allows for the derivation of fluence rate as a function of photonic distribution, absorption, and scattering effects across tissues such as the dura, skull, cerebrospinal fluid, and white and gray matter [6, 33].

Computational models of fluence rate as a logarithm using 630 nm, 700 nm, and 810 nm were examined using the Colin27 head atlas [6]. This head model allows for the calculation of fluence rate across divergent biological tissues with anisotropic conductivity. Analyses of penetration depth according to exposure to these NIR frequencies were evaluated at Cz. Cz is the central midline reference point according to the 10-20 International standardized placement system for electroencephalographic analyses [6].

Results indicate that applications of 700 nm and 810 nm had a higher penetration depth reflected by a higher fluence rate compared to 680 nm [6]. Furthermore, the comparison of absorption rate within gray matter was more significant for 810 nm compared to 700 nm and 680 nm. These results suggest that because 810 nm showed a higher fluence and absorption rate, this wavelength may show more efficacy in transcranial stimulatory interventions compared to lower wavelength frequencies within the NIR spectrum [6].

4. Mathematical considerations of phototherapy

Mathematical modeling of power density of tPBM applications must consider photometric units of luminance, wavelength of the light source, and the area by which the light is applied [34, 35]. Luminance, Lv, is calculated as a function of the energy emitted from a light source which accounts for the specific angle, direction, and scattering coefficients across the surface model by which the light source is applied [34, 35]. Thus, d^2 Φv represents the intensity of light energy according to units of time which is divided by dS. dS represents the area of the surface that the light is directed. dΩs denotes the differential solid angle by which the light source travels in a specified direction whereas θS accounts for the refraction and direction of light observed relative to the surface by which the light is applied [34, 35].

$$Lv = \frac{d^2 \Phi v}{dSd\Omega s cos\theta S} \tag{1}$$

Calculations of luminance and applications of PBM must consider Lambert's cosine law. Lambert's cosine law, also known as the cosine emission law, states that the intensity of light covaries according to the distribution area and its angle of incidence [28, 36–38]. As the power of light received by a surface depends on the cosine of the incidence angle, the power of a light source is significantly attenuated if the direction of light is applied orthogonal to the surface normal [28, 36–38].

Mathematical considerations of Lambert's cosine law and the clinical utility of tPBM require examination of implications related to the inverse square law of illuminance [39, 40]. The inverse square law of illuminance states that the light received by any surface is dependent upon the distance of the targeted surface in relation to the source of light. As this law indicates that the light received by a surface is inversely proportional to the square distance between the surface normal and a source of light; as distance increases, the power of the light significantly decreases. Eq. (2) indicates that illuminance, I, varies proportionately to the square distance, d, which is measured in meters and the intensity of the radiation is measured in candela [39, 40].

$$I \propto \frac{1}{d^2} \tag{2}$$

Mathematical modeling of the inverse square law may also apply geometric properties to evaluate light absorption within spherical models [41]. I represents the intensity of the light source at the surface of the sphere, whereas S represents the source strength divided by, $4\pi r^2$, the spherical area [41].

$$I = \frac{S}{4\pi r^2} \tag{3}$$

5. Quantum mechanics and pharmacodynamic implications of PBM

Quantum biology indicates that molecules derived from water can be converted to provide free energy to cells [23, 42–44]. Theoretical applications of electron transfer were developed that allowed for later technological advances such as femtosecond spectroscopy [23, 42–44]. The advent of femtosecond spectroscopy, which examines the dynamics of chemical reactions and the movement of atoms due to exposure to laser technology, has revolutionized non-invasive clinical medicine [23, 42–47].

Light water interactions are applicable to the treatment of complex diseases as water comprises 70% of the mass and 99% of the total molecules of the human body [23]. Because homeostatic regulation and biological signaling are mediated by energy and water interactions, physiological processes within humans can be considered as electrochemically and electrodynamically dependent. As muscle contraction, cardiovascular and neurological functioning are influenced by biochemical processes which facilitate electrical currents derived from ion gating and action potentials, considerations of tPBM and other LLLT interventions must be examined in relation to the clinicopathogenesis of divergent disorders [23].

Pharmacodynamic determinations of wavelength and dosage within LLLT modalities must be examined as increased water concentrations in target areas may increase the infrared energy absorption [48]. While limited research has established how scattering coefficients and the rate of light absorption can be altered due to water content within human skin, considerations of age, gender, and body mass suggest that clinical responses may be attenuated if variance across these factors is not accounted for. Furthermore, as altered water concentrations within a targeted area may attenuate clinical responsivity, clinicians that utilize these methodologies must also consider the depth of the targeted tissue [48].

6. Clinical implications of tPBM and neurodegenerative disorders

Because the clinicopathogenesis of neurodegenerative disorders often implicates alterations of neuronal communication between divergent Brodmann areas and functional connectivity networks, the conversion of photons from light therapies may improve cognitive functioning [12]. While comprehensive understanding of the factors that interact to induce the pathognomonic correlates of AD remain elusive, a prominent hypothesis of the etiopathogenesis of AD is related to the brain's inability to adequately produce cellular energy [12–16].

Current diagnostic criteria of AD delineate three stages of progression according to symptomatic presentation and neurological alterations [49, 50]. The preclinical stage of AD is characterized by a lack of significant clinical presentation of symptoms despite phenotypic alterations of volumetric and structural integrity, amyloid plaque buildup, and degeneration of neurons among other cellular changes. The second stage, mild cognitive impairment, is characterized by alterations in episodic memory retrieval, visual imagery performance, logical reasoning, and executive functioning that are not explained by educational level or age. Diagnosis of AD represents the last stage of neurodegeneration and is characterized by a loss of independence due to significant impairments across visuospatial, language, sensory processing, and memory domains [49, 50].

As the clinicopathogenic processes that underlie AD include diffuse neural degeneration and functional connectivity alterations, this suggests that dysregulation of the posterior dominant rhythm is a cardinal feature of severe disease progression [12]. Neurophysiological assessments within AD populations often indicate reduced signal complexity, reducing synchrony consistent with disease progression, and robust slowing across the electroencephalographic (EEG) frequency spectrum [12].

Quantitative EEG (qEEG) analyses were applied by Williams to evaluate functional connectivity alterations within a PAD sample after exposure to an active infrared treatment using 1070 nm or placebo treatment that mimicked the tPBM intervention [12]. Diagnosis of PAD was determined using guidelines put forth by the National Institute on Aging-Alzheimer's Association [50]. The sample (n=42) was composed of individuals between the ages of 40 to 85 which were randomly assigned to the active or control treatment group [12, 50]. All participants received 6 to 8 minute eyes open and eyes closed qEEG assessments prior to exposure to the tPBM or sham intervention and at the conclusion of the 8 week study [12, 50].

While statistically significant functional connectivity alterations were examined between pre and post measures for the tPBM and control group across the delta (1-4 Hz), theta (4-8 Hz), alpha (8-12 Hz) and beta (12-30 Hz) bands, the most significant alterations were focal to the EC tPBM group between 8 to 12 Hz [12]. Because AD is associated with diffuse neural degeneration that inherently affects the posterior dominant rhythm, it is interpreted that tPBM interventions may temporarily influence the posterior dominant rhythm via increases of ATP within subcortical structures [12].

These results are consistent with implications associated with neural packing density as the occipital regions have a greater metabolic demand than the frontal cortices [12, 51]. Implications of neurophysiological arrangement of cell distribution suggest that the number of cells in a specific area of the cerebral cortex is attributable to the amount of sensory processing and sensory integration demands of the specific region of interest. Thus, because the occipital regions have a significant metabolic demand where absorption of photons is associated with improvements in ROS, oxygen consumption, cerebral perfusion, the findings of this research suggest that the improvements in cognition may be attributable improvements in ATP synthesis and mitochondrial dysfunction [7, 12, 51–58].

7. Retinal disorders

Pharmacodynamic parameters and dosimetry of PBM in retinal disorders are not well understood [59]. Evaluations of clinical applicability of quantum dot light emitting diodes (QLEDs) have utilized narrow wavelength emissions of 670 nm while responsivity to other wavelengths have been limited [59–61]. Examination of responsivity to PBM at 670 nm and 830 nm were compared for age related macular degeneration in animal subjects [59]. Results suggest that reductions of cell death were observed in the animals that received 670 nm of PBM while no alterations in disease progression were reported for animals that received 830 nm [61].

Further examination of responsivity to 670 and 810 nm PBM were compared using ex vivo retina tissue cultivation [59]. The animal subjects were exposed to blue light irradiation to induce oxidative stress to the photoreceptors. As the photoreceptors contain high concentrations of mitochondria, assessments of improvement of the pathological effects after exposure to blue light allow for evaluations of molecular responses to near infrared and red light therapies [59].

Results indicate that exposure to 670 nm of red light and 810 nm of NIR increased activity of mitochondrial ATP [59]. Examination of oxidative phosphorylation (OXPHOS) activity after exposure to 670 nm and 810 nm was conducted using immunohistochemical staining of the retinal inner and outer segments. Mitochondrial pathways within the inner and outer layer of the retinal cells were damaged using blue light irradiation to elucidate specific alterations in OXPHOS complexes. Complexes I and II were emphasized in these analyses as they are implicated as the first two enzymes of the mitochondrial respiratory chain [59]. Alteration of the complex I+III+IV pathway approached statistical significance after exposure to 670 nm of red light and 810 nm of NIR as the oxygen consumption was restored to a normative range following mitochondrial induced apoptosis via blue light irradiation [59]. A similar trend was reported for the complex II+ III+ IV pathway after exposure to 670 nm of red light and 810 nm of NIR. Normalization of complex II after exposure to red light and NIR PBM suggests that improvements of ATP synthesis may be attributable to restoration of oxygen consumption within functional extramitochondrial pathways [59].

This study suggests that despite the robust hypothesis that CCO alterations are the primary mechanism that underlies responsivity to PBM in macular degenerative disorders, that functional extramitochondrial complexes may mediate this interaction [59]. As the results of this study did not show alterations in concentrations of CCO after exposure to PBM, it is suggested that CCO is mediated as a secondary effect of interactions between extramitochondrial complexes [59].

8. Epilepsy

Investigations of the clinical utility of tPBM in pediatric epileptic patients suggests that the addition of this intervention with Antiepileptic Drugs (AEDs) may yield better results compared to administration of AEDs alone [62, 63]. Applications of 808 nm tPBM were utilized in prepubescent rats that received low doses of valproic acid after exposure to Pentylenetetrazole [62, 63]. Pentylenetetrazole is a non-competitive antagonist of gamma-aminobutyric acid (GABA). This drug is utilized to study the GABA-ergic mechanisms that underlie epilepsy as it induces clonic convulsions [63].

The utilization of tPBM as an additive therapy reduced the seizure latency of Convulsive Status Epilepticus (CSE) [62]. Comparison of the seizure latency offset in animal models with Pentylenetetrazol induced CSE suggests that tPBM may serve as a seizure prophylaxis. While future research is required to establish dosage parameters of AEDs in combination with tPBM, the clinical utility of LLLTs may induce significant synergistic effects for epileptic populations that experience CSE, Refractory Status and Super Refractory Status Epilepticus [62].

9. Conclusion

tPBM is the utilization of LLLT which is thought to modulate mitochondrial activity, ATP synthesis, and biosignaling processes [13, 14]. While comprehensive understanding of the mechanisms that underlie this modality remain elusive, LLLT interventions have shown efficacy in pediatric epilepsy as adjunctive therapies to AEDs and across neurodegenerative disorders [12, 59, 62].

Considerations of quantum mechanics suggest that the clinical utility of tPBM is related to the ability of cells to derive free energy from water [23, 41–43]. As the advent of femtosecond spectroscopy has allowed for the study of the dynamic processes that underlie chemical reactions and molecular rearrangement after exposure to laser technology, the ability of LLLT therapies to induce photochemical alterations within mitochondrial pathways has become possible [12, 23, 42–47].

CCO is an endogenous photoreceptor which is implicated in electron transfer and metabolic processes [59]. This molecule is implicated as a targeted mechanism of tPBM as it is a photoreceptor that can absorb photons to increase ATP synthesis within the mitochondrial respiratory chain [19, 59]. As hypoxia is thought to underlie the etiopathogenesis of AD, the ability of tPBM to alter biological signaling via exposure to QLEDs is a significant advancement in clinical medicine [12, 16, 42–47].

Applications of quantum mechanics to the study of light therapy interventions suggest that the cosine emission law and the inverse square law of illuminance are relevant to the pharmacodynamic processes that underlie clinical responsivity to tPBM [28, 36–40]. The inverse square law of illuminance states that the light received by any surface is dependent upon the distance of the targeted surface in relation to the source of light [39, 40]. Considerations of the cosine emissions law indicate that the intensity of light covaries according to the distribution area of the light and its angle of incidence [28, 36–38].

As the pharmacodynamic parameters tPBM may vary according to water concentration levels across human tissues, considerations of wavelength, time course of exposure, and scattering effects must be examined prior to clinical applications of this intervention [48]. These factors suggest that individualized dose response curves may be required despite robust utilization of 670 to 1040 nm of near infrared and infrared light tPBM interventions [59–61].

Continued research of applications of quantum physics and photochemical responses in the development of precision medicine therapeutics is required. As the clinical efficacy of LLLTs is being examined as electromagnetic interventions that may curtail neurodegenerative processes, this suggests that exposure to NIR and IR therapies may modulate functional mitochondrial pathways [1, 7, 12, 16, 27, 28]. Current examinations of therapeutic responses to LLLTs suggest that the clinical utility and pharmacodynamics of these interventions may require multifactorial analyses which include, but are not limited to age, depth of the targeted tissue, and body mass [48]. Future directions of clinical research for LLLT interventions must seek to establish proper irradiation parameters and time course exposure to near infrared and infrared light therapeutics [28, 30, 31]. These implications suggest that clinical responsivity may not only vary according to the frequency of exposure to tPBM interventions but also according to the severity level and symptomatic profile of the disorder by which this intervention is applied.

Author details

Kristin S. Williams
Columbia University, New York, USA

*Address all correspondence to: kw2994@cumc.columbia.edu

References

[1] Anders J. Photobiomodulation. 2016. Available from: https://www.aslms.org/for-the-public/treatments-using-lasers-and-energy-based-devices/photobiomodulation

[2] Hochman L. Photobiomodulation therapy in veterinary medicine: a review. Topics in companion animal medicine. 2018;**33**(3):83-88

[3] Hamblin MR. Photobiomodulation or low-level laser therapy. Journal of Biophotonics. 2016;**9**(11-12): 1122-1124

[4] de Sousa M, Chacur M, Martins D, et al. Theoretical neuroscience. In: Hamblin M, Huang Y, editors. Photobiomodulation in the brain: low-level laser (light) therapy in neurology and neuroscience. London: Elsevier/Academic Press; 2019. pp. 9-19

[5] Karu T. Mitochondrial signaling in mammalian cells activated by red and near-IR radiation. Photochemistry and Photobiology. 2008;**84**(5):1091-1099

[6] Bhattacharya M, Dutta A. Computational modeling of the photon transport, tissue heating, and cytochrome C oxidase absorption during transcranial near-infrared stimulation. Brain Sciences. 2019;**9**(179):1-20

[7] Hong N. Photobiomodulation as a treatment for neurodegenerative disorders: Current and future trends. Biomedical Engineering Letters. 2019;**9**(3):359-366. DOI: 10.1007/s13534-019-00115-x

[8] Feldman H, Ferris S, Winblad B, et al. Effect of rivastigmine on delay to diagnosis of Alzheimer's disease from mild cognitive impairment: The InDDEx study. Lancet Neurol. 2007;**6**(6):501-512. DOI: 10.1016/S1474-4422(07)70109-6

[9] Lopez O. The growing burden of Alzheimer's disease. American Journal of Managed Care. 2011;**17**(13):S339-S345

[10] Petersen R, Thomas R, Grundman M, Bennett D, Doody R, Ferris S, et al. Vitamin E and donepezil for the treatment of mild cognitive impairment. New England Journal of Medicine. 2005;**352**(23):2379-2388. DOI: 10.1056/nejmoa050151

[11] Winblad B, Gauthier S, Scinto L, et al. Safety and efficacy of galantamine in subjects with mild cognitive impairment. Neurology. 2008;**70**(22):2024-2035. DOI: 10.1212/01.wnl.0000303815.69777.26

[12] Williams K. Quantitative Electroencephalographic Evaluations of Electrophysiological Biomarkers and Impaired Functional Networks of Older Adults Diagnosed with Dementia. The Journal of the Alzheimer's Association. 2022;**17**:e058610

[13] Mohandas E, Rajmohan V, Raghunath B. Neurobiology of Alzheimer's disease. Indian Journal of Psychiatry. 2009;**51**(1):55-61. DOI: 10.4103/0019-5545.44908

[14] Thal D, Grinberg L, Attems J. Vascular dementia: Different forms of vessel disorders contribute to the development of dementia in the elderly brain. Experimental Gerontology. 2012;**47**(11):816-824. DOI: 10.1016/j.exger.2012.05.023

[15] Facchin F, Canaider S, Tassinari R, Zannini C, Bianconi E, Taglioli V, et al. Physical energies to the rescue of

damaged tissues. World Journal of Stem Cells. 2019;**11**(6):297-321. DOI: 10.4252/wjsc.v11.i6.297

[16] Liebert A, Chow R, Bicknell B, Varigos E. Neuroprotective effects against POCD by photobiomodulation: Evidence from assembly/disassembly of the cytoskeleton. Journal of Experimental Neuroscience. 2016;**10**:1-19

[17] Liu Y, Fiskum G, Schubert D. Generation of reactive oxygen species by the mitochondrial electron transport chain. Journal of neurochemistry. 2002;**80**(5):780-787

[18] Turpaev K. Reactive oxygen species and regulation of gene expression. Biochemistry (Moscow). 2002;**67**(3):281-292

[19] de Freitas L, Hamblin M. Proposed mechanisms of photobiomodulation or low-level light therapy. IEEE Journal of Selected Topics in Quantum Electronics. 2016;**22**(3):1-37. DOI: 10.1109/JSTQE.2016.2561201

[20] Colombo E, Signore A, Aicardi S, Zekiy A, Utyuzh A, Benedicenti S, et al. Experimental and clinical applications of red and near-infrared photobiomodulation on endothelial dysfunction: A Review. Biomedicines. 2021;**9**(274):1-24

[21] Amaroli A, Ravera S, Parker S, Panfoli I, Benedicenti A, Benedicenti S. An 808-nm diode laser with a flat-top handpiece positively photobiomodulates mitochondria activities. Photomedicine and Laser Surgery. 2016;**34**(11):564-571

[22] Amaroli A, Pasquale C, Zekiy A, Utyuzh A, Benedicenti S, Signore A, et al. Photobiomodulation and oxidative stress: 980 nm diode laser light regulates mitochondrial activity and reactive oxygen species production. Oxidative Medicine and Cellular Longevity. 2021;**2021**:1-11

[23] Santana-Blank L, Rodríguez-Santana E, Santana-Rodríguez J, Santana-Rodríguez K, Reyes-Barrios H. Water as a photoacceptor, energy transducer, and rechargeable electrolytic bio-battery in photobiomodulation. In: Hamblin M, De Sousa M, Agrawal T, editors. Handbook of Low-Level Laser Therapy. Singapore: Pan Stanford Publishing Pte. Ltd; 2016. pp. 119-140

[24] Karu T. Primary and secondary mechanisms of action of visible-to-near IR radiation on cells. Journal of Photochemistry and Photobiology. 1999;**49**(1):1-17

[25] Schroeder P, Pohl C, Calles C, Marks C, Wild S, Krutmann J. Cellular response to infrared radiation involves retrograde mitochondrial signaling. Free Radical Biology and Medicine. 2007;**43**:128-135

[26] Muste J, Russell M, Singh R. Photobiomodulation Therapy for Age-Related Macular Degeneration and Diabetic Retinopathy: A Review. Clinical Ophthalmology (Auckland, NZ). 2021;**15**:3709-3720

[27] Hamblin M. Mechanisms and mitochondrial redox signaling in photobiomodulation. Photochemistry and Photobiology. 2018;**94**(2):199-212. DOI: 10.1111/php.12864

[28] Triana M, Restrepo A, Lanzafame R, Palomaki P, Dong Y. Quantum dot light-emitting diodes as light sources in photomedicine: photodynamic therapy and photobiomodulation. Journal of Physics: Materials. 2020;**3**(3):1-22

[29] Gonzalez-Lima F, Barrett D. Augmentation of cognitive brain functions with transcranial lasers.

Frontiers in Systems Neuroscience. 2014;**8**:1-4

[30] Chow R, Armati P, Laakso E, Bjordal J, Baxter G. Inhibitory effects of laser irradiation on peripheral mammalian nerves and relevance to analgesic effects: a systematic review. Photomedicine and Laser Surgery. 2011;**29**:365-381

[31] Lanzafame R, Stadler I, Kurtz A, Connelly R, Brondon P, Olson D. Reciprocity of exposure time and irradiance on energy density during photoradiation on wound healing in a murine pressure ulcer model. Lasers in Surgery and Medicine. 2007;**39**:534-542

[32] Lanzafame R. Light dosing and tissue penetration: it is complicated. Photobiomodulation, Photomedicine, and Laser Surgery. 2020;**38**(7):1-4

[33] L'Huillier J, Humeau A. Use of the finite element method to study photon-tissue interactions in biological media. In: Proceedings of the 17th IMACS Congress. Paris, (France); 2005. pp. 1-6

[34] Chaves J. Radiometry, Photometry, and Radiation Heat Transfer. In: Chaves J, editor. Introduction to Nonimaging Optics. 2nd ed. Boca Raton, FL: CRC Press; 2016. pp. 677-696

[35] Bernd D, Henriette M, Gross H. Radiometry. In: Gross H, editor. Handbook of optical systems: Metrology of optical components and systems. Vol. 5. Germany: Wiley-VCH; 2012. pp. 303-427

[36] Hyde E. Geometrical theory of radiating surfaces with discussion of light tubes. Bulletin of the Bureau of Standards. 1907;**3**(1):81-104

[37] Kuznetsov G, Nee A. Modelling of the conjugate natural convection in a closed system with the radiant heating source radiant energy distribution by Lambert's cosine law. Thermal Science. 2018;**22**(1 Part B):591-601

[38] dos Santos M-CL, de Lima V, Barbosa M, Dos Santos L, de Siqueira Rodrigues Fleury Rosa S, Tatmatsu-Rocha J. Photobiomodulation: systematic review and meta-analysis of the most used parameters in the resolution diabetic foot ulcers. Lasers in Medical Science. 2021;**36**(6):1129-1138

[39] Murdoch J. Inverse square law approximation of illuminance. Journal of the Illuminating Engineering Society. 1981;**10**(2):96-106

[40] Voudoukis N, Oikonomidis S. Inverse square law for light and radiation: A unifying educational approach. European Journal of Engineering and Technology Research. 2017;**2**(11):23-27

[41] Brownson J. Laws of Light. In: Brownson J, editor. Solar Energy Conversion Systems. Oxford: Academic Press; 2014. pp. 41-66

[42] Szent-Gyorgyi A. Biological oxidation and vitamins: Harvey Lecture, May 18, 1939. Bulletin of the New York Academy of Medicine. 1939;**15**:456-468

[43] Szent-Gyorgyi A. Bioenergetics. Science. 1956;**124**:3-64

[44] Szent-Gyorgyi A. Biology and pathology of water. Perspectives in Biology and Medicine. 1971;**14**:239-249

[45] Zewail A. Femtochemistry. Past, present, and future. Pure and Applied Chemistry. 2000;**72**(12):2219-2231

[46] Diddams S, Hollberg L, Mbele V. Molecular fingerprinting with the resolved modes of a femtosecond

laser frequency comb. Nature. 2007;**445**(7128):627-630

[47] Sibbett W, Lagatsky A, Brown C. The development and application of femtosecond laser systems. Optics Express. 2012;**20**(7):6989-7001

[48] Kim S, Shin S, Jeong S. Effects of tissue water content on the propagation of laser light during low-level laser therapy. Journal of Biomedical Optics. 2015;**20**(5):1-6

[49] McKhann G, Knopman D, Chertkow H, Hyman B, Jack C, Kawas C, et al. The diagnosis of dementia due to Alzheimer's disease: Recommendations from the National Institute on Aging-Alzheimer's Association workgroups on diagnostic guidelines for Alzheimer's disease. Alzheimer's & Dementia. 2011;7(3):263-269

[50] National Institute on Aging. Alzheimer's disease diagnostic guidelines. n.d. Available from: https://www.nia.nih.gov/health/alzheimers-disease-diagnostic-guidelines

[51] Young N, Collins C, Kaas J. Cell and neuron densities in the primary motor cortex of primates. Frontiers in Neural Circuits. 2013;**7**(30):1-11. DOI: 10.3389/fncir.2013.00030

[52] Begum R, Powner M, Hudson N, Hogg C, Jeffery G. Treatment with 670 nm light up regulates cytochrome c oxidase expression and reduces inflammation in an age-related macular degeneration model. PLOS One. 2013;**8**(2):1-11. DOI: 10.1371/journal.pone.0057828

[53] Chung H, Dai T, Sharma S, Huang Y, Carroll J, Hamblin M. The nuts and bolts of low-level laser (light) therapy. Annals of Biomedical Engineering. 2012;**40**(2):516-533. DOI: 10.1007/s10439-011-0454-7

[54] Desmet K, Paz D, Corry J, Eells J, Wong-Riley M, Henry M, et al. Clinical and experimental applications of NIR-LED photobiomodulation. Photomedicine and Laser Surgery. 2006;**24**(2):121-128. DOI: 10.1089/pho.2006.24.121

[55] Enengl J, Hamblin M, Dungel P. Photobiomodulation for Alzheimer's disease: Translating basic research to clinical application. Journal of Alzheimer's Disease. 2020;**75**(4):1073-1082. DOI: 10.3233/jad-191210

[56] Gkotsi D, Begum R, Salt T, Lascaratos G, Hogg C, Chau K, et al. Recharging mitochondrial batteries in old eyes. Near infra-red increases ATP. Experimental Eye Research. 2014;**122**:50-53

[57] Rojas J, Gonzalez-Lima F. Low-level light therapy of the eye and brain. Eye and Brain. 2011;**3**:49-67. DOI: 10.2147/eb.s21391

[58] Ying R, Liang H, Whelan H, Eells J, Wong-Riley M. Pretreatment with near-infrared light via light-emitting diode provides added benefit against rotenone- and MPP+-induced neurotoxicity. Brain Research. 2008;**1243**:167-173. DOI: 10.1016/j.brainres.2008.09.057

[59] Heinig N, Schumann U, Calzia D, Panfoli I, Ader M, Schmidt M, et al. Photobiomodulation mediates neuroprotection against blue light induced retinal photoreceptor degeneration. International Journal of Molecular Sciences. 2020;**21**(7):1-20

[60] Ivandic B, Ivandic T. Low-level laser therapy improves vision in patients with age-related macular degeneration. Photomedicine and Laser Surgery. 2008;**26**:241-245

[61] Giacci M, Wheeler L, Lovett S, Dishington E, Majda B, Bartlett C, et al.

Differential effects of 670 and 830 nm red near infrared irradiation therapy: a comparative study of optic nerve injury, retinal degeneration, traumatic brain and spinal cord injury. PLoS One. 2014;**9**(8):1-15

[62] Tsai C, Chang S, Chang H. Transcranial photobiomodulation add-on therapy to valproic acid for pentylenetetrazole-induced seizures in peripubertal rats. BMC Complementary Medicine and Therapies. 2022;**22**(81):1-15. DOI: 10.1186/s12906-022-03562-9

[63] Hansen S, Sperling B, Sánchez C. Anticonvulsant and antiepileptogenic effects of GABAA receptor ligands in pentylenetetrazole-kindled mice. Progress in Neuropsychopharmacology & Biological Psychiatry. 2004;**28**(1): 105-113. DOI: 10.1016/j.pnpbp.2003.09.026

Chapter 14

Virtual Reality in Stereometry Training

Penio Lebamovski

Abstract

A new stereoscopic system is presented in this chapter for training in stereometry. Virtual reality systems are two types: immersion and non-immersion. An example of a environments with immersion is the desktop computer in which the virtual world is displayed on a special stereoscopic monitor through 3D glasses. The HMD (Head Mounted Display) virtual helmet is an example of an immersion system. The new stereoscopic system can also implement both types of 3D technology. At the same time, the software can generate objects in the .obj file extension to print to a 3D printer or add to a virtual reality device. The 3D technology is one of the fastest-growing. Over the last decade, it has found application in almost all spheres of society, including education. In learning stereometry, a problem may arise in drawing the geometric figures, unlike planimetry, where the drawing 100% coincides with its original. The problem can be solved with the help of specialized 3D software.

Keywords: boundary method, extrude, stereometry, stereo system, virtual reality

1. Introduction

This chapter focuses on the application of 3D technology in teaching the discipline of stereometry. It is known that stereometry is a branch of Euclidean geometry, which mainly studies geometric figures in three-dimensional space. This course examines the properties, and determines the volumes and faces of the surfaces of geometric figures, such as cylinders, cones, truncated cone, sphere, prism and others. According to teachers and experts in mathematics, when studying this discipline, it is of particular importance for students to have a spatial imagination that allows the understanding of the studied geometric figures in three-dimensional space, i.e., in three dimensions: x, y, and z. What distinguishes 3D (volumetric) from 2D (flat) representation is that more information is acquired about geometric figures in three-dimensional space. Spatial thinking is an activity with the help of which it is possible to solve several practical and theoretical problems related to learning. With the use of 3D technology, the student can become part of the learning process, allowing him to actively participate in the study of geometric figures and discuss and discover new dependencies related to the studied material. The evolution of technology and the need to modernize teaching has led to the creation of specialized software applications to assist students in learning the discipline of stereometry. With the help of 3D simulations and information technologies, such as virtual reality (Virtual reality-VR) and augmented reality (Augmented reality-AR), students' cognitive activities are developed, as well as their interest in geometry (stereometry) increases [1, 2]. 3D modeling software applications, including VR and AR, help visualize complex and abstract shapes that are difficult for students who lack spatial imagination to understand. In the teaching process of mathematics (stereometry), the main goals of the teacher are: To demonstrate the possibilities of modern technologies by using 3D specialized systems.

- To familiarize students with how to work with 3D systems in modeling (generating) and processing geometric objects.
- To show the possibilities of VR/AR in solving a given geometric problem.

In the learning process, a problem may arise when drawing a stereometric figure, unlike planimetry, where the figures are 2D, and usually the drawing almost 100% matches its original. In the three-dimensional space, the figure transferred to a blackboard or white sheet does not match its real image, from where the problem that the teacher faces arises, namely that it is impossible to recreate the spatial figure completely. This problem can be solved using specialized 3D software to generate the geometric object (prism, pyramid, cube, etc.) and their corresponding sections and projections. Software with a suitable user interface can help the student "immerse" into virtual reality and absorb new knowledge of the studied discipline more easily.

The purpose is to perform an analysis of virtual reality systems and to present a new stereo system for stereometry training. The new system can be implemented by devices for VR with and without immersion.

2. Virtual reality

Virtual reality is a part of computer graphics. Today, it is possible for an ordinary user to become a part of VR through a suitable software product. The beginning of this field was given through computer games, which allow a person to experience things that are impossible to experience in real life [3]. Appropriate stereometry software will enable the student to enter this new research environment and interact with it instead of just looking at pictures on a monitor. The technology that enables such an experience is precisely virtual reality. In the current decade, it has become the most popular. VR primarily offers visual experiences on a computer screen or stereoscopic display. At the same time, VR can include additional multimedia, such as audio through headphones or a speaker and text. Interactivity in this virtual reality can be accomplished through the use of a keyboard, mouse, joystick or globe glove. The most used terms in the computer community are virtual reality, virtual environment, virtual world, artificial reality, etc. [3]. These real-time interactive graphics offers the user immersion in a three-dimensional world. Virtual reality systems can be grouped according to the level of immersion they offer the user. Mainly on immersion and non-immersion systems.

1. Non-immersive systems (desktop VR): Here, the immersion is relatively inferior than in other systems. It does not require special devices. Its other name is Window on World (WoW) [3]. The user can view the virtual environment through one or more computer screens. At the same time, he is allowed to interact but is

not immersed in it. These systems are relatively cheap and therefore are the most used. The effect they offer is satisfactory.

2. Semi-immersive (Fish Tunk VR)—this is an improved version of Desktop VR. Like HMDs, these systems support head tracking, which improves the sensation thanks to the motion parallax effect. They require the use of a monitor. But they do not use the sensors of the immersion systems.

3. Immersive systems—This is the best version of VR systems. These systems allow users to immerse themselves in the virtual environment fully. Such systems are HMD and CAVE. Unfortunately, however, they require a lot of equipment and technique and are therefore not widely used [3].

Main characteristics of immersion systems:

- 3D space navigation interface is provided, such as view, tour, interior view.
- Enables stereoscopic viewing, i.e. improves the perception of the sense of depth in space
- Realistic interaction is offered that allows manipulation

3. VR systems

VR can be enhanced by hearing technology. VR devices can be divided into three categories [4]:

- Desktop PC based
- VR all in one
- Mobile based

One of the most used virtual reality systems that offer the most excellent immersion effect is the virtual helmet. Compared to other systems with an immersive effect, they are relatively cheap; the CAVE system can be given as an example of an expensive one. The HMD virtual helmet provides a truly immersive VR experience through two head-mounted displays, while desktop-based helmets require the helmet to be connected to a powerful gaming PC. Each eye has its own display. Unfortunately, however, this category of virtual reality device is not widely used because it is quite expensive. An all-in-one is a mid-range option; these devices do not require a computer connection. The low-level mobile VR product relies on VR Head Set. The main advantage is that it has a relatively low price and is convenient to work with. Today, many schools in Europe and worldwide are already improving the educational process through innovative modern methods based on new technologies. VR is one of the most advanced technologies, and it is normal for it to be used in primary and secondary schools today. It allows students to grasp complex concepts and definitions more easily. The main goal is to support student-centered teaching. There are documented results from using VR in secondary school mathematics learning, with increased positive attitudes and interest in mathematics reported after using VR [5]. Another study concluded that students and teachers are open to a form of learning through VR and AR (Augmented Reality) [5]. In this way, students are helped to understand better geometric shapes that are difficult to represent. Virtual reality enables the user to use and interact with the virtual world similar to the real one [6]. VR finds application in many fields, such as flight simulation, motion simulation in physics and chemistry, etc. The output channels of virtual reality correspond to the human senses: vision and hearing. Vision is the most basic sense used. Stereoscopic vision is a primary human mechanism for depth perception. The techniques for stereoscopic vision are HMD, shutter, passive, and others. The main steps in implementing Virtual and Augmented Reality are.

1. Recall of facts and concepts
2. Understanding the new
3. Application of learning
4. Analysis: the relationship between concepts
5. Justification
6. Creativity

Let us summarize that the 3D scene can be observed through a virtual reality viewer, such as an HMD or a normal desktop monitor. The user can move around the virtual environment using a mouse and keyboard. In this scene, 3D geometric figures are visible, allowing us to understand mathematical functions that are generally difficult to draw.

4. Stereo technologies

Anaglyph technology is achieved by two identical, differently filtered images, respectively for each of the eyes [7]. The two images are placed on the screen with a slight offset from each other. No specific equipment is required here, just a pair of anaglyph glasses. This technology is most commonly seen in magazines and in the past when showing movies in movie theaters. 3D polarizing technology uses two images: for the left, and for right eye, they are superimposed on the screen before they pass through polarizing filters. Polarization is 45 for one eye and 135 for the other. There are two filters here, one for each eye [7]. The necessary equipment is two projectors aimed at the same screen; a polarizing filter is placed in front of each projector, allowing light to pass through. Shutter3D technology—with it, the frame alternates sequentially, first for one eye and then for the other eye; the frequency is 120 Hz by 60 Hz for each eye [7]. For this purpose, active shutter glasses are used, which darkens the glass of the eye, which should not receive information. 3D Display technology improves the perception of the 3D scene, from which the applications become more qualitative. Different 3D technologies are suitable for different applications, so it is important to know them thoroughly [4, 8]. Generally, these technologies are classified into two categories: stereoscopic and real 3D. The stereoscopic technique is based on binocular parallax by presenting separate images [4] to each eye. Motion parallax can also be simulated by adding a head tracking system. Unfortunately, not everyone can perceive 3D information through 3D Display technology, as (2–3% of the population are stereo blind) [4]. In turn, stereoscopic displays are divided into two subclasses: stereoscopic and autostereoscopic. With autostereoscopic displays, the user does not need to have special 3D glasses. Stereoscopic displays are based on blocking each eye from seeing the image corresponding to the other eye. This is achieved through glasses using various Display technologies. They are mainly used for simulation and training. Many of these 3D scenarios are impossible or too expensive to simulate. Because most of them are too risky to simulate or too expensive. To be as realistic as possible, it is necessary to use immersive 3D systems. Low-cost 3D stereo displays are used in middle school student learning. The aim is to increase the level of understanding on the part of the students.

5. Modern mathematics approaches for students

It is known that the main concepts related to the assimilation of knowledge in the discipline of stereometry are: axioms, definitions, theorems and their proofs, and, most importantly, the solution of tasks [9]. The role of the functions is as follows:

- To prepare.
- To acquire new knowledge.
- For control and evaluation.

Today, with the development of information technology, it can be said that the learning process is a function of objects: student, mathematical knowledge, teacher, and "virtual reality", with the teacher managing the learning process with the help of a computer. It is expected that with the developed StereoMV stereoscopic system, the efficiency of the learning process will increase. The material and applied tasks included in the system will be ordered by difficulty and complexity, starting from easier ones and ending with more complex tasks. Lessons can be solved both individually, and students can be divided into separate groups [9]. The course on the discipline of stereometry is a continuation of the course on planimetry. Very often, in the study of stereometry it is necessary to use various definitions and theorems of planimetry to clarify stereometric concepts, i.e., the solution of some stereometric problems is reduced to the solution of planimetry problems. The developed new stereo system makes it possible to build a spatial model of the studied object. At the same time, in the traditional way of learning, the drawings are mainly presented in a plane, which can make it difficult for the students and lead to confusion. The basic knowledge related to the study of stereometry in the middle course uses knowledge obtained already in the second school grade (8-year old students). This knowledge is related to planimetry and algebra, and students get to know them before the specific study of spatial objects. In the second grade, students study the types of triangles, measure the lengths of the sides of a triangle, square, and rectangle with a ruler, as well as study units of measurement (cm, dm, m). The teaching material of the third grade is related to the introduction of basic geometric knowledge, as students learn to use drawing tools (ruler and protractor), study-specific definitions, and the concept of an angle (straight, acute, and obtuse). Students learn to use the square grid, which is a propaedeutic of the Cartesian coordinate system introduced in the upper course. By measuring, they learn to distinguish which shape is a square and which is a rectangle. The fourth-grade learning material is all about learning and drawing a square and a rectangle on the square grid. The concept of the area of a geometric figure and the units of area measurement are introduced. After the fourth grade, the area of the studied figures from geometry is expanded, the properties of the figures are examined, and conclusions are formulated based on experience and observation. A habit of deductive thinking is formulated. The new geometric concepts of students at this age are: straight parallelepiped, cube, vertex, edge, wall, etc. The formula for volume and units of volume measurement are derived experimentally. One of the essential concepts in planimetry, which is the basis of stereometry in the study of a regular pyramid, is similar triangles and similar polygons. The system provides an opportunity to form initial knowledge of stereometry, implementing the material laid out in the textbook for the fifth grade. In connection with the teaching of geometry in IX–X grades, the problem of improving the methodology to develop spatial thinking takes on great importance. The development of this type of thinking in students is carried out through a unique system of tasks that meet the regularity of forming logical thinking. In the tenth-grade geometry course, students use cabinet projection and ways to depict polyhedra such as prisms and pyramids. Among the stereometric tasks of particular importance are the tasks of constructing a section of a polyhedron with a plane. In reality, these sections represent a polygon, which students are familiar with from the sixth-grade course and it is an essential element of the stereo system. The necessary knowledge related to areas are regular/irregular polygon and polyhedra. In the developed module of the system, there is a partial presentation of the material related to the above course on introductory sections of geometric bodies. In studying the topics associated with the cylinder and cone, the questions of depicting a circle and regular inscribed and circumscribed polygons are considered. The necessary propaedeutics for an upper course in stereometry are carried out through the course from I–III and IV–V classes, respectively. Up to this point, the measurement of sections in the planimetry course has been approached inductively. In this way, one goes outside the confines of the classroom by measuring objects with a meter. It is shown that there exists a mutual, unambiguous and reversible correspondence between the objects and the set of positive numbers [9]. The upper course students have the knowledge obtained from the elementary and intermediate courses about the volumes of geometric figures. and this knowledge should be expanded, ie. raised to a higher level. In Bulgarian schools, similar triangles and the signs of their similarity are first examined, then similar polygons are introduced. As it is known, two triangles are similar when there is a correspondence between their vertices, i.e., their angles are equal, and their sides are proportional. The Similar Polygon Theorem states, "If two polygons are divided by their diagonals into similar triangles, then they are also similar." Here the theorem of dependence between the perimeters and their corresponding sides is introduced. The formula for the volume of a pyramid can be obtained by the method of indivisibles of the Italian mathematician Cavalieri [9]. The pyramid's height is divided into a number of proportional parts, after which sections parallel to the base are drawn. There is a dependence between the area of the base and its parallel section. A polyhedron's section with a plane, is a polygon with sides intersecting the walls of the polyhedron with a plane.

6. Advantages of 3D software systems

In the theory of mathematical education, which deals with spatial and geometric imagination development, several experts and researchers are united in their opinion that this type of imagination is poorly developed in some students [10]. To help these students in the process of learning mathematics, in particular in the discipline of stereometry, there is a need to create software applications. These applications should enable teachers and students to visualize studied geometric figures, explore geometric relationships and concepts, and make and test assumptions in a dynamic learning environment by manipulating the studied objects, such as: constructing, dragging, rotating, and others, to train to be more effective, more accessible and enjoyable. The classical method of teaching with chalk, ruler, and compass ensures the understanding of geometric drawing in stereometry, the construction being done in the imagination but in a flat structure. To achieve better results in training in this discipline, some software applications have been created in recent years, such as CABRI 3D, GEOGEBRA, DALEST and others, the use of which opens up new opportunities for the educational process.

7. Methods and results

The stereoscopic system generates the following two types of objects studied in the discipline of stereometry:

- 3D edged geometric figures: polygons (regular), cube, parallelepiped, prism and pyramid;
- Rotational geometric bodies: cylinder, cone, sphere and torus.

There are two main techniques involved in generating a regular polygon in the stereoscopic system; the first is the traditional one that converts a 2D polygon into a 3D object. This is done by extruding a two-dimensional graphic into a three dimensional one. The extrusion technique creates a three-dimensional object from a two-dimensional that moves along a set trajectory. The trajectory can be a rotational or translational motion or a trajectory defined by an arbitrary curve. In the research on the generation of a quadrangular pyramid, a new boundary method is used, which is based on Cavalieri's Indivisible method and Newton's boundary method [11, 12]. The idea is as follows—the geometric figure must be divided into sections parallel to a given ruler. This approach can be applied to any regular polygon. Next, determine the relations of the parallel sections. In the case of four vertices, i.e., a square (regular polygon), the ratio of the segments is 1:1. Polygon vertices can grow without limit. Regular polygons generated in this way differ from traditional programming polygons because they do not need to be converted from a 2D to a 3D object by extruding. They are "mathematically more accurate" and more flexible than traditional ones. They have many advantages, the biggest one being that virtual reality devices can visualize them. At the same time, they can be exported to files with the extension .obj. It follows that the polygon is placed in the center of the coordinate axis and its vertices are calculated by the length of its side by the parameter **a**. To parametrically define a regular quadrangular pyramid, the values for each vertex of the three-dimensional space

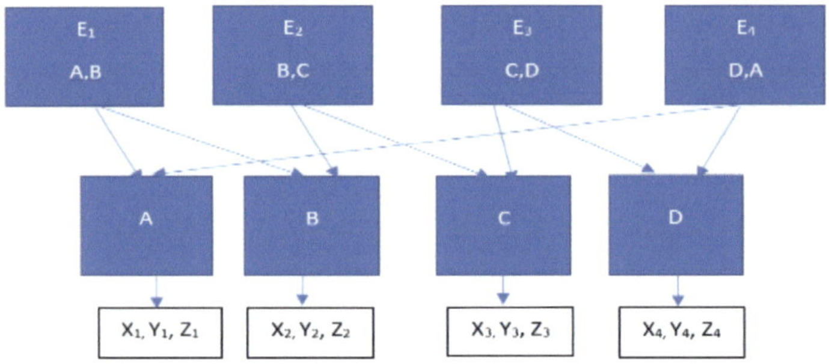

Figure 1.
Edge representation of the base of the pyramid.

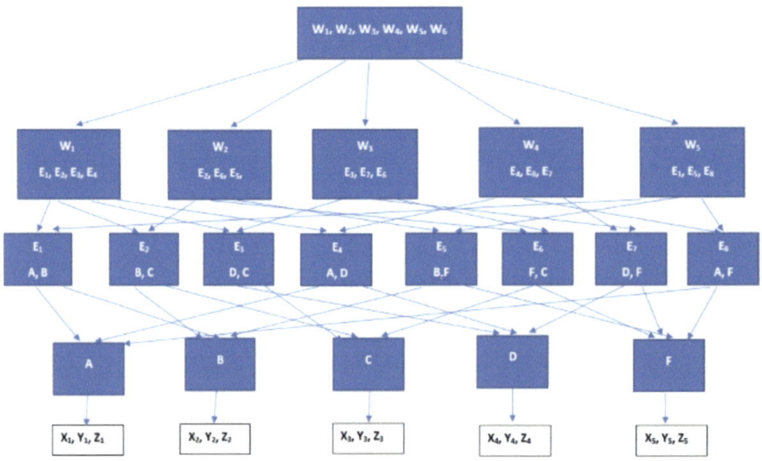

Figure 2.
Limit representation of a pyramid.

should be defined. The parameter "a" is the length of the side of the polygon that is the base of the pyramid. The parameter **h** is the height of the pyramid. When the value of vertex No. 5 is a number other than zero on the abscissa or ordinate, the final result will be a tilted pyramid. The system also enables the generation of a truncated pyramid. The methods for 3D geometric modeling and their schematic representation is done in two ways: edge and boundary. The edge representation (**Figure 1**) requires the following information:

The object's shape is represented by its edges (E1, E2,...).

1. Vertices metric information (A, B, C,...).

2. Partial topological information (X1, Y1, Z1,...).

The boundary representation of a regular quadrilateral pyramid presented in (**Figure 2**) requires:

1. List of all its walls (W1, W2,...)—W6 is the section of the pyramid

No. vertex	X	Y	Z	No. face	vertex
1	a/2	−a/2	0	1	1,2,3,4
2	a/2	a/2	0	2	1,2,5
3	−a/2	a/2	0	3	2,3,5
4	−a/2	−a/2	0	4	3,4,5
5	0	0	h	5	4,1,5

Table 1.
The parametric values of a pyramid.

No. vertex	X	Y	Z	No. face	vertex
1	1.5	−1.5	0	1	1,2,3,4
2	1.5	1.5	0	2	1,2,5
3	−1.5	1.5	0	3	2,3,5
4	−1.5	−1.5	0	4	3,4,5
5	0	0	4	5	4,1,5

Table 2.
Precise pyramid values.

2. The walls are described by the edges (E1, E2,...).

3. Edges are represented by vertices (A, B, C,...).

4. Vertices are represented by three-dimensional coordinates (X1, Y1, Z1,...).

For the parametric setting of a regular quadrangular pyramid (**Table 1**), values for the coordinates of each vertex of the three-dimensional space must be set. The pyramid's height is set with parameter **h**, and the length of the base with parameter **a**. When vertex number five values for the X or Y coordinates is a non-zero number, the result will be a tilted pyramid. The first four vertices form the base of the pyramid, and the fifth is the top of the pyramid. The peaks are crawling in the same direction in the counterclockwise case, starting from the IV quadrant (**Table 2**).

8. Regular/tilted/truncated pyramid

The algorithm for building a regular/ tilted /truncated pyramid (**Figures 3** and **4**) is as follows:

1. From the StereoMV object panel, select the pyramid shape. The base type is chosen, which is a regular polygon with 3–6 vertices.

2. The user sets the length of the main edge.

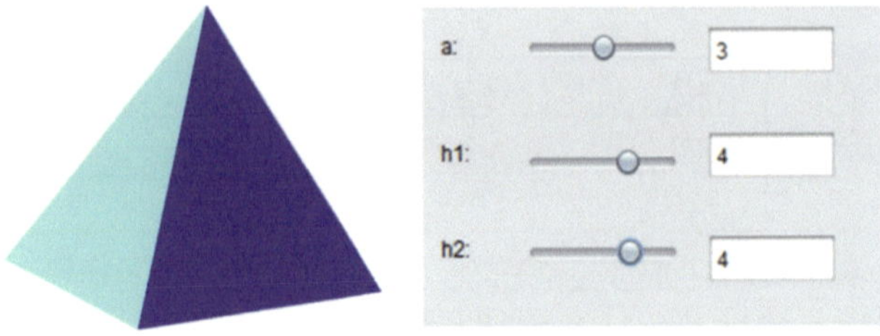

Figure 3.
Straight a regular pyramid.

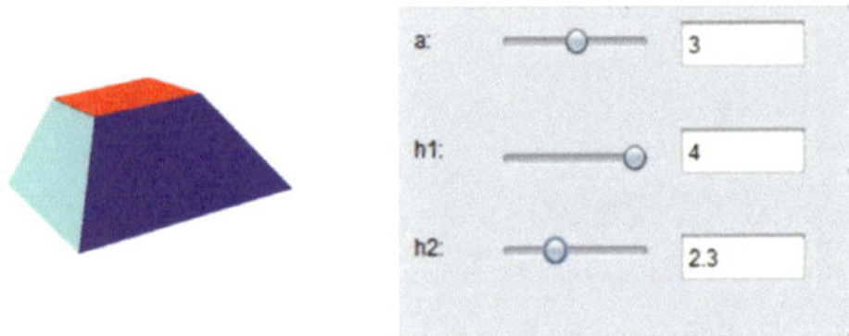

Figure 4.
Truncated pyramid.

3. The system builds a polyhedra along the third axis by setting a height value.

4. The value entered by the user for the abscissa/z coordinate is checked. If the value entered is a non-zero number, the final result will be a tilted pyramid; otherwise, a straight pyramid is generated.

5. The system also offers an additional option: generating a truncated pyramid. When selecting this option, the user must enter values for both heights.

The traditional way to generate a regular polygon is through trigonometry. It is characterized by the number of vertices and radius—its length. To generate a cone, the traditional method (**Figures 5** and **6**) is used. The following are set:

1. The base—a polygon with 100 vertices

2. The radius of the base is set.

3. The center of the circle rises along the third axis by extrude.

4. If the number on the abscissa or z coordinate is a non-zero number, the result will be an inclined cone. **Figure 5** shows the image of a straight cone and its dimensions. If the values for x and z are 0, then this Cone is straight. **Figure 6** shows the figure of a cone that is tilted to the left, and in its dimensions, as the value for x is −6 and z = 0. When the value of x is +6 and z = 0, the Cone will be inclined to the right.

The traditional method of generating a regular polygon is based on trigonometry and is characterized by the number of vertices and radius. To construct a prism and a pyramid through this method, the 3D graphics modeling technique is

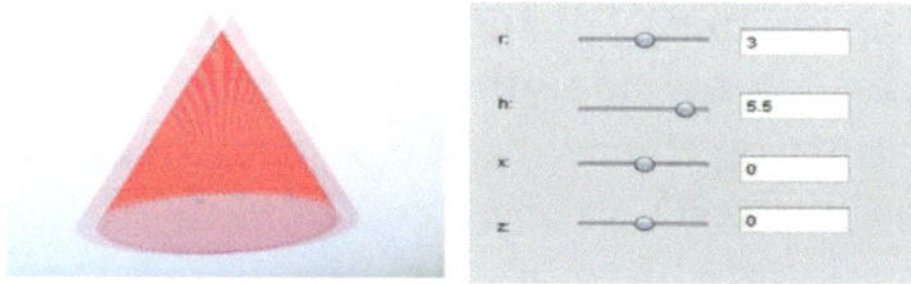

Figure 5.
A straight cone.

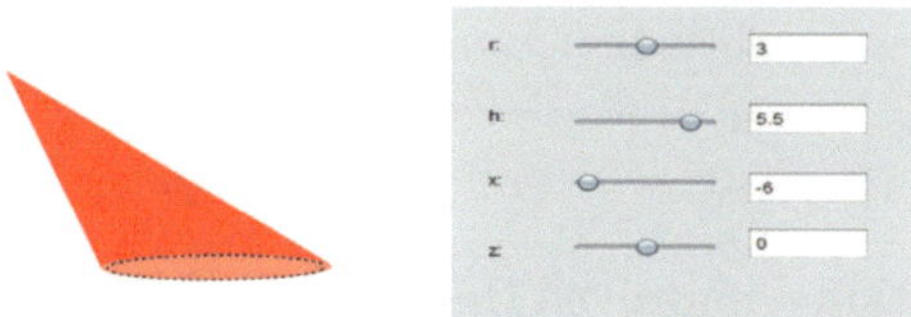

Figure 6.
Inclined cone.

used—extrusion, which is a big drawback. The new authoring method for generating polygons is characterized by the number of vertices and the side length of a regular polygon. This method gives a more accurate result than the traditional one because it does not need to be extruded. It uses only the ratio of parallel segments, not trigonometry, which differs from the traditional method. The author's method accurately results because it does not use trigonometry. A disadvantage of the new approach is that it requires a lot of computing power. Also, another limitation is that for now it is limited to 6 vertices. The new method can be applied to any regular polygon with unlimited vertices growth. The proof that the method works is that it allows visualization through a 3D virtual reality library and the ability to export to a .obj file.This paper uses the author's new boundary method to generate a regular quadrilateral pyramid. Values are set for the 4 vertices and the length of the base, respectively. Next, set a value for the height. While for the geometric body cone, the traditional method is used. Defines the base by a number of vertices, radius, and extrude height.

Before starting the process of 3D printing or adding a 3D model to a virtual/augmented reality device, it is necessary to create a file (.obj) containing the necessary information about the geometric object's spatial characteristics. This type of data is stored in text format and can be modified using a text editor (eg, notepad). These files contain the following information about the geometry of the object:

1. "v"—vertices of the geometric body;

2. "vn"—normal, which is a parameter for the light source;

3. "vt"—texture coordinate;

4. "f"—wall.

The stereoscopic system makes it possible to represent geometric objects in two ways—solid and wireframe mesh. The wireframe mesh representation will make it possible to understand the geometry of the three-dimensional model. While the solid gives a finished look to the geometric figure. These are its visual characteristics. It is preferable that before exporting the geometry object to a .obj file, it should be generated as a solid body. The Canvas3D class is a component of the AWT interface library and extends a two-dimensional object into a three-dimensional one by including the necessary 3D information. It represents the canvas where the three-dimensional objects are drawn. One of the modes for stereoscopic visualization is Mixed Immediate. The stereoscopic system uses this mode for stereo visualization through active glasses. The advantage of StereoMV is that it uses the Java3D library. And so it can be realized by a virtual reality device, such as:

1. Traditional Desktop

2. 3D active visualization—improved Desktop

3. 3D passive visualization—improved Desktop

4. HMD—an immersive system using a Desktop PC

5. CAVE—immersion system

In **Figure 7** presented: traditional, anaglyph, and stereo visualization of a objects. After exporting the geometry object to a .obj file. It can be added to an augmented or virtual reality device. The article uses the 3D viewer program in Mixed Reality mode using animation (**Figure 8**). In (**Figure 9**) HMD mode, each eye needs to have its own display. For this purpose, two canvas3D objects for the left and right eye need to be added.

Figure 7.
Traditional, anaglyph, and active visualization.

Figure 8.
Mixed reality 3D viewer.

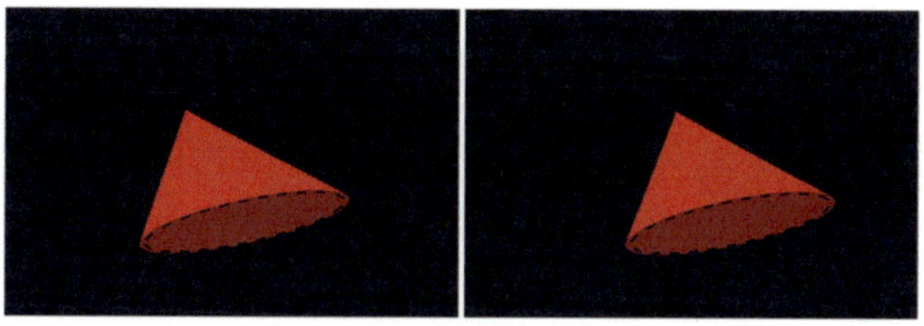

Figure 9.
Stereo cone—2 Canvas3D objects for HMD, left and right eye.

9. StereoMV

The author's software StereoMV (Stereo Math Vision), is the result of dissertation work on the topic "Stereoscopic Training System". It can be used to teach students about stereometry. It is aimed at the middle course of study. In the future, it will also include an upper course. The Java programming language and, more precisely, Java3D was used as a means of implementation. 3D active visualization of a non-immersive system was used at the beginning of the system development. After the thesis's successful defense, the software's development continued so that it could be implemented by an immersive system—HMD (virtual helmet). At the moment, the software also includes another direction, which is in the field of medicine. The goal is to create stressful situations through virtual reality that will take part in heart rate analysis by making recordings using a Holter device.

10. Conclusions

The software StereoMV proposed in this chapter can be used in teaching stereometry. It will be able to help develop spatial thinking in students who do not have it. VR systems are tracked, both with and without immersion. The stereo system can work with any VR device thanks to its Java3D library. The use of virtual and augmented reality in teaching mathematics, and in particular stereometry, is not new. There are documented results of the benefits of 3D technology in mathematics education. The conclusion is that VR helps students discover and track mathematical dependencies. At the same time, teachers are also open to this kind of training. The system's main advantage is that for the generation of 3D objects, a new author's boundary method is used with high accuracy compared to the traditional one which is using trigonometry. With this method, extruding the 2D polygons into 3D objects is unnecessary. This chapter uses the technique to generate a square (a polygon with four vertices). But the same rules can be applied to any regular polygon. The number of vertices can grow indefinitely. The chapter also presents modern educational approaches to teaching stereometry through the new author system.

Author details

Penio Lebamovski
Institute of Robotics, Bulgarian Academy of Sciences, Sofia, Bulgaria

*Address all correspondence to: p.lebamovski@abv.bg; penko_13a@abv.bg

References

[1] Gobbeti E, Scateni R. Virtual reality: Past, present and future. In: Riva G, Wiederhold BK, Molinari E, editors. Virtual Environments in Clinical Psychology and Neuro-Science. Amsterdam: IOS Press; 1998. pp. 3-20. DOI: 10.3233/978-1-160750-902-8-3

[2] Ross C, Gray T. The Effects of an Experiential Learning Course on Secondary Student Achievement and Motivation in Geometry, Dissertations. 2020. Available from: https://irl.umsl.edu/dissertation/933

[3] Mandal S. Brief introduction of virtual reality and its challenges. International Journal of Scientific & Engineering Research. 2013;**4**(4):304-309

[4] Mehrabi M, Peek E, Wuensce BC, Lutteroth C. Making 3D work: A classification of visual depth cues 3D display technologies and their applications. In: Proceedings of the Fourteenth Australasian User Interface Conference (AUIC2013). Adelaide, Australia. pp. 91-100

[5] Yi C. "Exploring the Effectiveness of Two Types of Virtual Reality HeadSets" for Teaching High School Mathematics, Science and Technology Education. 2021. DOI: 10.29333/ejmste/10996. ISSN:1305-8223

[6] Simonetti M, Perri D, Amato N, Gervassi O. Teaching Math with the help of Virtual Reality. arXiv:2111.01973v1. 2021

[7] [Internet]. 2022. Available from: Pro screens—Innovative Display Solutions. http://www.proscreens.eu/bg/3d-technologies

[8] Giraldi G, Silva R, Oliveira J. Introduction to Virtual Reality

[9] Ganchev I, Ninova U, Nikova V. Metodika na obuchenieto po matematika, UZU Neophit Rilski Book BG

[10] Rahman M, Puteh M. Learning trigonometry using GeoGebra learning module: Are under achieve pupils motivated? AIP Conference Proceedings. 2017;**1750**(1):39-42. DOI: 10.1063/1.4954586

[11] Lebamovski P, Petkov E. Usage of 3D technologies in stereometry training. In: CBU International Conference Proceedings. Vol. 1. 2020. pp. 139-146. DOI: 10.12955/pss.v1.61

[12] Lebamovski P. The Effect of 3D technologies in stereometry training. In: CBU International Conference Proceedings. Vol. 1. 2021. pp. 68-74. DOI: 10.12955/pns.v2.155